MANAGING FATIGUE IN TRANSPORTATION

Proceedings of
the 3rd Fatigue in Transportation Conference
Fremantle, Western Australia, 1998

Related books

Daganzo
Fundamentals of Transportation and Traffic Operations

Ettema & Timmermans
Activity based Appproaches to Travel Analysis

Gärling, Laitila & Westin
Theoretical Foundations Of Travel Choice Modeling

Hale
After the Event: From Accident to Organisational Learning

Hauer
Observational Before-After Studies in Road Safety

Stopher & Lee-Gosselin
Understanding Travel Behaviour in an Era of Change

Tilanus
Information Systems in Logistics and Transportation

MANAGING FATIGUE IN TRANSPORTATION

Proceedings of
the 3rd Fatigue in Transportation Conference
Fremantle, Western Australia, 1998

edited by

LAURENCE HARTLEY

Institute for Research in Safety and Transport
Murdoch University, Western Australia

United Kingdom – North America – Japan
India – Malaysia – China

Emerald Group Publishing Limited
Howard House, Wagon Lane, Bingley BD16 1WA, UK

Reprints and permission service
Contact: booksandseries@emeraldinsight.com

British Library Cataloguing in Publication Data
A catalogue record for this book is available from the British Library

ISBN: 978-0-08-043357-8

TABLE OF CONTENTS

Part III. Assessment of Fatigue

FOREWORD

This book presents a mixture of papers presented at the 3rd conference on Fatigue in Transportation held in Fremantle, Western Australia (WA) in 1998 and solicited papers from other authors. The conference was a memorable meeting with an outstanding team of scientists, practitioners, policy makers and industry representatives. This unique mixture of people who represent all aspects of the industry, has been the hallmark of past meetings. The focus of the 3rd meeting was on improving fatigue management practices as we steadily become a 24 hour society.

There is world wide interest in the better regulation of fatigue and not just in road transport. In most jurisdictions and industries there are prescriptive hours of service regulations which have been seen in the past as an appropriate mechanism for controlling fatigue, especially in road transport. However, they have recently come in for criticism because hours of work are not the only, and probably not the principle, cause of fatigue. Few hours spent asleep and the poor timing of work and rest are certainly very major causes of fatigue, but not the only ones. In Parts 1 and 2 of the book several authors review what we know about the causes and control of fatigue, and provide an update on the latest research findings about how we might be managing sleep and alertness as organisational resources in the not too distant future. These papers focus not only on road transport but maritime, air transport and military operations.

Although most chapters deal with road transport, the problems and the solutions are applicable across many safety critical industries. For example the management of sleep and alertness in a military organisation could inform better management of these resources in a road transport or maritime organisation. One lesson we have learned is that there is enormous variability in what freight drivers do, how they do it and how they deal with fatigue. Any regulatory framework has to cope with these facts, whilst at the same time setting standards for the industry.

Fundamental to improving the management of fatigue is its assessment. How can we identify and measure fatigue? In Part 3 of the book several authors address the issue of measurement and the shortcomings of controlling fatigue by hours of service regulations. In many road transport jurisdictions prescriptive hours of service of 10 or 12 hours of driving and related work, enforced by log books, is the method by which fatigue is controlled. This method of managing fatigue is held in wide disrepute, not only because of the ease with which enforcement can be corrupted, but also because it is believed that prescriptive hours of service may actually contribute to fatigue by setting targets for drivers to achieve, without permitting discretionary rest when needed. Part 3 raises the question of whether hours of service have served us well in managing fatigue, if not why not and what approaches might be better. These lessons are applicable to other industries too.

In an innovative approach to the better management of fatigue in Australian road transport Queensland Transport in conjunction with the Road Transport Forum developed an alternative compliance strategy to the hours of service regulations, termed the Fatigue Management Program. This is a proactive approach in which transport companies apply for exemption from the hours of service regulations on certain nominated routes and times if they can demonstrate they have good fatigue management practices in place in the company. Failure to comply with the contract between the company and the accreditation authority results in loss of accreditation and the company has to revert to operating under the hours of service regulation. This and other alternative strategies to manage fatigue are discussed in Part 4 of the book.

In Western Australia the Queensland approach to fatigue management cannot be adopted since there is no prescriptive hours of service regulation for road transport from which to be exempted. The Western Australian solution is to use the existing Occupational Safety and Health legislation. Specifically, the Occupational Safety and Health act provides for a Duty of Care by both employer and employee. To provide Duty of Care there must be a "*safe system of work... which is practicable and reasonable*". This Duty of Care extends to the *workplace*, which can be a transport environment; and it applies to *others in the workplace*, who can be other road users. The specification of a *safe system of work* is left to the industry, unions and Government to decide upon. The Standards for a safe system of work can be specified in an industry Code of Practice.

The Western Australian Code of Practice for Commercial Drivers described in Part 4, recognises that different sectors of the road transport industry and its drivers operate under different conditions and drivers differ in their capacity to manage their own fatigue. Several authors describe their research into individuals' differences including ageing, how this impacts on operational performance, and the need to address it in any regulatory framework. There can be no single formula for reducing fatigue in an industry as varied as the transport industry. The same will be true about other safety critical industries.

The strengths of this Occupational Safety and Health approach to better management of fatigue are that it is inclusive and applies to all drivers, subcontractors, their employers and customers. It is comprehensive and addresses the many causes of fatigue, whilst being flexible enough to permit different solutions to different demands on drivers. It provides a tool for companies to approach their customers and negotiate realistic freight schedules. It also provides a tool for employees to approach their employers and negotiate more realistic transport schedules and rosters. And the framework is transferable to other industries.

Dr. Laurence Hartley
Convenor.

CONTRIBUTORS

Ms Pauline Arnold, Murdoch University, South Street, Murdoch, WA, 6150, Australia
Thomas J. Balkin, Division of Neuropsychiatry, Walter Reed Army Institute of Research Washington, DC 20307-5100
S.D. Baulk, Seafarers International Research Centre, University of Wales, Cardiff, PO Box 901, Cardiff, Wales, CF1 3YG, United Kingdom
Dr Gregory Belenky, Director, Walter Reed Army Institute Of Research, Division Of Neuropsychiatry, Walter Reed Army Institute Of Research, Washington, DC 0307-5100 USA
Dr John Caldwell, Research Psychologist, US Army Aeromedical Research Laboratory, PO Box 620577, Fort Rucker, Alabama, 36362-0577, USA
Dr Lynn Caldwell, Research Psychologist, Sleep Disorders Center Of Alabama, Rt 1 Box 110, Jack, Alabama, 36346 USA
Mr James Dalziel, University Of Sydney, Department Of Psychology, A19, Sydney, NSW, 2006 Australia
A/Prof Drew Dawson, Director Centre For Sleep Research University Of South Australia, 5th Floor, CDRC, TQEH, Woodville Road, Woodville, SA 6011, Australia
Dr. P.A. Desmond, University of Minnesota, Minneapolis, USA
Prof David Dinges PhD, University Of Pennsylvania Health Systems, 1013 Blockley Hall, 423 Guardian Drive, Philadelphia, PA, 19104-6021, USA
Katharine Donkin, The Centre for Sleep Research, The Queen Elizabeth Hospital, Woodville Rd Woodville SA 5011, Australia.
Mr Adam Fletcher, The University Of South Australia, The Centre For Sleep Research, Level 5 CDRC The Queen Elizabeth Hospital, 11-23 Woodville Road, Woodville, SA, 5011, Australia.
Sandra Z. Fuller, Institute for Traffic Safety Management and Research, University at Albany, State University of New York
A/Prof Philippa Gander, Wellington School Of Medicine (Otago University), Dept Of Public Health, Wellington School Of Medicine, PO Box 7343, Wellington, New Zealand
Mark C. Hammer, Institute for Traffic Safety Management and Research, University at Albany, State University of New York
Assoc. Prof Laurence Hartley, Institute For Safety & Transport, Murdoch University, South Street, Muroch, WA, 6150, Australia
Dr Narelle Haworth, Senior Research Fellow, Accident Research Centre, Monash University, Clayton, VIC, 3168, Australia
Dr Ronald Heslegrave, Research Director, University Of Toronto, Dept. Of Psychiatry, Wellesley Central Hospital, 160 Wellesley Street, East Toronto, Ontario, M4Y 1J3, Canada
Mr Ron Knipling, Chief Research Division, U S D.O.T, USDOT/FHWA/HCS 30400, Seventh Street, Washington, DC 20590 USA

Nicole Lamond, The Centre for Sleep Research, The Queen Elizabeth Hospital, Woodville Rd Woodville SA 5011, Australia.
Jeffrey J. Lipsitz, M.D., Sleep Disorders Centre of Metropolitan Toronto, Canada
Dr. Anne T. McCartt, Institute for Traffic Safety Management and Research, University at Albany, State University of New York
Mr Alister Mckay, BP Oil New Zealand Limited
Mr Gary Mahon, Director (Road Use Management And Safety), Queensland Transport, PO Box 673, Fortitude Valley, QLD, 4006, Australia
Melissa M. Mallis, University Of Pennsylvania Health Systems, 1013 Blockley Hall, 423 Guardian Drive, Philadelphia, PA, 19104-6021, USA
Ms Michelle Millar, Junior Research Fellow, Wellington School Of Medicine (Otago University), Dept Of Public Health, Wellington School Of Medicine, PO Box 7343, Wellington, New Zealand
Dr. James C. Miller, Scripps Clinic and Research Foundation, La Jolla, California, 92037, USA
Prof. Merrill Mitler, Professor, Department Of Neuropharmacology, The Scripps Research Institute, 9834 Genesee Avenue, Suite 328, La Jolla, California, 92037, USA
Mr Barry Moore, Manager Economic Policy, National Road Transport Commission, PO Box 13105 Law Courts, VIC, 8010, Australia
Robert G. Norman, MS, Departments of Medicine/Division of Pulmonary and Critical Care Medicine, NYU/Bellevue Medical Centers, New York
Dr. Edward B. O'Malley , Departments of Medicine/Division of Pulmonary and Critical Care Medicine, NYU/Bellevue Medical Centers, New York
Mr Richard Phillips, Lecturer In Anatomy And Pathophysiology, University Of Tasmania, PO Box 1214, Launceston, TAS, 7250, Australia
Mr Lance Poore, Western Australian Department of Transport, 441, Murray St. West Perth, Western Australia, Australia.
David M. Rapoport, MD, Departments of Medicine/Division of Pulmonary and Critical Care Medicine, NYU/Bellevue Medical Centers, New York
Daniel P. Redmond, Division of Neuropsychiatry, Walter Reed Army Institute of Research Washington, DC 20307-5100
Ms Kathryn Reid, The University Of Adelaide, Centre For Sleep Research, The Queen Elizabeth Hospital, Woodville Road, Woodville, SA, 5011, Australia
Dr Louise Reyner, Lecturer In Sleep And Neurosciences, Loughborough University, Department Of Human Sciences, Loughborough University, Loughborough, Leics, LE11 3TU, United Kingdom
Dr William Rogers, Director Of Research, ATA Foundation, 2200 Mill Road, Alexandria, Virginia, 22314, USA
Mr Trevor Seal' BP Oil New Zealand Limited

Ted Schultz, Baker & Schultz, Santa Barbara, California
Helen C. Sing, Division of Neuropsychiatry, Walter Reed Army Institute of Research Washington, DC 20307-5100
Dr Alison Smiley, President, Human Factors North Inc., 118 Baldwin Street, Toronto, Ontario, M5T 1L6, Canada
Kingman P. Strohl, MD, Case Western Reserve University, Cleveland, Ohio
Dr Andrew Tattersall, Senior Lecturer, School Of Psychology, University Of Wales Cardiff, PO Box 901, Cardiff, Wales, CF1 3YG, United Kingdom
Prof Donald Tepas, Professor Of Industrial Psychology, University Of Connecticut, Ergonomics Laboratory, 406 Babbidge Road, Storrs, CT, 06269 1020, USA
Maria L. Thomas, Division of Neuropsychiatry, Walter Reed Army Institute of Research Washington, DC 20307-5100
David R. Thorne, Division of Neuropsychiatry, Walter Reed Army Institute of Research Washington, DC 20307-5100
Mr Sesto Vespa, Senior Development Engineer, Transport Canada, Transportation Development Centre, 800 Rene - Le Vesque Blvd, West 6th Floor, Montreal, Quebec, H3B 1X9, Canada
Dr Joyce Walsleben, Director - Sleep Disorder Center, New York University Medical Centre, Bellevue Hospital Center, 7N6 First Avenue And 27th Street, New York, NY 10016,USA
Dr David Waite, Wellington School Of Medicine (Otago University), Dept Of Public Health, Wellington School Of Medicine, PO Box 7343, Wellington, New Zealand
James K. Walsh, Ph.D. Sleep Medicine and Research Center, Chesterfield, MO
Nancy J. Wesensten, Division of Neuropsychiatry, Walter Reed Army Institute of Research Washington, DC 20307-5100
C. Dennis Wylie, B.A., Wylie & Associates, Goleta, CA

Part I.

The Scope of the Fatigue Problem: What Can Research Tell Us?

1

Fatigue Management: Lessons From Research

Dr. Alison Smiley, Human Factors North Inc., Toronto

Introduction

Over the last few decades, research has increasingly made clear that there is more to fatigue than simply the number of hours worked. The inadequacy of current hours of service regulations in providing for safe operation has become obvious. At the same time increasing global competition is forcing transportation companies to cut costs where possible. In such an environment, complicated red tape around hours of service that research might suggest is most unwelcome. As a consequence, the flexibility of fatigue management programs seems an ideal solution.

Research on fatigue in transportation offers many valuable lessons that managers of such programs and government regulators should heed. This paper reviews research which examines the relationship between the common causes of fatigue and its outcomes. The research is drawn from both the trucking and rail industries. Both truck and train operation depend on a single operator working long hours, paying continuous attention, in environments which are frequently monotonous, and at times of the day when performance is poor. These factors interact to create fatigue. The specific fatigue causes examined in this paper are long hours, time of day effects, and inadequate sleep: the outcomes examined are changes in operator performance and accident risk.

In reviewing the literature on fatigue, it is important to remember that not all variables can be controlled or examined in a single study. Every study has its weak points and no one study can offer the final answer. What is important is the overall picture that emerges.

FATIGUE AND ACCIDENT RISK

Research suggests that fatigue is implicated in a significant number of transportation accidents. In 1985, an American Automotive Association study looked at 221 serious truck accidents, serious being defined by the fact that the truck had to be towed from the scene of the accident. In-depth interviews were carried out with the truck drivers (if surviving) and with family. Log books and other receipts which could track actual driving time and routes were examined. For purposes of the study an accident was determined to be associated with fatigue if the driver action was consistent with falling asleep (e.g. drifting across lanes into the ditch) and if the driver had exceeded a duty period of 16 hours. Even with this conservative definition, it was estimated that fatigue was the primary cause in 40% of the accidents and a contributing cause in 60% of the total.

As a result of an extensive literature review, Haworth, Triggs and Grey (1988) concluded that the probable relationship between fatigue and accidents in articulated vehicles ranges between 5 and 10% of all crashes, about 20-30% of casualty crashes and about 25-35% of fatal crashes. They estimated that the contribution of fatigue may even reach 40-50% in particular types of crashes, for example fatal single vehicle semi-trailer crashes.

Comparisons show that the impact of fatigue as a contributing factor in truck accidents is similar to that of alcohol in all road accidents. The pattern is similar in that both alcohol and fatigue are increasingly likely to be contributing factors with increasing accident severity. Alcohol is a factor in only about 5% of all crashes, but about 20% of serious injury crashes and about 40% of fatal crashes (Fell, 1998). As with drivers who are fatigued, drivers who have consumed alcohol are at greater risk for single vehicle accidents, especially at night. Despite the parallels with alcohol, fatigue has not been viewed with the same level of concern. This may be related to the fact that fatigue is considerably more difficult to predict or to measure, and it is seen as more of a problem for commercial drivers than for the general public, the reverse of alcohol. The following sections review the main causes of fatigue and attempt to tease out the important interactions. These are the lessons from research that need to be understood and accounted for in any system of regulation or fatigue management.

Long Hours

The fatigue cause which has been most readily acknowledged in the past is long hours, and most countries regulate hours of driving, probably because of a "common sense" expectation that long hours lead to accidents. However, permitted daily driving time varies widely from 8 hours in the EEC to 15 hours in Alaska (McDonald, 1984), suggesting many factors other than science play a role in regulating limits.

Long Hours and Performance

The classic study of the impact of hours of work on performance and accident rates of truck and bus drivers was carried out by Mackie and Miller in 1978. They measured behaviour of a total of 12 truck drivers and 6 bus drivers over a week long period of driving. Conditions considered included regular and irregular schedules, sleeper cab operations, and light or moderately heavy cargo loading in addition to driving. Measures were multi-dimensional and included: vehicle control, critical incidents involving driver drowsiness, physiological changes, subjective ratings of fatigue, and test battery performance.

This very thorough study indicated that the effects of fatigue were evident long before the hours of service regulations had been exceeded. Fatigue effects showed up as significantly greater feelings of fatigue during the second half of all standard length (9.5 hour) trips, and after 6 hours on irregularly scheduled trips. Significant changes in steering patterns and in lane position variability showed up after 8 - 9 hours of driving on the regular schedule and after 4 - 5 hours on irregular schedules. (Steering and lane control variables are strongly affected by highway geometry. Therefore only equivalent inbound and outbound portions of trips could be compared. Thus, the estimate of the effect of long hours is limited by the comparisons which could be made.) Drivers engaged in sleeper operations showed earlier and/or greater signs of subjective fatigue and degraded performance as compared to drivers not so engaged.

Most recently, Mackie and Miller's work was extended by a study carried out jointly by the U.S. Federal Highway Administration and Transport Canada (Wylie et al., 1997). Two schedules of 10 hours driving were examined, one of which was a baseline regular daytime schedule and the second of which was according to the 21 hour daily cycle resulting from following U.S. regulations permitting 10 hours of driving, time for meals, refueling etc, and 8 hours off. With such a schedule, starting times advance by 3 hours each day. Two schedules of 13 hour driving (Canadian regulatory limits) were also examined, one primarily day driving, and the other primarily night. Schedules were compared on the basis of sleepiness (as judged from video recordings and EEG recordings), vehicle lane tracking, test batteries and subjective estimates. The 10 hour trips were carried out in the U.S. and the 13 hour trips in Canada.

Contrary to the findings of Mackie and Miller, this considerably larger study (80 drivers in total vs 18) did not show effects of hours of driving on performance. However, equivalent road sections were not compared for lane tracking effects on the different schedules. Mackie and Miller's study indicated that road geometry had a strong influence on lane tracking variability. Thus performance changes may have been hidden due to the different road geometrics. More analysis of these data is required. Wylie et al. (1997) found time of day to be more important than hours of work in determining drowsiness and poorer performance. In their conclusions, the authors state that " Night driving (0000 - 0600) was associated with worse performance on each of four important criterion variables (drowsiness, lane tracking, code substitution, and sleep length) whereas hours of driving and number of consecutive trips had little or no relationship to those criterion variables. **It was concluded that time of day was a far better predictor of decreased driving performance than time on task or cumulative number of trips."**

Long Hours and Accident Risk

The early studies of truck driving and hours of work seemed to indicate that most truck accidents occurred in the first few hours of driving. However, these studies failed to control for the fact that most truck trips were of short duration, and thus most accidents would be expected to occur during the first few hours. In 1972, Harris and Mackie analyzed 500 accidents which occurred over a one year period for a large common carrier engaged in long haul interstate operations. Exposure was taken into account in determining risk, by considering not only the number of hours prior to the accident occurring but also, the expected length of the trip. They found that 62% of the accidents occurred in the second half of the trip as opposed to 38% in the first half, irrespective of trip duration. This supports the premise that fatigue, which develops with time on task, is an important factor in accidents.

In their 1978 report, Mackie and Miller present analyses of 3 accident samples collected by the U.S. Bureau of Motor Carrier Safety: 406 dozing driver accidents, 226 single vehicle accidents and 116 rear end accidents. It is of interest that contrary to what might be expected, in only 7% of the single vehicle accidents had the driver been reported as dozing at the wheel. In all 3 samples, nearly twice as many accidents occurred in the second half of trips as occurred during the first half. Expected versus actual percentage of accidents were compared by driving time. **For the dozing driver accidents and the single vehicle accidents, the changeover to more accidents than expected from fewer accidents than expected occurred after about 5 hours of driving.**

In 1987 the Insurance Institute for Highway Safety in the U.S. used a case control approach to examine the relative risk associated with long hours of driving (Jones and Stein, 1987). This approach had been previously used with great success by Borkenstein et al (1964) to determine accident risk associated with blood alcohol level. For each large truck involved in a crash, three trucks were randomly selected from the traffic stream at the same time and place as the crash but one week later. A sample of 332 tractor-trailer crashes, each with 1, 2 or 3 case controls was extracted for analysis.

For tractor-trailers, the authors found that driving in excess of eight hours, drivers who violate logbook regulations, drivers aged 30 and under, and interstate carrier operations were associated with an increased risk of crash involvement. **In particular, the relative risk of crash involvement for drivers who reported a driving time in excess of 8 hours was almost twice that for drivers who had driven fewer hours.**

A Canadian study, by Saccomanno et al (1996), used several different databases to determine accident risk associated with different driving times. Databases included police accident reports, and a commercial vehicle survey of driver demographics, work hours, route etc. **Overall accident risk was higher for routes characterized by longer 85 percentile driving times starting at 9.5 hours.** There were more single vehicle accidents at night (assumed to be associated with fatigue) than during the day. Part of this effect is no doubt due to there being less traffic at night and therefore lesser opportunity to come into conflict with another vehicle. There was a higher proportion of such accidents on routes typified by long driving times. In remote regions, the nighttime single vehicle accident rates were particularly high - 13 times greater than for more populated areas in the daytime. For both daytime or nighttime conditions there were more single vehicle accidents at locations with high 85th percentile driving times. However, nighttime single vehicle accident rates were significantly higher than the daytime rates. **Thus it is clear that long hours, night time driving and driving in remote areas are associated with increased risk of single vehicle accidents due to fatigue.**

Weekly Hours, Performance and Accident Risk

As discussed above, a number of studies have examined the effect of long hours in a given day of driving. In contrast, only a few studies have examined the effect of several days of work on changes in performance or accident risk. Mackie and Miller (1978) observed 18 drivers over a 6 day period of driving. **They found cumulative fatigue effects on performance after 4 consecutive days of driving for both drivers on regular and irregular schedules. Despite these performance changes, drivers' subjective ratings did not reflect any evidence of cumulative fatigue.** For drivers engaged in sleeper operations there was also evidence of performance changes associated with cumulative fatigue, however this was strongly affected by time of day, with greatest changes occurring during night runs.

The author is aware of only one study concerning accident risk and cumulative fatigue. Jovanis and Kaneko (1990) report an analysis of carrier-supplied accident and non-accident data for a 6 month period in 1984. The data were obtained from a "pony express" type operation which operates coast to coast with no sleeper berths. Cluster analysis was used to identify 9 distinct patterns of driving hours over a 7 day period. The driving patterns of drivers who had an accident on the 8th day were compared to drivers who had no accident on the 8th day.

Accident risk on the eighth day was shown to be consistently higher for the 4 patterns involving infrequent driving the first 3-4 days followed by regular driving during the last 3-4 days, than the 4 patterns with the reverse arrangement. **This suggests that cumulative fatigue from driving over 3-4 days does occur, and leads to increased accident risk.**

Hours of Service Violations

Despite evidence that long hours are associated with increased accident risk and poorer performance, we know from various surveys that many drivers violate hours of work regulations. Braver et al. (1992) asked 1249 drivers in the U.S. about hours of service violations, and found that 73% of the interviewed tractor-trailer drivers were hours-of-service violators. Thirty-one percent reported driving more than the legal weekly limit (60 hours in 7 days, or 70 hours in 8 days). Over one quarter of the drivers reported working 100 hours or more per week.

In Australia, Williamson et al. (1992) reported that half the drivers admitted to breaking work-hour-regulations in at least half of their trips (56.6%). Approximately three-quarters of the drivers in this study rated fatigue as a substantial industry-wide problem **although only 35% rated fatigue as a substantial problem for them personally.**

Hertz (1991) attempted to obtain a more objective estimate of the percentage of violations. Drivers were interviewed at an inspection station and later observed arriving at sites in 2 cities 1200 miles from the initial inspection site (Hertz, 1991). Assumptions were made about average speed based on reports in the literature and fleet manager's estimates. Even with the most liberal estimate of the speed that could be achieved**, fully half of the drivers were in violation of the regulations.**

Summary: Long Hours

In summary, one major study (Mackie and Miller, 1978) which examined length of hours, irrespective of time of day, showed performance deficits in truck drivers after 8 - 9 hours on a regular schedule of daytime driving and after 4 - 5 hours on irregular schedules which included night driving. Another major study (Wylie et al, 1997) did not find lane tracking and steering changes associated with hours of driving. However, the effect of highway geometry, on lane tracking and steering, has not yet been accounted for in the analysis as it had been in the earlier study.

With respect to accident risk, studies showed increased risk of a dozing driver accident or a single-vehicle accident beginning at 5 hours (Mackie and Miller, 1978), and a doubling of relative risk for tractor trailer drivers after 8 hours of driving (Jones and Stein, 1987).

With respect to cumulative effects, both deterioration in performance and accident risk increases are demonstrated after 3-4 days of regular driving, but mainly for night driving. Despite performance changes, drivers' subjective estimates were not sensitive to cumulative effects.

One study showed that driver's perceptions of fatigue lagged the deterioration in performance by one hour . Another study showed that while 3/4 of drivers considered fatigue an industry-wide problem, only 1/3 considered it a problem for themselves.

TIME OF DAY

Current hours of work regulations in the trucking industry revolve around hours of driving and hours on duty. They are written as if the time of day at which the driving occurs is irrelevant. Yet research over the past 30 years into the effects of time of day on performance and accident risk show these to be as strong or even stronger than the effects of hours of work.

The basis of the time of day effect is our 24 hour physiological rhythm known as the circadian rhythm, whereby the body gears for action during the day and for recuperation at night. Many physiological functions, as well as mental alertness, follow a general pattern of rising during the day and falling during the night. Because of circadian rhythm, performance, particularly on vigilance tasks, which are similar in nature to much highway driving, is poorest in the early morning hours - 2:00 to 6:00 a.m., around the circadian low point, with a secondary low point after lunch, 2:00 to 4:00 p.m.

Our circadian rhythm is kept to a 24 hour cycle by time-givers: namely, the rising and setting of the sun, knowledge of clock time, and work time. When a person works at night, the circadian rhythm will adjust so that physiological activity is higher during the work period, and lower during sleep. However, this adjustment is slow, taking up to 10 days, and incomplete, because the usual time-giving cues of sun rising and setting and work time are in conflict with one another (Grandjean, 1988). The performance rhythms adjust only partially. Even after a number of days on shift, performance is still poorest in the early morning hours.

Time of Day and Subjective and Objective Sleepiness

One of the earliest studies on sleepiness and truck driving is that of Prokop and Prokop (1955). Figure 1 illustrates the frequency with which 500 truck drivers reported that they had experienced falling asleep by time of day. The greater sleepiness at the circadian low points, in the early morning hours and after lunch, is clearly evident.

In their on-road study of 18 drivers, Mackie and Miller (1978) found greater evidence of subjective fatigue and lowered physiological arousal in drivers who concluded nine to ten hour drives late at night than was evident for those same drivers after similar drives during the daytime. In addition, they recorded incidents where drivers complained of fatigue or slept through their rest breaks. **They found a larger number of these fatigue indicators during trips late at night or in the early morning hours.**

The strongest data on time of day and sleepiness come from the recent U.S.- Canadian joint study described above (Wylie et al., 1997). Continuous video recordings of the driver's face were sampled every half hour, and assessed for indications of drowsiness (eyelid closures, drooping eyelids, yawning, head nods etc.). In approximately 5% of these video samples, the driver was judged to be drowsy. **Of these drowsy episodes, 82% occurred during the hours of 1900 to 0700. Furthermore, most drivers were affected to some degree - about 2/3 were judged drowsy in at least one sample.**

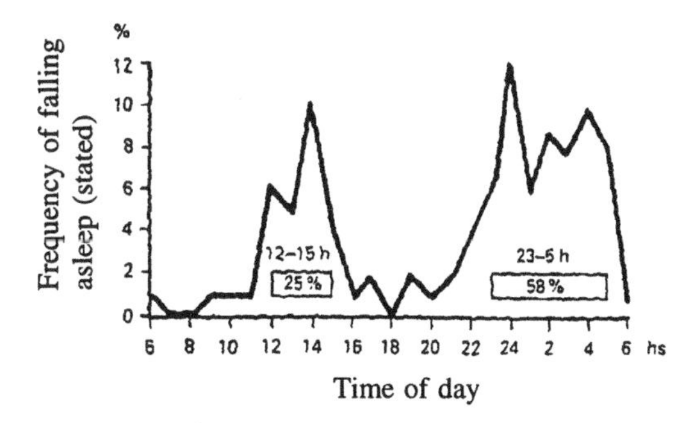

Figure 1 Frequency with which 500 truck drivers reported falling asleep at the wheel, in relation to the time of day (Prokop & Prokop, 1955 from Grandjean (1988).

Time of Day and Performance

Mackie and Miller (1978) found not only greater subjective sleepiness and lowered physiological arousal, but also deterioration in vehicle control during night as compared to day driving. In a simulator study, Gilberg, Kecklund and Akerstedt (1996) found similar effects. The performance of 9 professional truck drivers was examined using a truck simulator with sophisticated graphics and a motion-base. Drivers drove for three 30-minute periods. Night driving was carried out around the time of the circadian low point. Differences between day and night driving were small but statistically significant. Night driving was found to be slower, with a higher variability of both speed and lane position. In other conditions, the impact of a 30 minute nap or a 30 minute rest pause were compared. Neither the nap nor the rest pause had any effect.

Time of day effects were also examined in a study of railway engineers (Hildebrandt, Rohmert and Rutenfranz, 1975). The engineers were required to respond to light signals appearing randomly, on average once per minute. If the driver did not respond within 2.5 seconds, a loud warning hooter was sounded; if there was still no response within the next 2.5 seconds, an emergency brake was applied. Data were collected on performance of 1000 locomotive drivers over a total of 6304 work hours.

Analysis showed that soundings of the warning hooter peaked at 0300 and 1500, the latter peak being higher, no doubt due to the greater visibility of the light signals at night (see Figure 2). Peaks were more pronounced, indicating **greater fatigue for drivers who were in the 4th to the 6th hour of their shift, as compared to those in the 1st to the 3rd hour**.

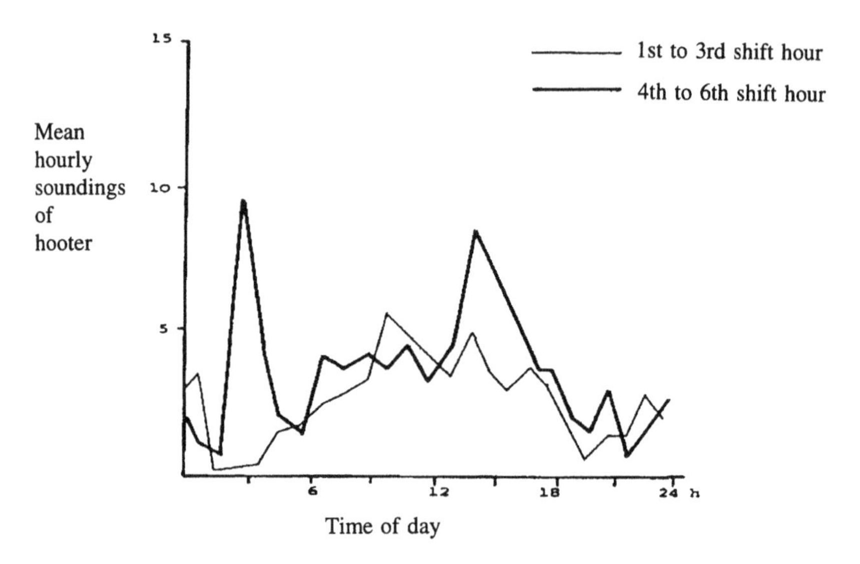

Figure 2: Daily course of the mean hourly frequency of sounding of the warning hooter with the first three and second three shift hours. After Hildebrandt et al. (1974). (McDonald (1984)).

Time of Day and Accident Risk

The Mackie and Miller (1978) analysis of 406 dozing driver accidents, described earlier, controlled for exposure by examining the ratios of percentage accidents to percentage trucks on the road by time of day. **An accident involving a dozing driver was found to be 7 times more likely to occur during the hours of midnight to 8:00 a.m. than in the other hours of the day, with the highest risk occurring between 4:00 and 6:00 a.m.** (88% of these dozing driver accidents were single vehicle accidents.)

Mackie and Miller (1978) also analyzed 226 single vehicle accidents and 116 rear end accidents. As shown in Figure 3, multi-vehicle accidents mainly reflect exposure: the higher the truck miles by time of day, the higher the number of such accidents. Single-vehicle accidents show a very different pattern, with the largest number occurring between 4:00 and 6:00 a.m. Despite only 7% of these single vehicle accidents having been reported as dozing driver accidents, there is a very clear time of day effect, suggesting sleepiness as a likely cause for many of these.

A study in Sweden, reported by Kecklund and Akerstedt (1995), also determined accident risk by time of day and controlled for exposure of cars and trucks separately. For trucks, the risk of a single vehicle accident increased during the night with a peak between 3:00 and 5:00 a.m. of 3.8 times the risk of an accident during the day (8:00 a.m. to 4:00 p.m.). Trucks also showed higher single vehicle accident risk during the weekend (peak of 6.0 between 4:00 and 6:00 a.m.). The authors provide a graph showing risks for car drivers as well (see Figure 4) showing a similar notable peak in risk in the early morning hours.

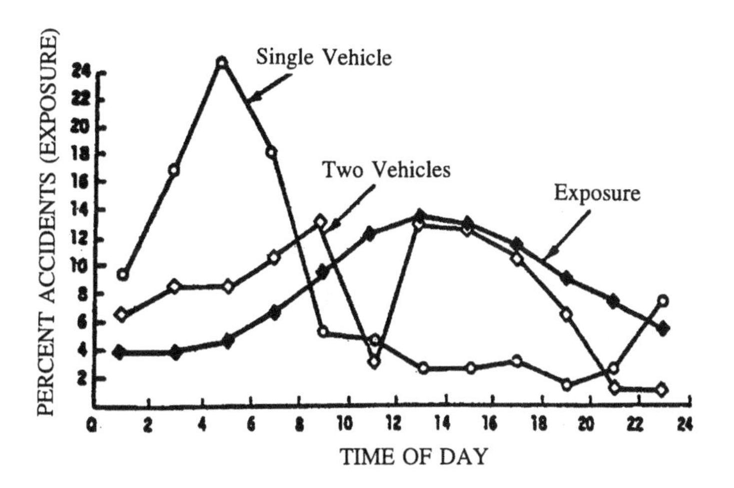

Figure 3: Percentage of accidents by time of day and type of accident for "dozing" drivers (N=493 accidents) (Mackie & Miller, 1978).

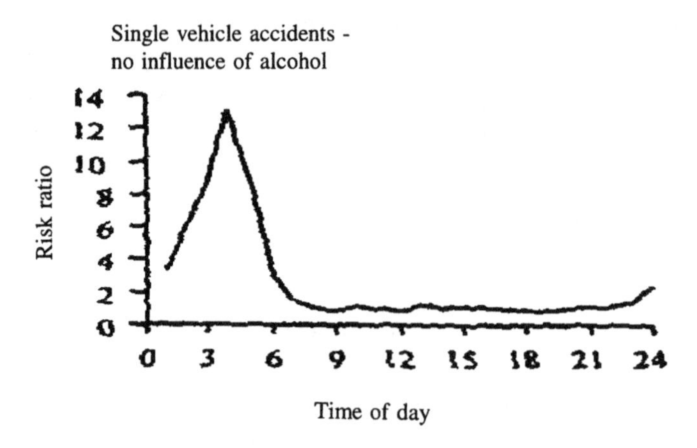

Figure 4: Relative risk of single vehicle accidents by time of day (Kecklund & Akerstedt, 1995)

Time of Day and Cumulative Effects

The interaction between cumulative fatigue and time of day was examined in the Jovanis and Kaneko (1990) study discussed above. Total hours of driving in a 7 day period varied between averages of 54 to 59 hours among the 9 patterns identified. **Drivers who began their trips near midnight and typically ended them around 10:00 a.m. faced a particularly increased crash risk after driving for several consecutive days**. In contrast, drivers who typically drove a regular daytime schedule (10:00 a.m. to 6:00 p.m.) showed little evidence of any effect due to continuous driving.

Summary: Time of Day

Research shows that in the early morning hours, drivers subjectively experience greater sleepiness, EEG measures demonstrate greater sleepiness, performance is worse and accidents, especially single vehicle accidents, are much more likely.

Unfortunately, it will not be easy to alleviate time of day effects through changes to regulations. The CANALERT study (1996) of Canadian railway engineers examined the effect of various fatigue countermeasures. These included scheduling that ensured drivers consistently started work either during the early part of the day or the later part of the day, opportunity for naps during the shift and improvements to the sleep facilities to promote better rest. The study concluded that "there are serious doubts as to the efficacy of any practicable hours of work regulation eliminating the risk of fatigue." The authors note that "some locomotive engineers were detected to have drowsiness episodes while operating locomotives shortly after commencing duty, especially at night, even after extended periods of prior rest off-duty of 24 hours or more." Nonetheless, the study found that modifications to work schedules that deal explicitly with time of day effects were effective in reducing, if not eliminating, objectively measured, and subjectively perceived fatigue. They also found that, not surprisingly, countermeasures to reduce fatigue were most effective for the drivers who actually used them.

SLEEP

Inadequate sleep obviously contributes to fatigue and occurs for various reasons. These include sleeping at a time of day when the body is not physiologically prepared to sleep, due to night work, cutting sleep time to meet a deadline, splitting sleep into two or more periods (e.g. as occurs with the use of sleeper berths) and a medical condition known as sleep apnea which is fairly common and can result in chronic sleepiness.

Sleep is a body function with a strong circadian component. Sleep lasts longest and is of the most recuperative quality if taken starting around 10:00 p.m.. If because of scheduling, sleep is taken during the daytime, it is significantly poorer in quality as measured by EEG recording, and shorter in duration. Figure 5, using data collected by Kogi (1985), clearly demonstrates that the time of day at which sleep starts dictates the amount of sleep that will be obtained. The most sleep is obtained when sleep is initiated in the evening around 9:00 p.m., with decreasing amounts obtained if sleep is initiated after 1:00 a.m. and the least amount if sleep is initiated in the afternoon. Sleep obtained at the optimum time of day typically lasts 7.5 to 8 hours.

There are a number of studies which have indicated that shiftwork affects ability to obtain sleep. Lille (1967) found that night shift workers obtained about 1.5 hours less per day than day shift workers, and slept up to 10 - 12 hours on days off. Drivers in Mackie and Miller's (1978) study reported obtaining 6.5 to 7 hours on regular daytime schedules. However, drivers with start times of 3:00 a.m. averaged 5.5 hours, and for 8:00 p.m. start times, only 4 hours. These short sleep lengths were found despite the fact that in this study drivers were always afforded the opportunity to obtain 8 hours of sleep. Drivers using sleeper berths averaged 4.5 hours on trips beginning at 8:00 a.m., but only 3 hours on trips beginning at midnight. The joint U.S.- Canada study (Wylie et al., 1997) recorded average total sleep lengths of 4.4 hours for drivers driving 13 hours at night, 40-85 minutes shorter than drivers on daytime or rotating schedules.

Given these findings it is not surprising that a study by Braver et al. (1992) suggests that sleepiness while driving is a widespread problem. **In a survey of tractor-trailer drivers at truck inspection stations, 19% admitted to falling asleep one or more times while driving during the past month**. Significantly more of the work-hour-rule violators than non-violators admitted to this problem.

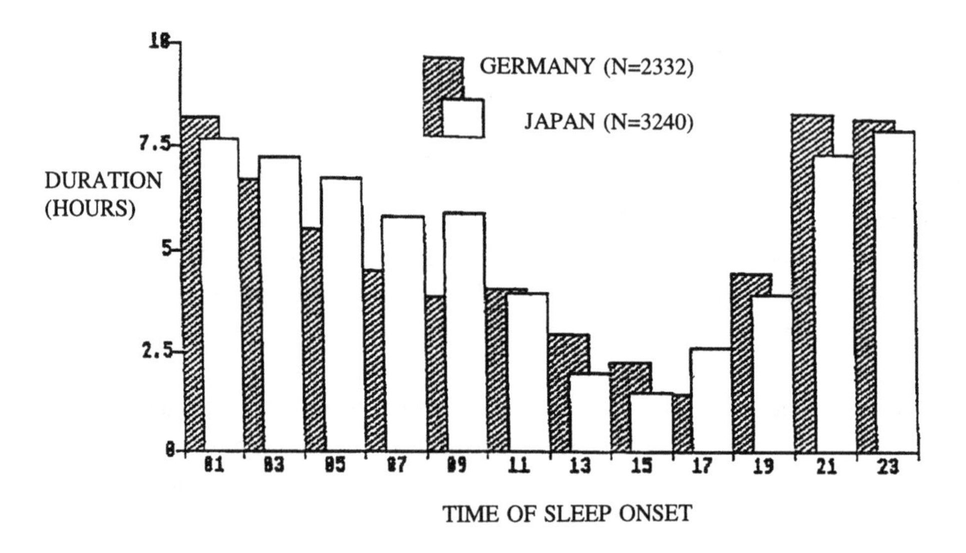

Figure 5: Hours of sleep obtained by time of sleep onset after Kogi 1985.

Sleep and Performance

While night and day driving performance have been compared, no studies could be found on the effect of the length of previous sleep on driving. However, laboratory studies allow precise measurement of the effects of sleep length on task performance. Even small reductions in sleep - by 2 hours - have been found to result in measurable changes in performance on tests of vigilance (Wilkinson et al., 1966). Thus one would expect a vigilance task such as highway driving to be affected by sleep deprivation as well. Sleeper berths are linked to shorter sleep and split sleep, which is inevitably poorer quality sleep. Mackie and Miller (1978) found driving performance deteriorated sooner in operations involving use of sleeper berths.

A recently recognized issue of concern for truck drivers is the medical condition of sleep apnea. This disorder is characterized by episodes when breathing ceases during sleep, resulting in chronic daytime sleepiness. The condition is fairly common, affecting about 4-5% of the population (Young et al., 1993). Furthermore, there are similarities between the demographics of the truck driver population and those who are most susceptible to sleep apnea, namely males in the age range of 30-55, and overweight persons. Indeed, a recent study by Stoohs et al (1994), using a liberal definition of sleep apnea, found 87% of a sample of long distance drivers were affected to some degree.

In tests of driving skills of sleep apnea patients, George, Boudreau and Smiley (1996) found that performance on a monotonous tracking and search task before treatment was similar to that of drivers dosed to 0.08% BAC, whereas after treatment, performance had returned to the same level as that of healthy controls. Similarly, Haraldsson et al. (1990a) who used a completely different treatment method, also found that the driving simulator performance of treated patients matched that of healthy controls. These studies would suggest that truck drivers with untreated sleep apnea are likely to perform more poorly than healthy drivers especially in monotonous driving environments, or at circadian low points.

Length of Sleep and Accidents

The U. S. National Transportation Safety Board (1995) studied 107 single-vehicle, nighttime accidents, where the driver survived and in which the previous 96 hours could be reconstructed. Of the total, 58% had fatigue as a probable cause, while the remainder were considered not fatigue-related. From the analysis the authors concluded that the most critical factors in predicting which nighttime accidents were fatigue-related were: the duration of the most recent sleep period, the amount of sleep in the past 24 hours, and split sleep patterns. The truck drivers in fatigue-related accidents were found to have obtained an average of 5.5 hours sleep in the last sleep period prior to the accident. This was 2.5 hours less than the drivers involved in non-fatigue-related accidents (8.0 hours). Truck drivers with split sleep patterns obtained about 8 hours sleep in a 24-hour time period; however, they obtained it in small segments, on average of 4 hours at a time. **The authors conclude that the data from**

this study indicate that driving at night with a sleep deficit is far more critical in terms of predicting fatigue-related accidents than simply nighttime driving.

The negative effect of split sleep is further supported by Hertz (1988) who examined the impact of the use of sleeper berths on accident causation. She compared sleeper berth use by 418 fatally injured tractor-trailer drivers to that of 15,692 non-injured drivers involved in property damage accidents. **Use of sleeper berths in two shifts increased the risk of a fatal crash by a factor of 3.** Univariate analysis was used to identify confounding factors, and then logistic regression was used to adjust for these confounding factors. Hertz found that the risk of an accident associated with sleeper berth use was as high for drivers driving alone as for drivers driving in a team. **In other words, the risk due to sleeper berth use does not appear to arise because of the disturbance of sleep due to the motion of the truck, but rather because of the splitting of sleep into two periods.**

Some studies of car drivers with sleep apnea have shown that they have a significantly increased overall risk of an accident as compared to healthy controls (George et al., 1987; Findley et al., 1988). Other studies, which have controlled for miles driven, have shown that the main increase in risk is for single-vehicle accidents. Haraldsson et al. (1990b) found that car drivers who were diagnosed with sleep apnea had a 9 times higher risk of a single vehicle accident as compared to healthy controls. Given the long hours driven by truck drivers, the risk of accidents associated with having a sleep apnea condition is likely to be considerably higher than for car drivers.

Sleep and Subjective Measures

Unfortunately research indicates a lack of awareness among drivers of their own sleep needs. The NTSB (1995) study of single vehicle heavy truck accidents which occurred at night was cited above. Many of the truck drivers who were involved in fatigue-related accidents did not recognize that they were in need of sleep and believed that they were rested when they were not. The authors note that **"about 80 percent of the drivers involved in fatigue-related accidents rated the quality of their last sleep before the accident as good or excellent".**

LESSONS LEARNED

The lessons learned about fatigue effects from the research outlined above can be summarized as follows:

1. Accident risk for dozing driver accidents and single vehicle accidents starts to increase after 5 hours of driving. One study shows that overall crash risk doubles after 8 hours driving. Another study shows that overall crash risk increases after 4 days of driving. A study which accounted for the effects of highway geometry on lane tracking and steering found performance changes associated with time on task occur after 8 - 9 hours driving on regular schedules and earlier (after 4 - 5 hours) on irregular schedules where nighttime driving is involved.

2. The early morning circadian low period is associated with poorer performance, even when drivers start work in a rested condition. The majority of drivers exhibit signs of sleepiness in this time period. One study showed that single vehicle accident risk is 3.8 times higher in the 3:00 to 5:00 a.m. period, as compared to daytime driving. Another study showed 7 times higher risk of single vehicle accidents in the midnight to 7:00 p.m. period as compared to the remainder of the day.

3. Sleep deficits increase the risk of a fatigue related accident at night. Split sleep taken in sleeper berths is associated with 3 times the risk of a fatal accident. The medical condition of sleep apnea is associated in some studies with significantly increased overall crash risk, and in other studies with a significantly higher risk of single vehicle accidents. Performance of drivers with untreated sleep apnea is shown to be worse than legally impaired drivers. However, treatment has been shown to raise performance levels of patients back to those of healthy drivers. Studies suggest that a significant number of truck drivers may be afflicted by sleep apnea.

4. Subjective estimates of alertness and sleep quality are poorly correlated with objective reality, particularly with respect to cumulative effects of fatigue and adequacy of previous rest.

APPLYING THE LESSONS FROM RESEARCH

Scientists have the easy job of determining performance and accident risk changes associated with various fatigue producing factors. Regulators and company managers, in the case of fatigue management programs, have the difficult job of determining how much deterioration is acceptable, and where to "draw the line". One similar area where policies are well developed is the area of driving under the influence of alcohol. It is instructive to look at some of the findings in this area, to determine how one should regard changes in performance and accident risk.

First, one must be wary of assuming that because performance changes due to long hours, or time of day are small, these changes are not of concern. For example, one on road study of alcohol effects showed that lane position variability increased a mere 8 cm at 0.12% as compared to 0.00% blood alcohol concentration (BAC) (Louwerens, Gloerich, de Vries, Brookhuis and O'Hanlon, 1987). In another on road study, detection of obstacles by the side of the road was 71 m. for sober subjects and 62 m. for subjects at 0.10% BAC, equivalent to 1/2 second difference in response time at the speed travelled (Laurell, McLean and Kloeden, 1990). At first glance these changes seem rather trivial in magnitude. Nonetheless, these modest changes in performance are associated with sizable increases in risk of an accident. Hurst's reanalysis (1973) of Borkenstein's Grand Rapids study showed that the risk of a collision at 0.08% BAC was double or more than found at 0% BAC. Most countries which regulate alcohol and driving have limits of 0.08% BAC or less. In other words, a doubling of risk is considered unacceptable when that risk arises from consumption of alcohol.

Similarly, we should not be sanguine about the doubling of risk of a tractor-trailer accident after 8 hours of driving, or the tripling of risk of a fatal tractor-trailer accident when a sleeper berth is used. Furthermore, when a truck accident occurs, there is a public safety issue. A U.S. study found that 70% of all large truck crashes, and 65% of all fatal crashes, involve other vehicles - mostly passenger cars. When cars and trucks crash, car occupants are 35 times more likely to be killed than truck occupants (Insurance Institute for Highway Safety, 1985). However, as with alcohol, the most frequent type of accident associated with fatigue is a nighttime, single-vehicle accident, in which it is the truck driver who is at risk of injury or death.

While more studies are certainly needed, research to date provides ample guidance for regulators, managers, and drivers. Suggested guidelines are:

1. Driving during the circadian low point of the day (2:00 - 6:00 a.m.) should be avoided.

2. If driving must be done during this period, it is critical that drivers be well-rested and that they not enter this time period towards the end of a long trip.

3. Rest is most recuperative if taken starting 9:00 p.m. to 1:00 a.m., and if taken in a lengthy sleep period, rather than split as in a sleeper berth.

4. Since daytime sleep is shorter and poorer quality than night sleep, drivers working at night should avoid successive night shifts, because the inevitable result is that drivers do not obtain sufficient rest before a second night shift.

5. Drivers should be screened for the medical condition of sleep apnea and treated.

6. Drivers are poor judges of their own condition - others should be involved in the determination of whether a driver is adequately rested.

With respect to the question of limits on hours, it is clear that there is a strong interaction between hours and time of day. Although the recent large U.S. - Canada study did not find performance changes associated with time on task, an earlier study did. Furthermore a number of studies have shown increased accident risk associated with hours worked. More epidemiological studies are needed to separate an accident risk per kilometre associated with hours driven and that associated with time of day. Such data would provide the basis for determining the appropriate hourly limits for day driving and for night driving. In the meantime, managers, regulators and drivers should be concerned about evidence already available. One study has shown increased risk of dozing driver and single vehicle accidents after 5 hours of driving. Another study has shown a significant increase in overall crash risk after 9.5 hours driving. A third study has shown a doubling of accident risk after 8 hours. This increase in risk is equivalent to that associated with alcohol at 0.08% BAC. In all cases increases in accident risk occur well before hours of driving would be exceeded in many jurisdictions.

With respect to cumulative days of work, only one study was found and it shows increased risk after 4 days of driving, but mainly for drivers driving at night. At a minimum 4 days should be considered the limit for night driving, until further epidemiological studies can be carried out.

References

AAA Foundation for Traffic Safety. (1985) *A Report on the Determination and Evaluation of the Role of Fatigue in Heavy Truck Accidents,* Report to the AAA Foundation for Traffic Safety, 8111 Gatehouse Road, Falls Church, Virginia, 22047, October,

Borkenstein, R.F., Crowther, R.F., Shumate, R.P., Zeil, W.B., and Zylman, R. (1964) *The role of the drinking driver in traffic accidents.* Bloomington: Indiana University Press..

Braver, E.R., Preusser, C.W., Preusser, D.F., Baum, H.M., Beilock, R., Ulmer, R. (1992) Long hours and fatigue: a survey of tractor-trailer drivers. *Journal of Public Health Policy*, *13*, 341-66..

Brown, I. Driver fatigue. (1994) *Human Factors,* 36(2), 298-314..

CANALERT'95 - *Alertness Assurance in the Canadian Railways.* (1996) Phase II report, Circadian Technologies, Inc.,.

Fell, J. (1998) Personal communication.

Findley, L.J., Unverzagt, M.E., and Suratt, P.M. (1988) Automobile accidents involving patients with obstructive sleep apnea. *American Review of Respiratory Diseases*, 138, 337-40..

George, C.F., Nickerson, P.W., Hanly, P.J., Millar, T.W., and Kryger, M.H. (1987) Sleep apnea patients have more automobile accidents. *Lancet 1* (8556):447..

George, C.F.P., Boudreau, A.C., and Smiley, A. (1996) Simulated driving performance in patients with obstructive sleep apnea. *American Journal of Respiratory and Critical Care Medicine.* Vol. 154, Issue 1, 175-181, July,.

Gillberg, M., Kecklund, G., and Akerstedt, T. (1996) Sleepiness and performance of professional drivers in a truck simulator - comparisons between day and night driving. *Journal of Sleep Research* 5, 12-15..

Grandjean, E. (1988) *Fitting the task to the man: a textbook of occupational ergonomics.* London: Taylor and Francis..

Harris, W., and Mackie, R.R. (1972) *A study of the relationships among fatigue, hours of service, and safety of operations of truck and bus driving.* Report No. BMCS-RD-71-2 prepared for US DOT, FTA, Bureau of Motor Carrier Safety, Washington, DC..

Haraldsson, P.O., Carenfelt, C., Diderichsen, F., Nygren, A., and Tingvall, C. (1990b) Clinical Symptoms of Sleep Apnea Syndrome and Automobile Accidents. *ORL* 52:57 - 62..

Haraldsson, P.O., Carenfelt, C., and Persson, H.E. (1990a) Simulated long-term driving performance before and after uvulopalatopharyngoplasty. *ORL J Otorhinolaryngol Relat Spec.*, 53: 106-110..

Haworth, N.L., Triggs, T.J., and Grey, E.M. (1988) *Driver Fatigue: Concepts, measurements, and crash countermeasures.* Department of Transport and Communications, Federal Office of Road Safety. Report #CR-72..

Hertz, R.P. (1988) Tractor-trailer driver fatality: The role of nonconsecutive rest in a sleeper berth. *Accident Analysis and Prevention, 20,* 431-439..

Hertz, R.P. (1991) Hours of service violations among tractor-trailer drivers. *Accident Analysis & Prevention*, 23(1), 29-36..

Hildebrandt, G., Rohmert, W. and Rutenfranz, J. (1975) The influence of fatigue and rest period on the circadian variation of error frequency in shift workers (engine drivers). In *Experimental Studies in Shift Work*, W.P. Colquhoun, S. Folkard, P. Knauth, and J. Rutenfranz, eds., pp. 174-187. Westdeutscher Verlag, Opladen, West Germany..

Insurance Institute for Highway Safety. (1985) Big Trucks and Highway Safety. Washington, U.S.A..

Jones, I.S. and Stein, H.S. (1987) Effect of driver hours-of-service on tractor-trailer crash involvement. Arlington, VA.; *Insurance Institute for Highway Safety..*

Jovanis, P.P., and Kaneko, T. (1990) *Exploratory analysis of motor carrier accident risk and daily driving patterns.* Transportation Research Group, University of California at Davis, Research Report UCD-TRG-RR-90-10..

Kecklund, G. and Akerstedt, T. (1995) Time of day and Swedish road accidents. *Shiftwork International Newsletter*, Vol. 12(1), pg. 31..

Kogi, K. (1985) Introduction to the problems of shiftwork. In S. Folkard & T.H. Monk (Eds.) *Hours of work: temporal factors in work scheduling*, pp. 165-184. Chichester: John Wiley & Sons Ltd..

Lille, F. (1967) Le sommeil de jour d'un groupe de travailleurs de nuit. *Travail Humain,* Vol. 30, pp. 85-97..

Louwerens, J.W., Gloerich, A.B.M., de Vries, G., Brookhuis, K.A., O'Hanlon, J.F. (1987) The relationship between drivers' blood alcohol concentration (BAC) and actual driving performance during high speed travel. In P.C. Noordzij (Ed.), *Proceedings of the 10th International Conference on Alcohol, Drugs, and Traffic Safety.* Amsterdam: Elsevier..

McDonald, N. (1984) *Fatigue, safety and the truck driver.* Taylor & Francis, London and Philadelphia..

Mackie, R.R. and Miller, J.C. (1978) *Effects of hours of service regularity of schedules, and cargo loading on truck and bus driver fatigue.* U.S. Department of Transport Report No. HS-803 799..

National Transportation Safety Board. (1995) *Safety study: Factors that affect fatigue in heavy truck accidents.* NTSB PB95-917001 - SS-95/01..

Prokop, O. and Prokop, L. (1955) Ermüdung und Einschlafen am Steuer [Fatigue and falling asleep in driving]. *Deutsche Zeitschrift für Gerichtliche Medizin*, 44, 343-355..

Saccomanno, F.F., Shortreed, J.H. and Yu, M. (1996) Effect of driver fatigue on commercial vehicle accidents. In *Truck Safety: Perceptions and Reality*, F. Saccomanno and J. Shortreed (eds.), Institute for Risk Research, University of Waterloo, Waterloo, Canada..

Stoohs, R.A., Guilleminault, C., Itoi, A. and Dement, W.C. (1994) Traffic accidents in commercial long-haul truck drivers: the influence of sleep disordered breathing and obesity. *Sleep*, 7(7), 619-623..

Wilkinson, R.T., Edwards, R.S., and Haines, E. (1966) Performance following a night of reduced sleep. *Psychonomic Science,* 5, 471-472..

Williamson, A.M., Feyer, A., Coumarelos, C., and Jenkins, T. (1992) *Strategies to combat fatigue in the long distance road transport industry: Stage 1: The industry perspective.* CR-108. Transport and Communications, Federal Office of Road Safety, Canberra, Australia..

Wylie, C.D., Shultz, T., Miller, J.C., and Mitler, M.M. (1997) *Commercial Motor Vehicle Driver Rest Periods and Recovery of Performance.* Transportation Development Centre, Safety and Security, Transport Canada, Montreal, Quebec. TP 12850E. April,.

Young, T., Palta, M., Dempsey, J., Skatrud, J., Weber, S., Badr, S. (1993) The occurrence of sleep-disordered breathing among middle aged adults. *New England Journal of Medicine*, 328, 1230-1235..

2

It's Not Just Hours Of Work: Ask The Drivers

Pauline K. Arnold & Laurence R. Hartley,Institute for Research in Safety and Transport, Murdoch University, Western Australia

Introduction

As concern about fatigue as contributor to road crashes has grown considerable efforts have been made, both in Australia and internationally, to estimate the size of the problem. Despite these efforts, it is as yet unclear to what extent fatigue compromises road user safety. Estimates of the size of the problem of fatigue in truck crashes vary widely, from as low as two or three percent to as much as sixty percent. Despite this uncertainty, however, evidence of the seriousness of crash outcomes, especially for involved parties other than truck drivers themselves, suggests that fatigue in the road transport industry merits attention. In both laboratory and field studies, a few variables have been consistently found to be related to performance changes and subjective reports associated with fatigue.

Time of Day and Time Spent Working

There is evidence that crash risk is dependent upon the time of day at which driving is being done (McDonald, 1984; Hamelin, 1987; Haworth, Heffernan & Horne, 1989; Sweatman, Ogden, Haworth, Vulcan & Pearson, 1990; Haworth & Rechnitzer, 1993). Findings such as these led Mitler et al. (1988) to conclude that catastrophic events arising from human performance are more likely to occur in the early hours of the morning (1.00 a.m. to 8.00 a.m.), with a second period of elevated vulnerability, compared to other times, early to mid afternoon (2.00 p.m. to 6.00 p.m.).

Meta analyses undertaken by Folkard (1997) confirmed the presence of circadian effects in crashes, with a peak at about 3.00 a.m.. However, Folkard concluded that this rhythm could not account for all the variation in crash risk, with residual peaks in crashes that are difficult to account for with this factor alone. He proposed that time on task may account for these residual peaks. Folkard found that, after circadian effects have been taken into account, crash risk peaks first after 2 to 4 hours on task, declines and then rises again to reach a peak equal to the 2 to 4 hour peak only after 12 hours on task.

There is evidence that crash risk increases with time on task for truck drivers, though research outcomes are very variable. As well as with time of day, Hamelin (1987) showed that the risk of being involved in a crash increases with the number of hours spent driving, particularly after eleven hours on the road. More conservatively, the Insurance Institute for Highway Safety (1987) reported that driving in excess of eight hours results in a near doubling of crash risk compared to the risk associated with less than two hours of driving. Summala and Mikkola (1994) found that truck drivers who had driven in excess of 10 hours were more likely to be at fault in a crash than not at fault. Kaneko and Jovanis (1992) found crash risk for the truck drivers studied remained fairly constant during the first four hours of driving. Beyond the fourth hour risk fluctuated, but increased overall, peaking at nine hours. These findings should be considered, however, in light of confounding factors such as the 10 to 12 hour driving limits imposed by hours of service regulations applying to drivers in these studies, and changes in the driving environment which may be associated with arrival at an urban destination.

Freund & Vespa (1997) found that hours spent driving was not a strong or consistent predictor of fatigue in their heavy goods vehicle drivers. They did report, however, that self-report ratings of fatigue were related to time on task, suggesting that feelings of fatigue increase with increasing task duration even if there are no noticeable performance decrements. Freund and Vespa did find some evidence of cumulative fatigue across several days of driving, though it was inconsistent. Vigilance performance during the last days on all four of the work schedules studied declined, and drivers rated themselves more fatigued after multiple trips.

Many researchers have concluded that time of day is the most consistent predictor of fatigue; a few others have reported that time spent working has a greater effect; for example, Lin, Jovanis and Yang's (1994) study based on the operational records of a national trucking company. However, Tepas (1994) warned that when work is within reasonable limits, how physically or mentally tired a worker feels is not correlated with the length of their workday.

Sleep Loss

Sleep is a vital and indispensable physiological function, though the specific amount required varies between individuals. An individual who obtains less than their required amount of sleep over several nights acquires a sleep debt. Substantial sleep debts can result in uncontrolled sleep episodes lasting up to several minutes and severe performance decrements (Rosekind et al., 1994). Others studies have produced minimal disruption of performance but have revealed changes in mood and subjective fatigue (for example, Friedman et al., 1977). With increased sleepiness, individuals may become indifferent to the outcome of their performance, and report fewer positive emotions, more negative emotions and a worsening of mood (Rosekind et al., 1994).

Studies show that individuals with an extensive sleep deficit can perform brief tasks, up to about 15 minutes, depending on the difficulty of the task. The ability to sustain performance is greatly reduced, however, as the task becomes more demanding. Performance of tasks that are long in duration, familiar, monotonous, uninteresting and complex deteriorates more than performance of short, novel, interesting and simple tasks (McDonald, 1984; Brown, 1994). Professional driving in rural and remote regions of Western Australia is typified by regular (perhaps weekly trips) for long distances travelled along generally straight, flat roads with little traffic and little variation in scenery. Thus, it has all of the characteristics identified by McDonald and Brown as most problematic for maintaining performance following long hours of work and loss of sleep.

Some research supports the notion that sleep may be the most important factor in fatigue. Neville, Bisson, French, Bol & Storm (1994) examined fatigue experienced by air crew during Operation Desert Storm. Neville et al. did not find the predicted relationship between fatigue scores and cumulative flying time over 30 days. Rather, they reported that fatigue was related to recent flying and sleep times (within the previous 48 hours), and that sleep time was the more strongly associated of the two.

Sweeney et al. (1995) reconstructed activities over the 96 hours prior to crashing for 107 heavy truck drivers. More than half of the crashes (58%) were identified by the researchers as fatigue related, based on drivers' self report or physical evidence at the crash scene. The principle factors related to crash type were: the length of the last sleep period prior to the crash; the total hours of sleep obtained during the prior 24 hours; and the presence of split sleep patterns in the prior 96 hours. A second group of measures that played a lesser role in predicting group membership included: exceeded hours of service limits; hours of driving in the past 24 hours; and hours driving since last sleep. The variables which were least predictive included: duration of most recent duty period; hours on duty and hours driving in prior 48 hours; irregular sleep schedules; and hours since last sleep. Drivers in fatigue related crashes had obtained, on average, 5.5 hours sleep in their last major sleep period prior to crashing. In contrast, drivers whose crashes were attributed to other causes averaged 8 hours of sleep. The fatigue crash drivers obtained, on average, 6.9 hours of sleep in the 24 hours prior to crashing, while the non-fatigue crash drivers averaged 9.3 hours. Thus, the research conducted by Sweeney et al. suggests that the measures that best discriminate between fatigue and non-fatigue crash drivers are those to do with sleep, rather than measures of work. They concluded that drivers involved in fatigue crashes do not obtain adequate sleep.

Research in Western Australia on Fatigue in the Transport Industry

As elsewhere, there is concern in Australia about the part played by fatigue in truck crashes. Fatigue has been found to be a significant contributor to all Australian road crashes, including crashes involving heavy goods vehicles (e.g. Haworth et al., 1989; Ryan & Spittle, 1995; Sweatman et al., 1990; Tyson, 1992). As in several other highly motorised countries, most Australian states employ regulation of driving and related working hours as a mechanism for controlling fatigue in the transport industry. These Australian states restrict truck driving hours to 11 or 12 per 24 hour period, mandate a half hour break within every 5½. hours of driving, and also mandate a 5 or 6 hour continuous rest period in every 24 hour work day. These regulations are enforced by requiring drivers to maintain a logbook of driving hours that must be produced on demand from a police officer. However, regulations regarding driving and related working hours are generally not enforced in Western Australia (though they do exist, they are applied in very limited circumstances).

Within the last decade national road safety bodies have attempted to unify road user and transport regulations throughout Australia. It was suggested that driving hours restrictions for heavy transport drivers should be implemented in Western Australia and the Northern Territory where they are not currently enforced. Government and industry bodies in these states argued that local conditions imposed dramatically different demands on the transport industry, and that the driving hours restrictions operating in the more populous eastern seaboard states of Australian would not suit these conditions. However, little information about current working conditions of truck drivers in Western Australia was available to support these claims.

Information about current work practices was needed to understand transport operations in the unregulated environment, and assess the possible impact of the imposition of the uniform driving hours regulations in WA. Prior to the survey reported here, the only information available came from a national survey of drivers working mainly in Australian states that have prescriptive driving and working hours (90% of the sample) (Williamson, Feyer, Coumarelos & Jenkins, 1992). It seemed likely, however, that work practices are different when driving hours are self regulated. So the research reported here obtained information from truck drivers and transport companies in an Australian state, Western Australia, which does not enforce restrictions on driving hours.

The aim of this study was to obtain information about hours of work and sleep from drivers operating without restrictions on driving hours (referred to as unregulated drivers). It also aimed to have both drivers and companies identify the causes of fatigue experienced by Western Australian drivers, and the countermeasures used to manage it. Where possible these findings are compared with those of Williamson et al's drivers (referred to as regulated drivers). A full report of the findings of the study are presented in Hartley, Arnold, Penna, Hochstadt, Corry and Feyer, (1996) and Arnold, Hartley, Penna, Hochstadt, Corry & Feyer, (1996a & b).

SURVEY METHODOLOGY

During a seven day period, 1249 truck drivers were invited to participate in the survey. Two hundred and one of these drivers had already been interviewed at other locations. Of the remaining 1048 drivers, 638 (60.9%) agreed to be interviewed. Since these drivers were not working prescribed hours they are referred to as unregulated drivers. The absence of information about the numbers of truck drivers operating in the industry in Western Australia make it impossible to determine whether this sample was representative of the industry as a whole. However, the Agricultural Protection Board of Western Australia do monitor vehicles travelling on the major highway connecting eastern and western Australia. From these data, it is estimated that 30% of the drivers who travelled along this highway during the survey period were interviewed.

Data were collected by teams of research assistants using a standardised driver's questionnaire, at six road houses (road side businesses which serve petrol and refreshments, and provide restroom facilities) in WA. Road houses were chosen so as to provide a sample of drivers from each major long distance transport route within the state. Drivers were asked to provide details about their driving and non-driving work schedules and the amount of sleep they had obtained in the past week. This was achieved by obtaining an hour by hour record of activities over the 24 hours prior to the interview, and by obtaining a global estimate of the number of hours spent doing these things over the 7 days prior to the interview. Drivers were asked about the frequency of fatigue related events such as nodding off whilst driving, crashes and near misses. They were also asked how frequently they experienced fatigue, and how frequently it was experienced by, or posed a danger for, others. They were not provided with any description of fatigue.

Management representatives of transport companies operating in WA were also interviewed using a second standardised company questionnaire. Eighty eight companies were contacted by telephone, 84 agreed to participate and nominated someone knowledgeable about their transport operations for the interview. As there is no reliable account of the number of transport companies operating in the state, it is impossible to report what proportion these 84 companies represent. Rather, interviews were solicited from a spread of companies representing large, medium and small, city and country operations with a variety of freight types identified in a traffic count of trucks on major highways. Interviews were conducted by pairs of research assistants, one of whom asked the standardised questions while the other recorded responses. Company representatives were asked how frequently their drivers experienced fatigue, and how great a problem it was for the transport industry.

SURVEY RESULTS

Daily Hours of Driving, Working and Sleep

Estimates of daily driving hours across the heavy transport industry in Western Australian were obtained using two methods. The first relied upon drivers' reports of their activities over the previous 24 hours; that is, it was retrospective. The second calculated daily driving hours on the basis of drivers' predictions about the completion time of their trip; this method was prospective.

Table 1 Percentage of unregulated drivers exceeding 14 hours driving & non-driving work in 24 hours

Activities	Estimates		Average Estimate
	Retrospective over previous 24 hours	*Prospective until end of current trip*	
Drivers exceeding 14 hours driving in 24 hours	33%	43%	38%
Drivers exceeding 14 hours driving & non-driving work in 24 hours	47%	55%	51%

The suggested national hours of service regulations would limit truck drivers' hours of service to 14 hours per day. Table 1 provides estimates of the proportions of drivers who would have exceeded this limit during the week of data collection. The data showed that 33% of drivers had driven in excess of 14 hours in the 24 hours prior to being interviewed (the retrospective estimate of hours of work per day). In contrast, the prospective estimate suggested that 43% of drivers would have driven more than 14 hours by the time they reached their destination. These data suggest that about 38% of the drivers who were interviewed drove more than 14 hours in day.

However, hours of driving alone do not take into account all of the work that drivers may do in a working day. The addition of driving plus other non driving work such as loading and unloading revealed that, based on the retrospective estimate, 47% of drivers had worked for more than 14 hours in the 24 hours prior to their interviews. The prospective estimate shows that 55% of drivers would have worked in excess of 14 hours by the time they reached their destination. These data suggest that about 50% of the drivers who were interviewed worked more than 14 hours in day.

Indeed 29% of drivers had worked more than 16 hours in the 24 hours prior to their interview. The projected estimate showed about 37% would have done more than 16 hours of work by the time they reached their destination. These figures are of some concern because they suggest that one third of drivers do not get sufficient time for rest and sleep in a working day (assuming most people need about 8 hours, and have other things to do besides work and sleep).

Drivers were also asked about their hours of sleep prior to their current trip. About 20% had had less than 6 hours of sleep before the current trip. (The mean was 8.25 hours; this is similar to the 7.5 hours of sleep before departure obtained by Williamson et al's (1992) mainly regulated drivers.) The concern about drivers' lack of opportunity to obtain adequate sleep and rest is validated by the average daily hours of sleep reported over the week preceding the interviews. Only working days were included in analyses of hours of sleep obtained by drivers, and sleep preceding days off work was excluded. Consistently, about 1 in 20 drivers (5%) reported having not slept on at least one workday of the preceding 7 days. A further 7.5% reported less than 4 hours of sleep on at least one workday. Thus, one in eight drivers (12.5%) obtained less than 4 hours of sleep on one or more of their workdays in the week preceding their interview. These data also show that about 30% of drivers obtained less than 6 hours of sleep on at least one working day within that week.

Weekly Hours of Driving and Working

The suggested national hours of service regulations would limit truck drivers' hours of service to 72 hours per week. The distribution of hours driven per week was obtained from the data collected for the seven days prior to the interview. About one third of drivers drove between 40 and 60 hours in the week preceding their interview. Another one in five drove 60 to 80 hours. Nearly 2% of respondents drove in excess of 100 hours in the week preceding their interview. About one in six (17.5%) drove more than 72 hours in the week.

Again, hours of driving alone does not take into account all of the work that drivers may do in a working week. The data show that about 5% of drivers worked in excess of 100 hours in the seven days preceding their interview. More than one in ten (11%) did both driving and non driving work for more than 90 hours per week, and nearly one third of drivers (30%) worked in excess of 72 hours.

The national survey conducted by Williamson et al. (1992) reported that about 35% of their regulated drivers exceeded 72 hours of driving per week. In so far as the present data from an unregulated state and Williamson et al's data are comparable, a greater proportion of regulated truck drivers' work long hours each week. (Williamson et al. did not report daily driving hours.)

Hazardous Events During Driving

This section deals with how often potentially hazardous events, that may be fatigue related, occur to drivers. These events include nodding off, near misses and accidents. Thirty five drivers (5.5%) reported they had had an accident in the 9 months since the beginning of the year in which the study was conducted, 1995. These drivers were asked whether their accidents were related to fatigue and four (11.4%) said they were. (This is comparable to Maycock's (1995) data for U.K. car drivers, of whom 15% cited tiredness as a contributor to their accidents on major motorways.)

Drivers were also asked whether they had experienced any hazardous events resulting from fatigue on their present trip. None of the 35 drivers who reported accidents also reported having experienced a fatigue related event on their current trip. Of the 638 drivers, another 33 (5.2%) stated that a potentially dangerous event had occurred. The most common events were nodding off (or falling asleep whilst driving) and near misses, and together these accounted for 60.6% of all reported events (see table 2).

Table 2 Fatigue-related events reported on present journey

Events	Frequency	Percentage
Nodded off	11	33.3
Near miss	9	27.3
Ran off road	3	9.1
Collision	2	6.1
Other	8	24.2
Total	33	

Modelling of Hazardous Events

The data related to hazardous events were submitted to segmentation modelling using CHAID. Fatigue related events on the current trip, accidents over the past nine months and a combination of these two variables were used as dependent variables. Variables included as potential predictors were:

- Driving operations characteristics (9 variables related to route driven, employment status, freight type, loading / unloading and schedule)
- Hours of driving over previous week (mean hours driving per work day, and longest hours driving on any day)
- Hours of non-driving work over previous week (mean hours non-driving work per work day, and longest non-driving work driving on any day)
- Hours of driving plus non-driving work over previous week (mean hours driving plus non-driving work per work day, and longest driving plus non-driving work driving on any day)
- Hours of sleeping (hours sleeping prior to current trip, mean hours sleeping on work days and days preceding work days over the past week, and shortest hours sleeping on any work day and day preceding any work day over the past week)
- Number of days in the past week containing driving, non-driving work, and both driving and non-driving work.

Hours of driving, non-driving work and sleeping for *yesterday* were included in the analyses of events occurring on the current journey because of the time frame of reporting, but were not included in analyses of variables with a longer frame of reference (for example, events occurring since beginning of the year).

Segmentation modelling identified three groups of drivers as having experienced more fatigue related events during their current journey compared to others. Overall these events were reported by about 5% of drivers, but much greater proportions of these three groups reported such events. Mean hours of sleep before and on work days was the best predictor of these events: CHAID divided the sample into three initial groups - those who had no more than 6 hours sleep on average, 6 to 10 hours, and more than 10 hours. Figure 1 shows the tree diagram of the significant predictor variables identified by CHAID.

Figure 1 CHAID Segmentation Model for Fatigue Related Events on the Current Journey

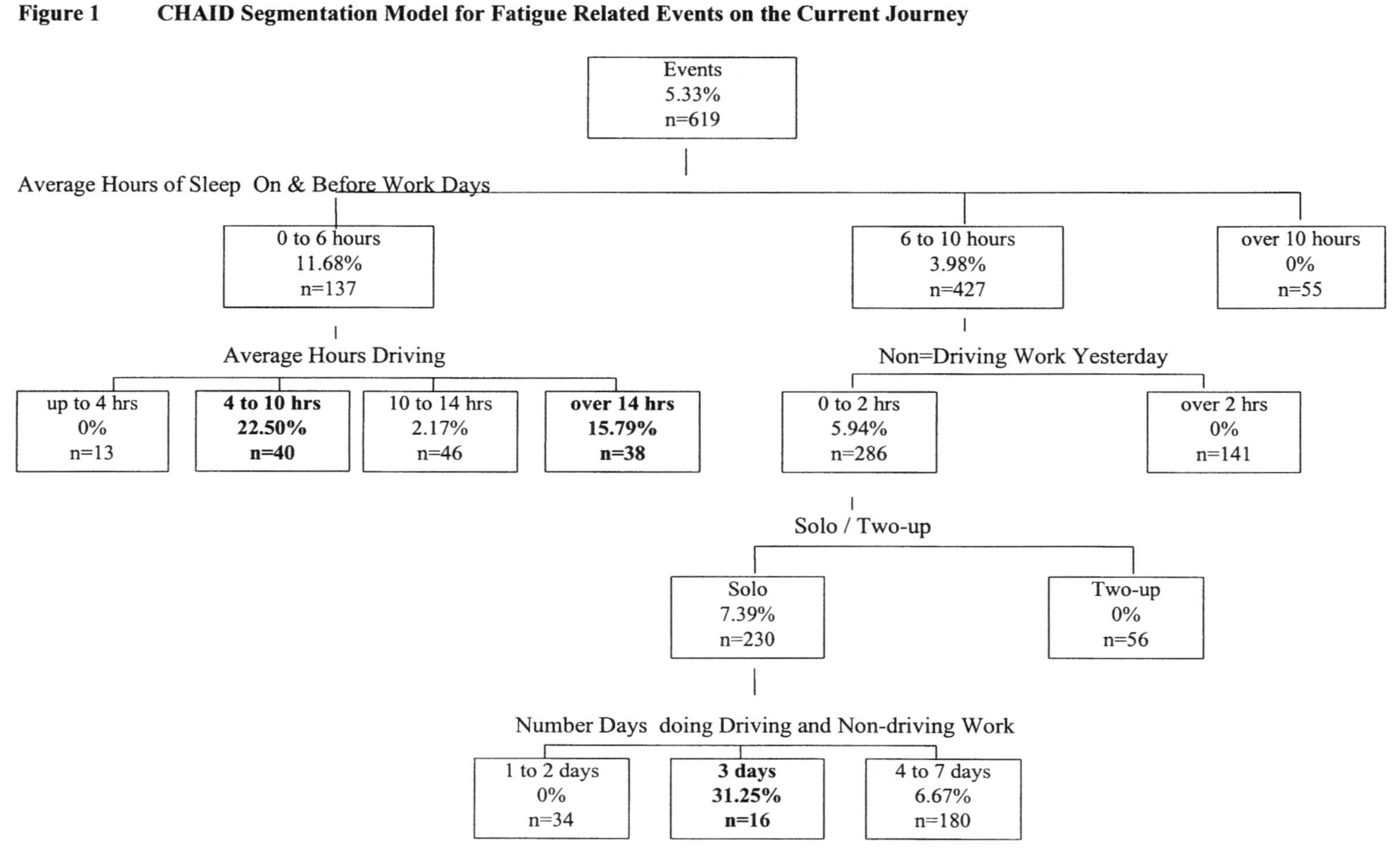

The segmentation modelling showed that the following groups reported fatigue related events on the current journey at a greater rate than other drivers:

- solo drivers who had slept 6 to 10 hours per night in the week prior to their current trip, and who had worked for only three days in that week, and did less than two hours non driving work prior to the trip - 5.9 times more events than other drivers;
- drivers who had slept less than 6 hours per night and driven 4 to 10 hours per working day in the week prior to their current trip - 4.2 times more events than other drivers;
- drivers who had slept less than 6 hours per night and driven more than 14 hours per working day in the week prior to their current trip - 3.0 times more events than other drivers.

Similarly, the absence / occurrence of accidents since the beginning of the year was submitted to segmentation modelling using CHAID and logistical regression. No adequate predictors were found in the CHAID analysis; the small numbers of accidents reported constitute a problem for analyses. To overcome this problem, it was considered reasonable to combine accidents and fatigue related events on the current trip, as the latter may well have become accidents if the driving environment was not so forgiving on the Western Australian roads used by the surveyed drivers (for example, they are mostly low traffic density). However, it must be kept in mind during discussion of these results that this new variable is related to the one previously described. Overall, occurrence of hazardous events was reported by 10.7% of drivers (N=68).

As it did with fatigue related events on this trip, CHAID identified average hours slept on or before workdays in the previous week as the best predictor for hazardous events, dividing the sample into two initial groups - those who had no more than 6 hours sleep on average and those who had more than 6 hours (see figure 2).

The segmentation modelling showed that the following groups reported hazardous events at a greater rate than other drivers:

- drivers who had, on average, no more than 6 hours sleep per night in the week prior to their current trip and were interviewed as they drove to the Northwest of Western Australia - 2.4 times more events than other drivers;
- drivers who had more than 6 hours sleep per night, drove solo, and did non-driving work for 6 hours or less on working days - 1.5 times more events than other drivers.

Figure 2 CHAID Segmentation Model for Hazardous Events (Combined Events on This Journey & the During Previous Nine Months)

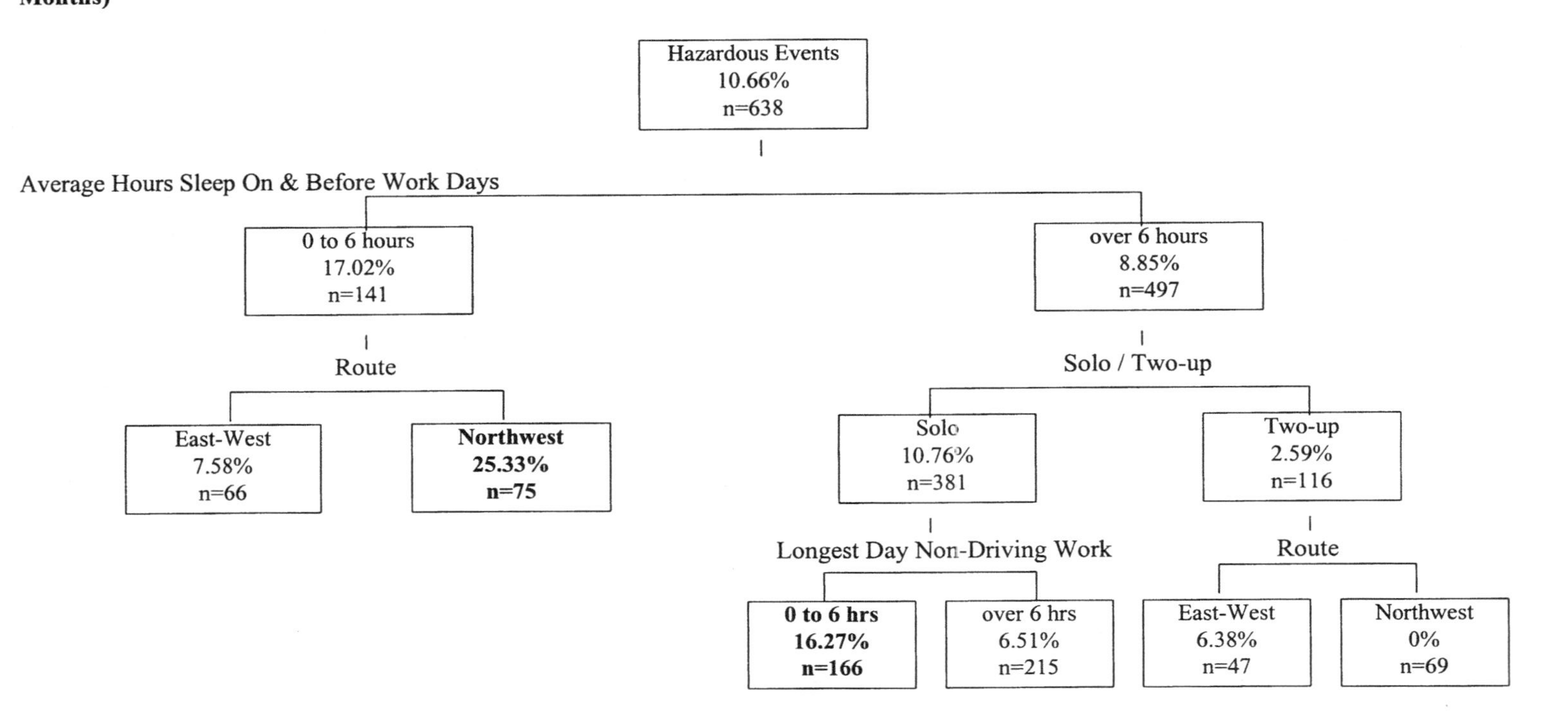

Frequency of Nodding Off and Near Misses Since the Beginning of the Year

Drivers were asked to indicate the frequency with which they had either nodded off or fallen asleep whilst driving, in the 9 months preceding the study. Two thirds of the drivers (66.7%) indicated they had not fallen asleep whilst driving in that time (see table 3). A smaller proportion (10.7%) indicated they had fallen asleep occasionally. About one in thirty five drivers (2.8%) reported that it had happened often or very frequently.

Table 3 Frequency of nodding off & near misses from January to September, 1995

	Nodding Off		Near Misses	
Frequency	***Frequency***	***Percentage***	***Frequency***	***Percentage***
Very frequently	9	1.4	12	1.9
Often	9	1.4	24	3.8
Occasionally	68	10.7	68	10.7
Rarely	105	16.5	113	17.7
Never	425	66.6	402	63.0
No Response	22	3.4	19	3.0
Total	638		638	

Drivers were also asked about how often they had had near misses in the 9 months since the beginning of the year. Again, almost two thirds of drivers (62.5%) said they had not had a near miss whilst driving during the year of the study (table 3). One in ten (10.3%) said they had had occasional near misses. About one in twenty drivers (5.8%) reported they often or very frequently had near misses.

The report by about 35% of drivers that they had near misses in the last nine months should be considered in context of the frequency with which all drivers encounter near misses. Though empirically based estimates of the frequency of near misses are unavailable, anecdotal evidence suggests almost all drivers experience near misses occasionally, and some drivers experience them regu!arly. Indeed, it seems likely that the data gathered about the frequency of nodding off and near misses events over the previous nine months under represents their true frequency. Evidence of this is apparent in the comparison between the two questions about fatigue related events on the current trip and the frequency of nodding off in the last 9 months. One of the nine drivers who reported having nodded off on their present trip also reported that he never nodded off or fell asleep whilst driving. This anomaly highlights the shortcomings of self-report data.

Modelling of Nodding Off and Near Misses

The data about nodding off and near misses were submitted to segmentation modelling. Variables included as potential predictors were the same as those used for analyses of accidents and fatigue related events on the current trip. Nodding off and near misses were treated as ordinal variables in their analyses.

Segmentation modelling identified three groups of drivers who reported higher ratings of the frequency of nodding off while driving compared to other drivers. Overall, the mean rating for frequency of nodding off was 1.5, indicating a rating between never and rarely (see figure 3 for the tree diagram produced). As it did with hazardous events, CHAID identified average hours slept on or before workdays in the previous week as the best predictor, dividing the sample into three initial groups - those who had no more than 6 hours sleep, 6 to 8 hours and more than 8 hours.

The segmentation modelling showed that the following groups gave higher ratings of the frequency of nodding off while driving than other drivers:

- drivers who had slept less than 6 hours per night in the week prior to their current trip, and did driving and non driving work for 20 to 24 hours on at least one day in the previous week - 2.3 times higher ratings than other drivers;
- drivers who had slept less than 6 hours per night and did non-driving work for 14 or more hours on at least one day in the previous week - 1.5 times higher ratings than other drivers;
- drivers who had slept 6 to 8 hours per night, and had worked in excess of 18 hours on at least one work day - 1.5 times higher ratings than other drivers.

Segmentation modelling identified two groups of drivers who reported higher frequencies of near misses compared to other drivers. Overall, the mean rating for frequency of near misses was 1.6, an average rating between never and rarely (see figure 4). CHAID identified route driven as the best predictor of ratings of the frequency of near misses over the past nine months, and amongst drivers travelling to the Northwest identified two segments whose ratings were greater than the sample average.

Figure 3 CHAID Segmentation Model for Frequency of Nodding Off Since the Beginning of the Year

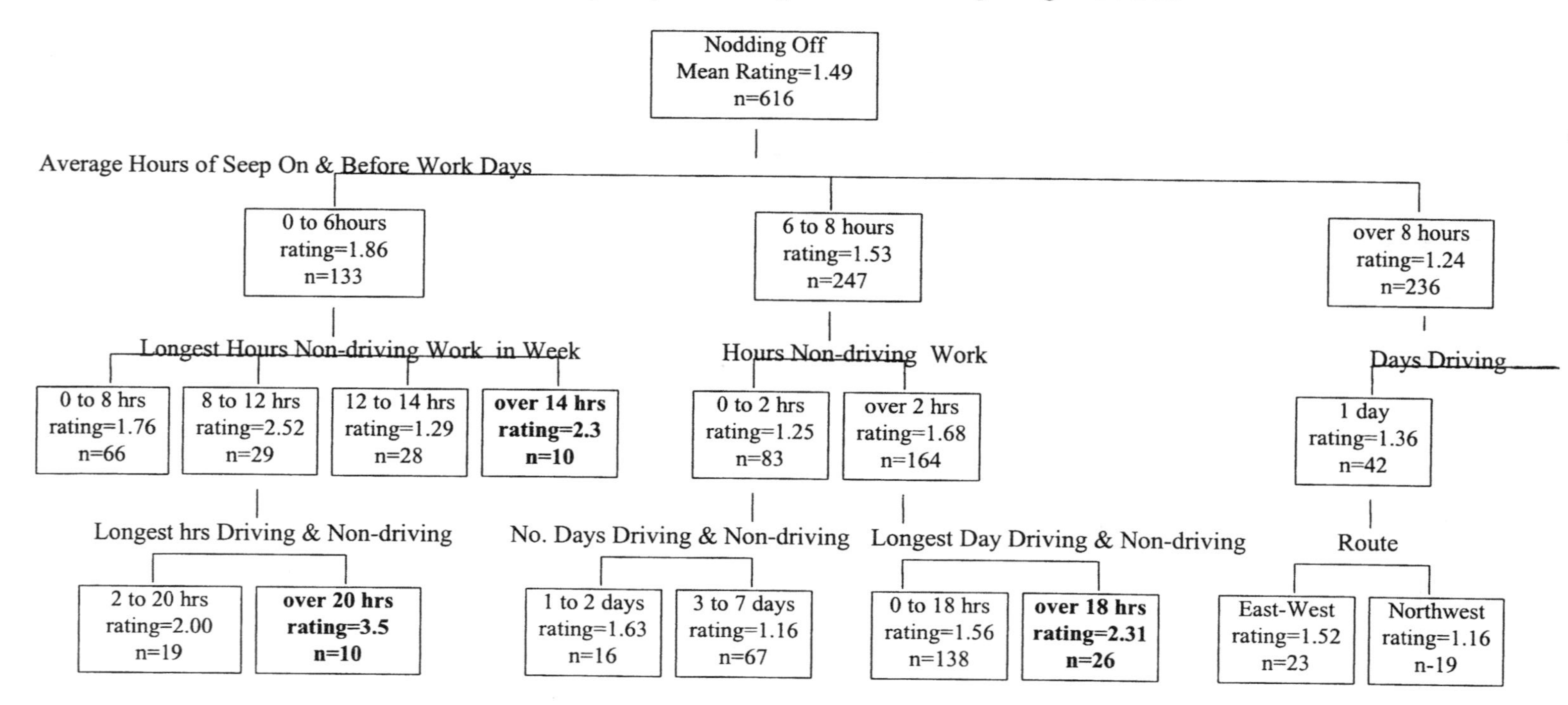

Figure 3 continued CHAID Segmentation Model for Frequency of Nodding Off Since the Beginning of the Year

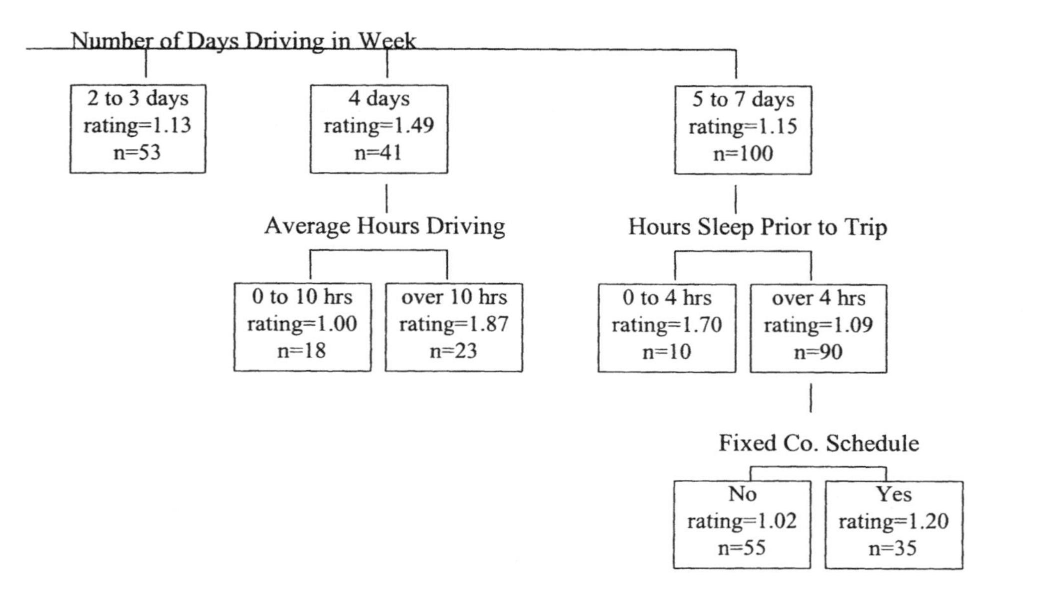

Figure 4 **CHAID Segmentation Model for Frequency of Near Misses Since the Beginning of the Year**

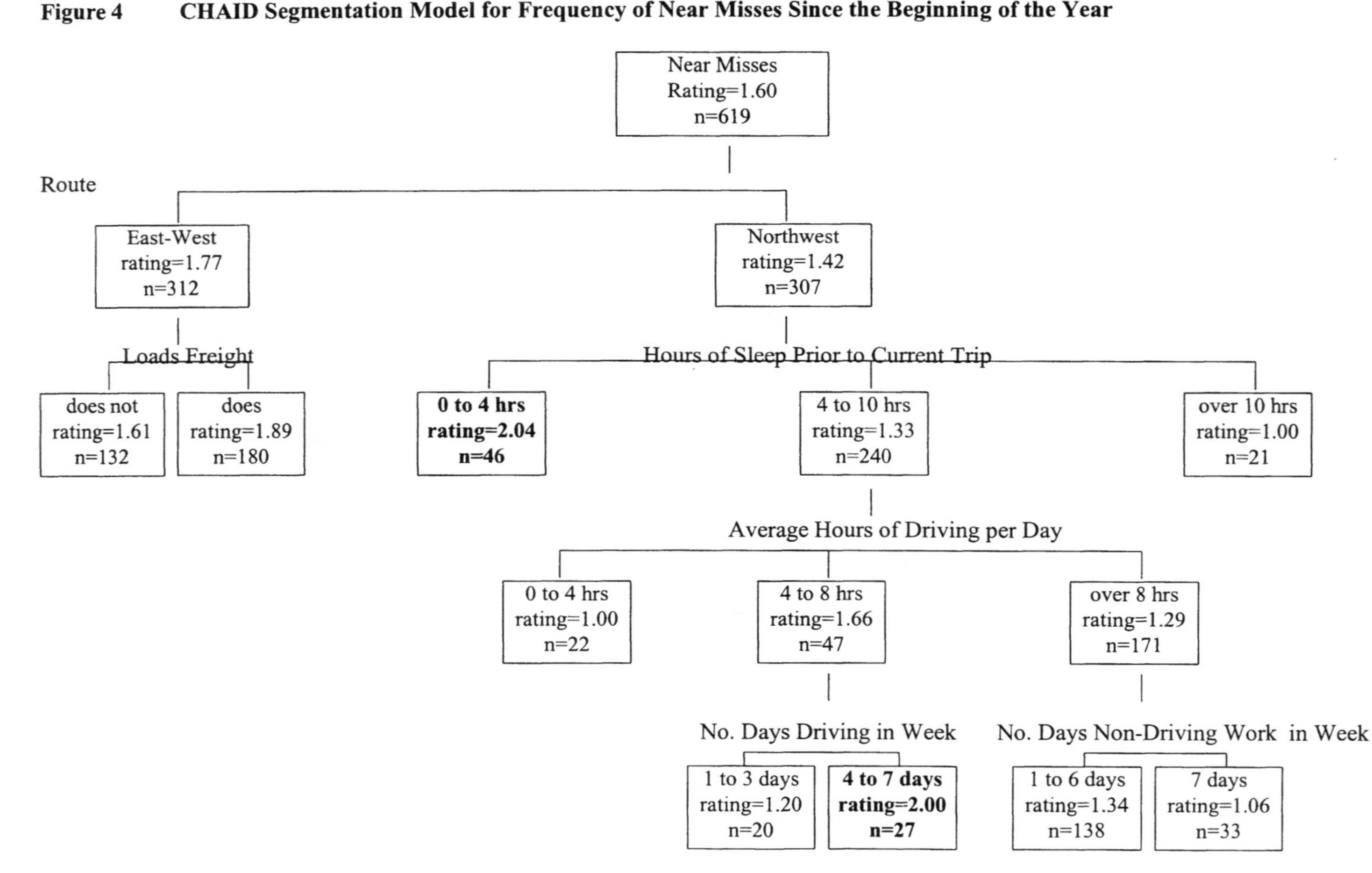

The segmentation modelling showed that the following groups gave higher ratings of the frequency of near misses while driving than other drivers:

- drivers to the Northwest who had slept less than 4 hours prior to departure on their present journey - 1.3 times higher ratings than other drivers;
- drivers to the Northwest who had slept 4 to 10 hours prior to departure, and on average drove 4 to 8 hours for 4 to 7 days in the previous week - 1.3 times higher ratings than other drivers.

Use of Alcohol and Drugs to Manage Fatigue & Their Modelling

Drivers were asked whether they used any of a list of strategies provided by the interviewer to lessen fatigue or tiredness. Drinking alcohol and taking drugs were two of the items in that list. Almost one in twenty drivers (4.5%) said they drank alcohol to lessen their fatigue. One in six drivers (16.5%) said they took pills or drugs. Overall, 19.9% of drivers reported the use of either alcohol or drugs, or both, to control fatigue.

The data about alcohol and drug use were submitted to segmentation modelling. Variables included as potential predictors were the same as those used for analyses of accidents and fatigue related events on the current trip.

Segmentation modelling identified two groups of drivers who reported more use of alcohol to lessen fatigue compared to other drivers. Overall, alcohol use was reported by about 4.6% of drivers, but much greater proportions of these two groups reported its use (see figure 5 for the tree diagram produced). CHAID identified shortest hours of sleep on or before any workday in the previous week as the best predictor, dividing the sample into two, those who had 6 hours or less and those who had more.

The segmentation modelling showed that the following groups reported higher incidence of alcohol use to lessen fatigue while driving than other drivers:

- Northwest drivers who had slept only 2 to 4 hours on or before at least one night in the week prior to their current trip - 2.3 times higher incidence than other drivers;
- East-West drivers who had slept less than 6 hours on or before at least one night in the week prior to their current trip - 2.0 times higher incidence than other drivers.

Segmentation modelling identified four groups of drivers who reported more use of drugs to lessen fatigue compared to other drivers. Overall, drug use was reported by about 16.5% of drivers, but much greater proportions of these four groups reported its use (see figure 6 for the

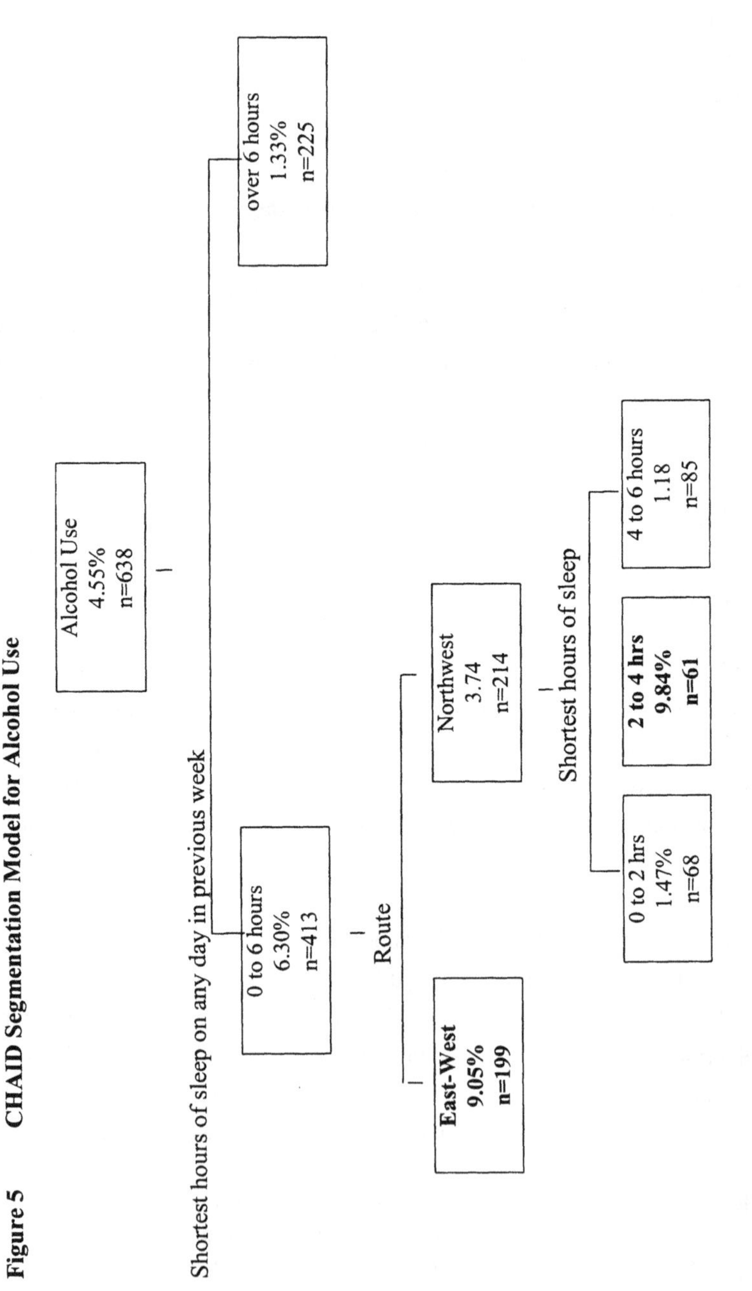

Figure 5 CHAID Segmentation Model for Alcohol Use

Figure 6 CHAID Segmentation Model for Drug Use

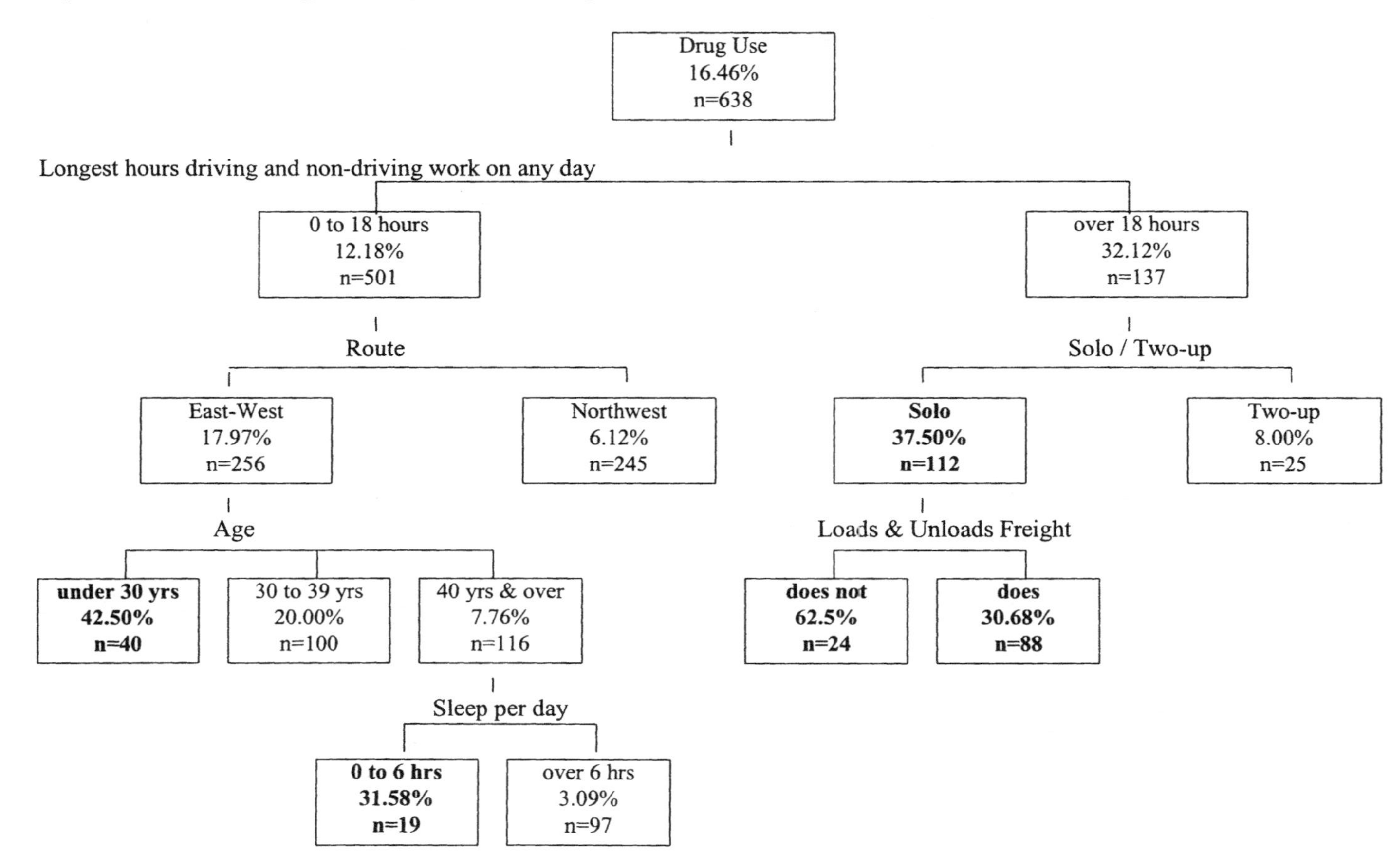

tree diagram). CHAID identified longest hours of driving plus non-driving work on any workday in the previous week as the best predictor, dividing the sample into two, those who had done up to 18 hours of work and those who had done more.

The segmentation modelling showed that the following groups reported higher incidence of drug use to lessen fatigue while driving compared to other drivers:

- Solo drivers whose longest workday in the previous week was more than 18 hours and who did not assist with loading and unloading their freight - 3.8 times higher incidence than other drivers;
- East-West drivers aged under 30 years whose longest workday in the previous week was less than 18 hours - 2.6 times higher incidence than other drivers;
- East-West drivers aged 40 years and over whose longest workday in the previous week was less than 18 hours and who had, on average, less than 6 hours sleep on or before work days - 1.9 times higher incidence than other drivers;
- Solo drivers whose longest workday in the previous week was more than 18 hours and who did assist with loading and unloading their freight - 1.9 times higher incidence than other drivers.

Contributors to, & Strategies Used to Manage, Fatigue

Drivers were asked to identify the three main causes of their fatigue, and company representatives were asked to identify three main causes of fatigue for their drivers. The principal causes of fatigue identified by drivers and company representatives were, driving long hours, loading, delays in loading, lack of sleep, tight schedules and dawn driving (see table 5). There were some differences between the causes identified by drivers and company representatives. While, nearly 70% of company managers thought that long hours of driving are a main contributor to fatigue, less than 40 % of drivers named long hours. In contrast, more drivers blamed both loading the truck and delays in loading for their fatigue while few company representatives identified these as causes.

About half the company representatives thought lack of sleep contributed to their drivers' fatigue, while about one third of drivers identified this as a contributor to their own fatigue. One explanation for this difference may lie in respondents' different experience of the problem. While for company representatives lack of sleep may seem an obvious contributor, drivers' more direct experience of fatigue while driving may make them more likely to identify things which cause them to lose sleep.

Table 5 Contributors to fatigue reported by unregulated drivers (N = 638) and company representatives (N = 84) (percentages in parentheses)

Contributors	Drivers		Companies	
Driving long Hours	244	(38.2)	58	(69.0)
Loading/Unloading	213	(33.4)	18	(21.4)
Delays in loading	206	(32.4)	6	(7.1)
Lack of sleep	206	(32.4)	41	(48.8)
Over tight delivery schedules	135	(21.2)	30	(35.7)
Driving between 2-5 am	135	(21.2)	16	(19.0)
Breakdowns	97	(15.2)	8	(9.5)
Irregular trip schedules	63	(9.9)	11	(13.1)
Poor rest in truck	57	(8.9)	9	(10.7)
Irregular rest hours on road	52	(8.2)	6	(7.1)
Inexperience	36	(5.6)	16	(19.0)

Williamson et al. (1992) also found driving long hours, lack of sleep, loading and dawn driving were principal causes of fatigue in regulated drivers, though long hours and lack of sleep were nominated more frequently by regulated drivers compared to unregulated drivers. Prescriptive driving hours appears to have done little to diminish the most common causes of fatigue in states with this regulation.

Drivers were asked what strategies they used to manage their driving fatigue. Many said they obtained a good night's sleep prior to departure, or could pull over to the side of the road to rest when they felt tired (see table 6). It is noteworthy, however, that about 10% of the unregulated drivers reported they are unable to get regular sleep, a good night's sleep prior to trips or pull over to rest on the roadside during trips. Thus, these drivers appear to be denied use of the most natural countermeasure to fatigue, sleep. Drinking caffeine beverages was reported by more than two thirds of unregulated drivers and almost half reported use of nicotine products. One in six reported using drugs to manage their fatigue. About the same proportion of Williamson et al's (1992) drivers reported caffeine and nicotine as useful. However, the regulated drivers reported taking drugs about twice as frequently as unregulated drivers, and those taking drugs reported it to be the most effective countermeasure they used.

Table 6 Countermeasures drivers used to combat fatigue

Countermeasures	Frequency	Percentage
Pull over when tired	521	81.7
Drink caffeine beverages	437	68.5
Good night's sleep before departure	398	62.4
Smoke cigarettes, chew nicotine gum	279	43.7
Keep fit & healthy	269	42.2
Sleep regular hours	205	32.1
Eat lollies, chocolates	199	31.2
Take pills or drugs	104	16.3
Drink alcohol	29	4.5
Other	32	5.0
Nothing	3	0.5
Number of Drivers	638	

Drivers who were employed by companies and the company representatives were asked what countermeasures to fatigue companies might use, but were not currently doing so. Many drivers and company representatives thought there was nothing more their company should be doing to minimise fatigue (35.0% and 42.9% respectively). However, the remaining drivers endorsed a number of strategies they thought their company should be doing. These included giving them more time to do trips, providing more assistance with loading and unloading, and giving drivers greater control over their schedules. Most commonly, company representatives thought their company should be providing more education about fatigue. All other suggested strategies received little support from the company representatives.

Table 7 Strategies that drivers (N = 348) and company representatives (N = 84) suggest companies should, but do not, use to minimise fatigue (percentages in parentheses)

Countermeasures	Driver Responses		Company Responses	
Nothing	122	(35.0)	36	(42.9)
More time to do trip	66	(19.0)	6	(7.1)
Provide help with loading / unloading	65	(18.7)	1	(1.2)
Provide education	53	(15.2)	24	(28.6)
Give driver control over schedule	49	(14.1)	3	(3.6)
Allow driver more time off	48	(13.8)	4	(4.8)
Enough sleep time on trip	38	(10.9)	4	(4.8)
Use shared driving arrangements	18	(5.2)	3	(3.6)
Minimise night driving	16	(4.6)	5	(6.0)
Other	12	(3.4)	9	(10.7)

Allowing drivers more time off work and providing enough time for sleep were endorsed by relatively few drivers (13.8% and 10.9%), suggesting drivers believe they already have both. In contrast, these suggested strategies were endorsed by even fewer company representatives

(4.8%), suggesting that either they believed their company already gave drivers adequate time off work and provided for sleep, or these strategies were things their company could not afford to do.

Discussion

The survey findings regarding hours worked per day show that a large proportion of unregulated truck drivers in this study work in excess of 14 hours in typical working days. Conversely, this suggests that many drivers obtain less than 10 hours off between work shifts. Concern about these conditions of work is reinforced by the finding that one in eight drivers get less than four hours sleep on one or more of their working days during a week. However, the hours these unregulated drivers spend working and sleeping are comparable with those of drivers in states which regulate driving hours (Hartley et al., 1996; Williamson et al., 1992). This suggests that regulations to limit driving hours do not appear to reduce the time spent at work in the road transport industry.

Average hours of sleep over the week prior to being interviewed was the strongest predictor for self reports of the frequency of hazardous events and nodding off while driving over the previous nine months. Shortest hours of sleep was the best predictor for self reported alcohol use as a means of managing fatigue. Duration of sleep was also a predictor for frequency of near misses and use of drugs to manage fatigue. The findings here are consistent with those obtained by Neville et al. (1994) and Sweeney et al. (1995). These researchers concluded that the measures that best discriminate between fatigue and non-fatigue operators are those to do with sleep, rather than measures of work duration or timing.

These findings regarding sleep are interesting but problematic in that the data obtained in this study were self reports of work activities and event frequencies. Inconsistencies in some drivers' reports provided some evidence of these problems within these data. Drivers were asked to recall their work history over seven days; some drivers' records become increasing inaccurate over the seven days, necessitating their removal from the data set. It is also likely that the frequencies of events such as near misses and nodding off were understated by drivers as, in the first case, drivers may simply have forgotten minor events and, in the second case, some drivers may not be aware of sleepy episodes which have no adverse consequences. Furthermore, though drivers were assured that the information they provided would be kept confidential, they are likely to have been reticent to respond honestly to questions which had legal implications, such as for drug taking and extremely long hours of continuous driving.

Thus, caution must be exercised in discussing these results. It is, nevertheless, interesting to note the consistent presence of a measure of sleep as a stronger predictor than other variables.

Despite reservations, these findings support the view that because sleep cannot be mandated by imposition of regulations, the alternative is to ensure drivers have the *opportunity* to obtain adequate sleep and educate them on the importance of making the most of that opportunity. The success of making such provisions will depend very much upon drivers' appreciation of the importance of sleep. Some of the data collected here suggest both drivers and companies do not adequately consider sleep as an important factor in fatigue causation when it is missed. About half the company representatives thought lack of sleep contributed to fatigue, while about one third of drivers thought so. Relative to other causes identified by both drivers and companies, neither group accorded it the importance the modelling analyses suggest it deserves. Drivers do seem to recognise the importance of sleep as a countermeasure to fatigue and many reported that they tried to get a good night's sleep prior to journeys. Yet, the data regarding hours of sleep during the week before being interviewed and prior to the current trip show that up to one half of drivers are unsuccessful in obtaining 7 hours of sleep prior to working days.

As important as it seems to be, lack of sleep was not the only cause of fatigue identified by drivers. The breadth of causes of fatigue identified by drivers highlights the necessity for a multi-faceted fatigue management system in the road transport industry. Effective fatigue management requires attention to all causes of fatigue, not just driving hours and opportunities for sleep. Further, the disparity between drivers' and companies' views about causes and effective countermeasures will need to be resolved for, unless there is agreement on what factors need controlling and how it should be done, effective management will be hard to achieve.

These drivers and company managers rated themselves and their own drivers as experiencing problems with fatigue much less frequently than other companies' drivers (reported in Arnold et al., 1996a). Similar findings were reported by Williamson et al. (1992). These findings suggest that initiatives to raise awareness of the fatigue problem will have difficulty changing many drivers' and employers' views about personal vulnerability even though there may be increased awareness of the problem in general. Initiatives to raise awareness of fatigue should, in addition, address the problem of drivers' perceived personal invulnerability to fatigue. The transport industry Code of Practice currently being developed in Western Australia aims to raise awareness in all industry participants.

Acknowledgments: This research study was commissioned by the Western Australian Road Transport Industry Advisory Council and funded by the Traffic Board of Western Australia.

REFERENCES

Arnold, P.K., Hartley, L.R., Penna, F., Hochstadt, D., Corry, A. & Feyer, A-M. (1996a) *Fatigue in the WA transport industry: The drivers' perspective.* Report Number 118. Institute for Research in Safety & Transport, Perth, Western Australia.

Arnold, P.K., Hartley, L.R., Penna, F., Hochstadt, D., Corry, A. & Feyer, A-M. (1996b) *Fatigue in the WA transport industry: The company perspective.* Report Number 119. Institute for Research in Safety & Transport, Perth, Western Australia.

Brown, I. (1994) Driver fatigue. *Human Factors,* **36**, 298-314.

Folkard S. (1997) Black times: Temporal determinants of transport safety. *Accident Analysis and Prevention,* **29**, 417-430.

Freund, D.M. & Vespa, S. (1997) The driver fatigue and alertness study: From research concept to safety practice. *Proceedings of the international large truck safety symposium.* University of Tennessee Transportation Centre, Knoxville.

Friedman, J., Globus, G., Huntley, A., Nulaney, D., Naitoh, P. & Johnson, L.C. (1977) Performance and mood during and after gradual sleep reduction. *Psychophysiology,* **14**, 245-250.

Hamelin, P. (1987) Lorry drivers' time habits in work and their involvement in traffic accidents. *Ergonomics,* **30**, 1323-1333.

Hartley, L.R., Arnold, P.K., Penna, F., Hochstadt, D., Corry, A. & Feyer, A-M. (1996) *Fatigue in the WA transport industry: The principle and comparative findings.* Report Number 117. Institute for Research in Safety & Transport, Perth, Western Australia.

Haworth, N.L.; Heffernan, C.J. & Horne, E.J. (1989) *Fatigue in truck crashes.* Monash University Accident Research Centre, Melbourne.

Haworth, N.L. & Rechnitzer, G. (1993) *Description of fatal crashes involving various causal variables.* Report number CR119. Federal Office of Road Safety, Canberra.

Insurance Institute for Highway Safety (1987) More than 8 hours behind the wheel? Twice the risk. *Status Report,* **22**, 1-2.

Kaneko, T. & Jovanis, P.P. (1992) Multiday driving patterns and motor carrier accident risk: a disaggregate analysis. *Accident Analysis and Prevention,* **24**, 437-456.

Lin, T-D., Jovanis, P.P. & Yang, C-Z. (1994) Time of day models of motor carrier accident risk. *Transportation Research Record,* 1467. National Academy Press, Washington.

Maycock, G. (1995) *Driver sleepiness as a factor in car and HGV accidents.* TRL Report 169. Transport Research Laboratory, Crowthorne (UK).

McDonald, N. (1984) *Fatigue, safety and the truck driver.* Taylor & Francis, London.

Mitler, M.M., Carskadon, M.A., Czeisler, C.A., Dement, W.C., Dinges, D.F. & Graeber, R.C. (1988) Catastrophes, sleep and public policy: Consensus report. *Sleep,* **11**, 100-109.

Neville, K.J., Bisson, R.U., French, J. Bol, P.A. & Storm, W.F. (1994) Subjective fatigue in C-141 aircrews during Operation Desert Storm. *Human Factors,* **36**, 339-349.

Rosekind, M.R., Co, E.L., Johnson, J.M., Smith, R.M., Weldon, K.J., Miller, D.L., Gregory, K.B., Gander, P.H. & Lebacqz. J.V. (1994) Alertness management in long-haul flight operations. *Proceedings of the thirty-ninth corporate aviation safety seminar.* Flight Safety Foundation, St Louis, Missouri.

Ryan, G.A. & Spittle, J. (1995) *The frequency of fatigue in truck crashes.* In Proceedings of the National Road Safety Research and Enforcement Conference: Promaco Conventions, Perth, WA.

Summala, H. & Mikkola, T. (1994) Fatal accidents among car and truck drivers: Effects of fatigue, age and alcohol consumption. *Human Factors,* **36**, 315-326.

Sweatman, P.F., Ogden, K.J., Haworth, N., Vulcan, A.P. & Pearson, R.A. (1990) *NSW Heavy Vehicle Crash Study Final Technical Report.* Report number CR 92, CR5/90. Federal Office of Road Safety, Canberra.

Tepas, D.I. (1994) The special relevance of research on shiftworkers employed in manufacturing to work schedule problems in the transport industry. *Proceedings of the 12th triennial congress of the International Ergonomics Association.*

Tyson, A.H. (1992) *Articulated truck crashes in South Australia 1978-1987: A discussion of countermeasures.* Office of Road Safety, South Australian Department of Road Transport, Adelaide, SA.

Williamson, A.M.; Feyer, A.M.; Coumarelos, C.; Jenkins, T. (1992) *Strategies to combat fatigue in the long distance road transport industry.* Federal Office of Road Safety, Canberra.

3

SLEEP SURVEY OF COMMUTERS ON A LARGE US RAIL SYSTEM

Joyce A. Walsleben, RN, Ph.D[1]*., Robert G. Norman, MS* [1]*, Ronald D. Novak, Ph.D., M.P.H.*[2]*, Edward B. O'Malley , Ph.D.*[1]*, David M. Rapoport, MD*[1]*, Kingman P. Strohl, MD*[2]
Departments of Medicine/Division of Pulmonary and Critical Care Medicine, NYU/Bellevue Medical Centers, New York [1]*and Case Western Reserve University, Cleveland, Ohio*[2]

INTRODUCTION

The issue of fatigue in transportation and the work place is of particular interest at this conference. Much of the focus is toward the driver, be they a trucker, air or ship pilot. However, an unexamined group of work place participants, particularly in urban/metropolitan areas is the daily mass transit commuter. We note that in many areas of the world, a large proportion of employed adults commute daily to and from work by means which may restrict nocturnal sleep time. While restricting nocturnal sleep times, some modes, particularly the train, offer the opportunity to add sleep through napping. Further, there have been a number of surveys of adult workers which show a high prevalence of sleep disorders and poor sleep habits in the general population (The National Commission on Sleep Disorders 1993, The Gallup Organization 1997, Young et al 1993) as well as an association between sleep disorders and cardiovascular disease (Hla et al 1994). Little is known about the impact of commuting on sleep, sleep and napping behavior, symptoms of sleep disorders and the association of commuting with general health. To address these issues, we designed a questionnaire to assess general health, sleep-related health habits and symptoms of sleep disorders among a large cohort of commuters in a US suburban area serviced by the Long Island Rail Road (LIRR). The LIRR transports over 100,000 persons daily on each of the morning or evening commutes. We report the prevalence of difficulties with sleep and wakefulness and their correlations with commuting times, general health and potential sleep disorders.

METHODS

Twenty-two volunteers distributed postage-paid, mail-back questionnaires (N=21,000) at 15 different stations, encompassing rides of 30-90 minutes during the westbound morning rush hour (6-9 AM) over 6 consecutive weekdays in March of 1994. Questions regarding sleep were selected and formatted to match those of previous community-based studies and covered a range of sleep issues (Haraldsson et al 1992, Kapuniai et al 1988, Maislin et al 1995, Stradling et al 1991). Answers pertaining to sleep quality and symptom prevalence were arranged using a 5 point Likert frequency scale (1=Never, 2=Rarely, 3=Moderately, 4=Very often, 5=Always) and a "Don't Know" category. The "don't know" category was coded as missing data. When appropriate, categories "never" and "sometimes" were combined and considered not present or negative responses and "moderately", "very often" and "always" were considered present or positive. Additional demographic and general medical information was gathered. Responses could be anonymous. Commuters were asked to complete and return the questionnaire. The project received NYU Institutional Review Board and LIRR management approval.

Data from each questionnaire were entered into a Paradox database (Borland, Scotts Valley CA) and verified. We employed the following definitions: A long commute: over 75 minutes in length; a short commute: under 45 minutes in length; excessive daytime sleepiness (EDS): a positive report (at least moderate presence) of difficulty staying awake when necessary (as in working, driving or talking on the telephone) or difficulty staying awake in the movies, theater or show; respiratory-related symptoms (RRS): "snoring", "gasping" and/or reports of "stop breathing" during sleep; suspected sleep disordered breathing (SDB): a positive report of any two or more respiratory variables; simple snoring: a positive report of "snoring" and a negative report for "gasping" and "stop breathing"; body mass index (BMI): wt(Kg)/ht(M^2) (Khosla et al 1967); obesity: a BMI over 27 (Partinen 1994) ; socioeconomic status (SES): categorized by the Hollingshead Scale using level of education and job title to form social levels I-V with I being the highest (Hollongshead and Redlich 1958) .

Statistical modeling of the data was performed using CSS Statistica 3.1 (Statsoft, Tulsa, OK). Statistical significance of categorical associations was analyzed by Chi Square. 95% Confidence Intervals for the difference between means and t-tests were used to compare groups for interval data. Adjusted odds ratios, with 95 % confidence intervals, were calculated from logistic regression models for evaluation of predictor variables on hypertension. For each analysis, subjects with missing data for that analysis were deleted. A p. level of 0.05 was considered significant; however, p. values are provided to allow adjustment if desired.

RESULTS

Based on the LIRR statistics of use for the stations targeted for this project, there was an available pool of approximately 31,300 commuters. Of the 21,000 questionnaires distributed,

4715 (22%) of these questionnaires were returned. The demographic breakdown of the respondents (59.7% male, 40.3% female, 31.9% between ages 35-49) compares favorably to previously gathered LIRR statistics for same stations (59.5% male, 40.5% female; 40% between ages 35-49). Respondent's mean educational level was 16 years with 26% of the sample having greater than 16 years of education. Mean SES was 2.2. Mean BMI was 25.32 $\pm$4.3 with 38.6 % males and 16.1% females reporting a BMI>27. There were only slight but statistically significant educational and SES differences between long and short commuters. Long commuters were less educated (Short=16.48 years versus long=15.61 years; t=6.55; p<.0001), and had lower SES (Short = 2 versus Long = 2.28; t=-7.72; p<.0001). Additionally, long commuters were bigger (mean BMI value for men with short commutes was 25.98 versus men with long commutes - 27.54). For women, the mean BMI for those with short commutes was 23.15 versus 23.76 for those with long commutes.

One third (37%) of the sample complained of at least moderate difficulty falling or staying asleep. More women than men reported this symptom (40.6% females versus 35.2% males, Chi^2 14.2; p<.001). In response to questions regarding EDS, 34.8% of respondents reported at least moderate difficulty staying awake (while working, driving, talking on the telephone and in the theater). There was no difference across gender on this response. Furthermore, there were no significant differences noted between long and short commuters on responses to these questions.

2.8% of respondents reported seeking professional help for difficulty falling or staying asleep. Overall, respondents reported the use of various remedies to "improve" sleep: alcohol (12.3%); sedatives (5.8%); tranquilizers (4.1%); antidepressants (2.2%); and relaxation therapy (11.2%). Significantly more short commuters used tranquilizers (5.5% versus 2.7%; p<.02) and sedatives (7.4% versus 3.6%; p<.005) when compared to long commuters. Only 0.09% of respondents sought professional help for excessive daytime sleepiness. Yet, 4.9% of the respondents answered positively to the question of whether they had ever used "stimulant drugs" to improve alertness. There was no difference between long and short commuters on this variable. Overall, only 0.09% of respondents reported having been diagnosed with a sleep disorder.

Respondents tended to sleep significantly less on week nights than on weekend nights (95%CI for difference 74.3-79.8). Women reported sleeping significantly longer than men during both week day (95% CI for difference 12.7-19.3) and weekend nights (95% CI 21.7-30.8).
See figure 1.

Length of commute was negatively associated with length of week night sleep. Respondents with short commutes reported a mean sleep time of 428($\pm$51) minutes compared to those with long commutes who reported a mean of 389(±55) minutes (95% CI 24.3-36.2). On weekends, long commuters slept a mean of 97 minutes longer than they did on weeknights, while short commuters slept 64 minutes longer. There was no statistical difference between mean weekend sleep times of these groups; 492(±72) for the short commuters versus 495(±79) for the long commuters (95% CI -5.2 -11). However, the short commuters slept a mean of 144 minutes longer per week than long commuters. There were no differences in this behavior according to sex or age.

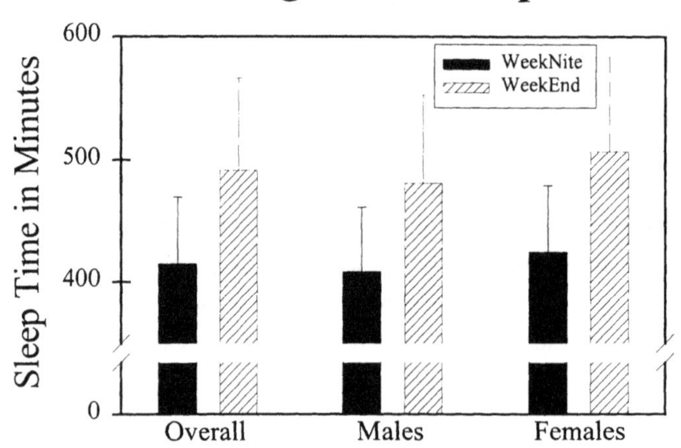

Figure 1

A total of 70% of 4683 respondents, reported at least moderate napping during their commute. Among nappers, 40.9% reported napping on both morning and evening rides. As shown in Table 1, significantly more women than men reported napping on both rides. Similarly, more women napped on the morning ride than men. More long commuters napped on the AM or PM rides and on both the AM and PM rides.

Table 1

	Male	Female	Chi^2	Short commute	Long commute	Chi^2	EDS	No EDS	Chi^2
AM Nap	45.5%	53.8%	31.8*	27.9%	72.1%	240.6	53.9%	45.9%	26.6*
PM Nap	62.5%	61.5%	.49	51.1%	69.6%	46.6*	68.1%	58.8%	38.9*
AM or PM Nap	69.7%	70.6%	.38	56.0%	82.1%	97.7*	75.6%	66.8%	37.0*
AM and PM	38.2%	44.8%	20.3*	22.9%	57.2%	176.4	46.4%	37.8%	32.1*

*p=<.00001

Those respondents who napped slept significantly less during their work week and appeared to compensate by increased sleep time during the weekend. Nappers reported sleeping a mean of 408±52.63 minutes during the weeknights while non-nappers reported 431±55.3 minutes (95% CI for difference 19.5-26.5). Over the weekend nights, there was no difference. Nappers slept 496.3±75.4 minutes versus 489±75.1 minutes for non-nappers (95% CI for difference -0.9-8.9).

Respondents who reported at least moderate "EDS", compared to those with little or no EDS, were more likely to nap on either ride and on both morning and evening rides. Interestingly, among respondents reporting no "EDS", 66.8% reported at least moderate napping on either the morning or evening commutes and 37.8% reported napping on both rides.

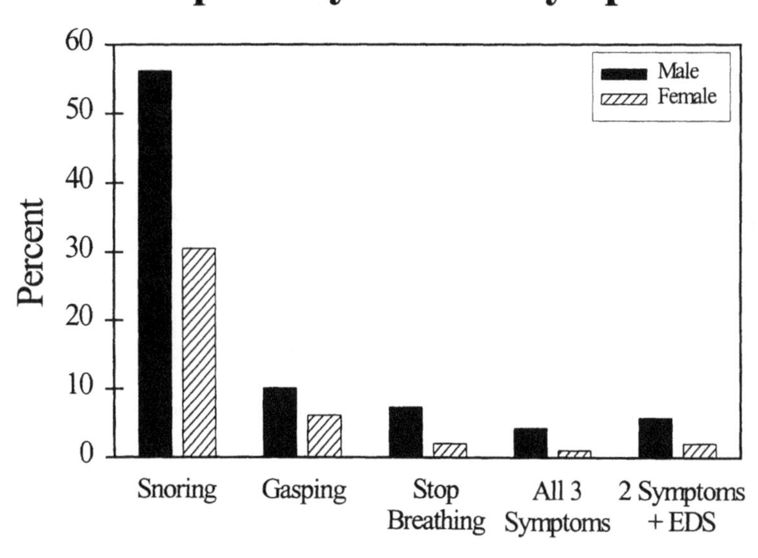

Figure 2

Over four thousand respondents (N=4171) answered the questions regarding respiratory-related symptoms. Positive reports by gender are shown in Figure 2. More males than females reported the presence of snoring, gasping and/or stopping breathing during sleep. Of those reporting two or more of the respiratory related symptoms (8.6%) 11.4% were male and 4.5% female. Furthermore, 2.9% of the sample reported all three symptoms (4.2% males/1% females, Chi square 260.8; p<0.00001). There was no significant association between the prevalence of napping and reports of respiratory symptoms. However, 4.2% (5.7% male/2% female) of the respondents reported at least two respiratory symptoms and at least moderate EDS.

Among the 4048 respondents who answered questions on BMI and respiratory symptoms, it was clear that the prevalence of obesity increased as the number of reported respiratory related symptoms increased (Table 2). Of the respondents who denied any respiratory symptoms, 80.7% were non-obese. Conversely, of those respondents reporting all three respiratory symptoms, 58.3% were obese (Overall Chi square 215.7;p<0.00001).

Similarly, self reported EDS also increased as the number of reported respiratory-related symptoms increased (Table 2). Thirty five percent of these respondents also reported one respiratory -related symptom versus 52.9 % who reported three symptoms.

Table 2

Respiratory Related Symptoms	Percent with Self reported EDS n=1393	Percent with BMI ≥27 n=1170
No symptoms	30.1%	19.3%
1 of 3 symptoms	35%	37.6%
2 of 3 symptoms	46.8%	40.7%
3 of 3 symptoms	52.9%	58.3%
Chi ^{2}	52.14 p<.00001	215.7 p<.00001

Of the 4564 respondents who answered questions regarding hypertension and diabetes, 398 cases of hypertension were reported. More respondents with hypertension reported at least moderate snoring than would be expected by chance (74% versus 26% Chi square 112;p=<0.0001).

When respondents were divided into simple snorers (those who reported only the symptom of snoring) and suspected SDB (those reporting at least two of the respiratory symptoms) and compared to those reporting no respiratory symptoms ("normals"), there was an association between reported respiratory variables and EDS, hypertension, diabetes, obesity and gender (Table 3).

TABLE 3

PERCENT	"NORMALS" n=2152(52.9%)	SIMPLE SNORERS n=1560 (38.4%)	SUSPECTED SDB n=354 (8.7%)	Chi^2	p value
EDS	30.0%	34.6%	48.9%	50.08	p<.00001
HYPERTENSION	4.5%	12.4%	16.9%	105.2	p<.00001
DIABETES	1.3%	2.7%	4.2%	17.21	p<.00005
OBESITY	19.3%	38.9%	46.5%	218	p<.00001
MALES	48.2%	71.7%	79.3%	270	p<.00001

Because of the known association of multiple factors on hypertension, we looked at the adjusted odds ratios for diabetes, obesity, age by decade, length of commute, gender, alcohol, EDS, and smoking in simple snorers and suspected apneics. Both simple snorers and suspected apneics have an increased risk of hypertension when other factors are held constant. Simple snorers have an adjusted odds ratio of 2.21 (1.54-3.18) and suspected apneics have an increased adjusted odds ratio of 2.20 (1.36-3.57) (Table 4). Only gender, alcohol, EDS and smoking were non-significant. Based on this model, it appears there may also be an increased risk of hypertension associated with a long commute.

TABLE 4

	Adjusted Odds Ratio of Hypertension	95% Confidence Interval for Odds
SIMPLE SNORE	2.21	1.54-3.18
SUSPECTED APNEA	2.20	1.36-3.57
DIABETES	7.42	4.58-12.02
OBESITY	2.65	1.87-3.75
AGE (Decade)	1.98	1.71-2.30
LONG COMMUTE	1.69	1.20-2.38
GENDER	1.34	0.92-1.97
ALCOHOL	1.11	0.81-1.53
EDS	0.74	0.53-1.03
SMOKING	0.91	0.60-1.40

Discussion

We were able to distribute questionnaires to 67% of the available commuters on the LIRR which is the largest commuter rail line in the country, servicing over 100,000 daily commuters between the end of Long Island and New York City. Our respondents were representative of the available adult population. Half of these suburban adults report significant difficulties with sleep and wakefulness. One third reported excessive daytime sleepiness but few sought professional help. It may be that commuters accept sleepiness as part of suburban life where the realities of a long commute and early work start times could impact on their ability to schedule adequate nightly sleep time. respondents with long commutes appear to sleep less during the week nights and attempt to "catch up " with a greater increase in weekend sleep than do short commuters. Additionally, they nap more during commutes than short commuters. The result appears to indicate significant sleep deprivation across week nights affects the respondent's ability to remain fully awake during the day. The large percentage reporting difficulty with sleep and wakefulness which is uninvestigated, untreated or self treated suggests that people with sleepiness may not recognize their "difficulties" as abnormal. Alternatively, one could wonder if there is some positive feedback interacting with the symptom of sleepiness. For instance, recent work from Dawson's laboratory in Australia correlates performance after significant sleep deprivation to performance with blood alcohol levels consistent with intoxication (Dawson et al, 1997). Perhaps respondents receive some pleasant intoxication-like feedback from their chronic levels of sleepiness. If so, they may not seek to correct the situation. Unfortunately, the large percentage of respondents reporting and apparently accepting difficulty with sleep and wakefulness may also indicate there is a general lack of public awareness regarding proper sleep habits and expectations of normal sleep.

It is interesting to note that even those respondents not complaining of EDS frequently nap. This could suggest the nap is valuable in increasing daytime levels of alertness. However, nappers still attempted to "catch up" on sleep over the weekend. It may be that the sleep obtained on the moving train is not refreshing and cannot be factored into total sleep time over the week. Alternatively, nappers who do not report EDS may be reflective of unrecognized EDS and sleep disorders in the adult working public. Unfortunately, it is difficult to incorporate the role of motivation into the reported incidence of napping. Sleep physiologists would argue that manifestations of EDS such as napping would not occur in the presence of adequate sleep, and that the rested individual would remain alert even under unstimulating circumstances.

Snoring and EDS are important symptoms. In the LIRR cohort, 48% of respondents report at least moderate snoring. This is somewhat higher than in community based epidemiological studies (Lugaresi et al 1975, Schmidt-Nowara et al 1990). However, the estimate of the prevalence of suspected sleep disordered breathing in this cohort, based on the definition of reports of at least two "respiratory" symptoms (snoring, gasps or stop breathing), is 8.6% (11.4% male/4.5% female) and is lower from that of previous studies(Young et al 1993, Hla et al 1994). For example, in the Wisconsin Cohort, Young et al. note that 9% of middle-aged women and 24% of middle-aged men will have an Apnea/hypopnea Index (AHI) >5. However, Young et al report that 4% of men and 2% of women will have an AHI >5 with symptoms such as EDS. The LIRR cohort is similar. We report the combination of at least moderate reports of EDS, snoring, gasping and/or stop breathing among 4.2% of the cohort (5.7% males and 2% of females). The combination of these symptoms suggest significant sleep apnea in these respondents.

Symptoms suggestive of sleep apnea were associated with increased reports of excessive daytime sleepiness, hypertension, diabetes and obesity. Links between sleep apnea, diabetes, hypertension and cardiovascular disease have been noted in the literature (Katsumada et al 1991, Strohl et al 1996). Our results also suggest a higher risk for "hypertension" in the population defined as simple snorers and those defined as suspected apneics when compared to "normals" and adjusted for age, gender, obesity, diabetes, smoking, alcohol, EDS and length of commute. This is similar to the findings in 1060 subjects from the Wisconsin Sleep Cohort Study where an association between hypertension and sleep apnea was independent of obesity, age and gender (Young et al 1997). Our data also suggest an independent risk of developing hypertension for those with long commutes compared to short commutes. This may indicate the additional stress or lack of consistent sleep allowed the long commuter. This is consistent with recent reports from Japan which note shifts in heart rate variability among long commuters (Kageyama et al 1997). The authors suggest that the variability may induce cardiovascular dysfunctions related to the onset of cardiovascular disease. While this is an interesting finding in our cohort, further research will be needed to clarify this association.

The impact of the consequences of EDS remains a significant risk to society in terms of loss of productivity and safety. This risk may increase in suburban areas as the numbers of long distance commuters increase. From a social perspective, based on previous gathered data, the large percentage of LIRR responders who report significant daytime sleepiness raises the issue of serious consequences to society from sleep deprived commuters (Mitler 1988). Specifically, there could be concern raised about highway safety particularly in suburban areas during the commuter's morning and evening travel to and from the train. A recent survey regarding drowsy drivers in New York State, surveyed 1000 licensed drivers in NYS (McCartt 1995).

Fifty-five percent had driven while drowsy over the last year; 24% reported falling asleep at the wheel. Fewer hours of sleep at night and greater difficulty staying awake during the day were variables predictive of fall asleep incidents while driving. Furthermore, 33% of those questioned reported snoring during sleep. Snorers were more likely to have fallen asleep at the wheel than non-snorers (34.4% versus 18.9% ;OR 2.25; 95% CI= 1.66-3.06).

We recognize several inherent limitations in this study. First, our sample may be biased with an over representation of people who are concerned about sleep and sleepiness. While the sample respondent demographics were representative of the available pool of commuters, and even though the topic of sleep and sleepiness are popular topics of discussions to commuters (author's personal observation), "non responders" may have ignored the questionnaire or considered it irrelevant. Second, we were not able to do test-retest reliability in our sample. Third, our questionnaire was not validated to physiologic measures. However, we were careful to use items from questionnaires validated in other populations in these analyses and sorted responses using categories similar to those validated studies. Additionally, there is face validity of self reported napping during the commute.

CONCLUSION

This study polled a representative group of suburban rail road commuters and examined results as they applied to those with short commutes (<45 minutes) versus long commutes (>75 minutes). Over 50% of respondents reported significant difficulty with sleep and wakefulness. Respondents with long commutes reported less sleep and more daytime sleepiness overall. It is surprising that few respondents attempted to improve sleep times. It may be that respondents accept the price of sleep loss for the perceived improvements in quality of life in suburbia. This study also suggests that a large portion of society may have unrecognized and consequently untreated sleep disorders. Over 8% of this population report symptoms and conditions which in combination suggest a presence of sleep disordered breathing, a treatable medical condition. Failure to seek professional help may reflect the public's failure to recognize the significance of symptoms related to sleep disorders. This is especially true for the group of long commuters who are at risk to be significantly sleep deprived by their imposed sleep schedules. Public education and improved awareness among health professionals regarding sleep needs, the risks of sleep loss and sleep disorders may encourage evaluation of the personal impact of increased length of commute on health.

References

Dawson D, Reid K. (1997) Fatigue, alcohol and performance impairment. Nature; 388:235.

Haraldsson P-O, Carenfelt C, Knutsson E, Persson HE, Rinder J. (1992) Preliminary report:validity of symptom analysis and daytime polysomnography in diagnosis of sleep apnea. Sleep 15(3):261-263.

Hla KM, Young TB, Bidwell T, Palta M, Skatrud JB, Dempsey J. (1994) Sleep apnea and hypertension. Ann Intern Med. 120:382-388.

Hollingshead AB, Redlich EC. (1958) Index of social position. In (Eds) Hollingshead and Redlich. Social Class and Mental Illness. J. Wiley and Sons, Inc, New York.

Kageyama T, Nishikido N, Kobayashi T, Kurokawa Y, Kabuto M.(1997) Commuting, overtime, and cardiac autonomic activity in Toyko. Lancet 350:

Kapuniai LE, Andrew DJ, Crowell DH, Pearce JW. (1988) Identifying sleep apnea from self-reports. SLEEP 11(5): 430-436.

Katsumada K, Okada T, Miyao M, Katsumata Y. (1991) High incidence of sleep apnea syndrome in a male diabetic population. Diab Res Clin Prac 13:45-52.

Khosla T, Lowe FR. (1967) Indices of obesity derived from body weight and height. Br J Prev Soc Med 21: 122-128.

Kump K, Whalen C, Tishler PV, et al. (1994) Assessment of the validity and utility of a sleep-symptom questionnaire. Am J Respir Crit Care Med 150:735-741.

Lugaresi E, Coccagno G, Farneti M, Mantovani M. Cirignotta F. (1975) Snoring. Electroenceph and Clin Neurophysiol 39:59-64

Maislin G, Pack AI, Kribbs NB, et al. (1995) A survey screen for prediction of apnea. Sleep 18(3):158-166.

McCartt A. Ribner SA, Pack AI, Hammer MC. The scope and nature of the drowsy driver problem in New York State. Presented at the 39th Annual meeting of the Association for the advancement of Automotive Medicine; October 16-18, 1995, Chicago, Illinois.

Mitler M, Carskadon M, Czeisler C, Dement W, Dinges D, Graeber R. (1988) Catastrophes, sleep and public policy:consensus report. Sleep 11:100-9.

National Commission on Sleep Disorders Research Report. V 1. (1993) Executive Summary and Executive Report. Bethesda, Maryland: National Institutes of Health.

Partinen M. (1994) Epidemiology of sleep disorders. In Kryger MH, Roth T, Dement WC (Eds) Principles and Practice of Sleep Medicine. 2nd edition. Philadelphia:WB Saunders 437-452.

Schmidt-Nowara WW, Coultas DB, Wiggins CL, Skipper BE, Samet JM. (1990) Snoring in an

Hispanic-American population: risk factors and association with hypertension and other morbidity. Arch Intern Med. 50:597-601.

Stradling JR, Crosby JH. (1991) Predictors and prevalence of obstructive sleep apnea and snoring in 1,001 middle-aged men. Thorax 46:85-90.

Strohl KP, Redline S. (1996) Recognition of sleep apnea. Am J Respir Crit Care Med 154:279-289.

The Gallup Organization, (1997) Sleep In America II. (Published by) The Gallup Organization, Princeton, NJ,1997.

Young T, Palta M, Dempsey J, Skatrud J. Weber S, Badr S. (1993) Occurrence of sleep disordered breathing among middle aged adults. NEJM 328: 1230-1235.

Young T, Peppard P, Palta M, Hla M, Finn L, Morgan B, Skatrud J. (1997) Population-based study of sleep and disordered breathing as a risk factor for hypertension. Arch Intern Med157:1746-1752.

4

DRIVER FATIGUE: PERFORMANCE AND STATE CHANGES

P.A. Desmond, University of Minnesota, Minneapolis, USA

INTRODUCTION

Driver fatigue presents us with a serious problem. In the United States, the National Highway Traffic Safety Administration (NHTSA, 1990) estimated that driver drowsiness was a factor in 57,000 crashes as determined from the Police Accident Report (PAR). Maycock (1995) found that 7% of motor vehicle accidents could be attributed to fatigue in a sample of 4600 male car drivers in the United Kingdom. The hazardous consequences of driver fatigue have been highlighted by recent research which has shown that prolonged driving may give rise to a range of subjective symptoms of tiredness, which may be related to performance decrements (e.g. Summala, Salmi, Mikkola and Sinkkonen, 1996). While the subjective symptoms of fatigue are commonplace, studies of driving performance have failed to produce a consistent picture of how fatigue affects the component skills of driving (see McDonald, 1984). For example, there are some studies which show that prolonged driving results in increased lane drifts (e.g. O'Hanlon and Kelley, 1974) while others have failed to replicate this finding (e.g. O'Hanlon, 1971). Some of the inconsistency in the pattern of results might be attributed to the fact that many studies have focused on the duration of driving performance without considering other factors which may contribute to fatigue (McDonald, 1984). Recent research (e.g. Desmond & Matthews, 1997a) has emphasized the role of task demands in driver fatigue, and has characterized the driver as an adaptive operator who can alter his or her effort under varying task demands. In the first part of this chapter I will describe two theoretical explanations for effects of task demands. Next, research evidence that demonstrates how fatigue effects vary with the demands of the driving task will be reviewed.

Despite the prevalence of changes in the individual's subjective state under fatiguing conditions, a comprehensive understanding of the driver's subjective experience of fatigue has escaped researchers' attention (McDonald, 1984). It is essential that we develop a thorough understanding of how different types of fatiguing driving conditions affect the driver's subjective state change. The research reviewed in the present chapter will demonstrate the complexity of fatigue, and will provide a profile of the subjective symptoms that characterize fatigue. A common feeling that is associated with fatigue is an aversion to the investment of further effort in the task at hand (Holding, 1983). Thus, motivation seems to be important component of fatigue reactions. Several studies will be described here that have investigated the role of motivation in fatigue-related impairments in driving performance.

Future transportation developments such as the Intelligent Vehicle Highway System promise us improved driver safety through the automation of many aspects of the driving task. In an automated driving environment the driver may be forced to exercise sustained monitoring of the status of an automated system. Vigilance research has highlighted some of the negative subjective reactions, such as fatigue and boredom, that are associated with such tasks (e.g. Scerbo, 1998). It is necessary to understand how automation will affect the driver's subjective state and performance. In the final part of this chapter, I will review some preliminary work that has attempted to address this issue.

Theoretical Perspectives on the Interaction Between Task Demands and Fatigue

Resource theories of attention (e.g. Wickens, 1992) and dynamic models of stress and sustained performance (e.g. Hancock & Warm, 1989) represent two opposing theoretical perspectives on the interaction between task demands and fatigue effects. Desmond and Matthews (1997a) provide a detailed account of these theories, so only a brief resume is necessary here. Resource theories propose that individuals have a limited supply of resources for attentional processing (Wickens, 1992). In order to achieve successful control of the vehicle, the driver may be required to allocate his or her resources. Resource theory proposes that as an increase in task demands occurs, so a decrease in attentional resources will take place. Several studies have shown decrements in secondary task performance as the task demands of driving increase, supporting resource theory (e.g. Harms, 1991). It is possible that fatigue might act to deplete resource availability. There is some evidence to support this proposal; Matthews (1992) showed performance decrements consistent with resource theory in a study of attention in states of tiredness. If fatigue reduces attentional resources, we can predict that fatigue-related impairments in driving should be exacerbated as task demands increase. Adaptive models provide a contrasting approach to performance impairment. There is some evidence to suggest that individuals' performance efficiency adapts to workload. Kahneman (1973) has proposed that resource availability may vary adaptively with task demands. Studies of complex tasks such as air traffic control (e.g. Desmond & Hoyes, 1996) show that operators can sometimes maintain a constant level of performance as workload increases. In contrast to the view that fatigue represents a general impairment of function, adaptive models propose that fatigue and stress lead to specific changes in the strategies adopted by individuals. Hancock and Warm (1989) have proposed an adaptive model of stress and fatigue effects on human performance. The model accounts for stress effects on performance at both low and high levels of stress and is linked to recent theories of attention (Wickens, 1987). A potential danger of fatigue is that matching effort to task demands may be impaired because fatigue reduces the range or efficiency of strategies available for regulation of effort. Recent simulator studies of driver stress have demonstrated that stress impairs performance mostly during low workload conditions, consistent with Matthews, Sparkes and Bygrave's (1996) effort-regulation hypothesis. Thus, adaptive models predict that fatigue should impair performance when workload is low.

TASK-INDUCED FATIGUE EFFECTS IN DRIVING

In a recent study of simulated driving performance, Desmond and Matthews (1996, in press) tested the predictions of resource theory and adaptive models for the variation of fatigue effects with task demands. The authors developed a fatigue-induction technique for examining task-induced fatigue. The technique allowed the assessment of two types of task-induced performance decrements: task-specific impairments arising from sustained performance of the same task; and after-effects or transfer effects arising from generalization of fatigue across various levels of workload (see Craig and Cooper, 1992). In the context of driving we can imagine task-specific fatigue occurring during a period of busy freeway driving. However, an important question is whether the fatigue associated with the previous period of demanding freeway driving carries over to performance of a drive along a quiet country road where workload is reduced. The fatigue-induction technique in Desmond and Matthews' study produces fatigue through a demanding secondary-task, during which task-specific fatigue effects were assessed. After effects of fatigue induction were assessed by having drivers proceed with single-task driving. Task demands were manipulated by requiring drivers to follow straight and curved road sections. Two contrasting hypotheses were tested in the study: resource theory predicts impairment in driving performance on curved but not straight road sections, but adaptive models predict the opposite.

In the Desmond and Matthews study, 80 drivers performed both a control and fatiguing drive on separate occasions. Drivers' lateral control of the vehicle was assessed on straight and curved road, before and after performance of the fatigue-induction technique. In the control drive, drivers were required to perform the primary task of driving without fatigue induction. A variety of subjective measures were administered before and after both drives to assess mood, fatigue, motivation, cognitive interference (thought intrusion), perceived control and active coping. The findings from the subjective data showed that the fatigue-induction technique produced a variety of subjective changes in drivers. As well as experiencing increased physical and perceptual fatigue and boredom, drivers also experienced more stressful reactions such as increased tension, anger, depression, and cognitive interference. An important finding was that task motivation was shown to decrease following the fatiguing drive, a finding that is consistent with previous studies of fatigue in contexts other than driving (e.g. Holding, 1983). Moreover, active coping was also found to decrease following the fatigue drive. During fatigue induction drivers' lateral control of the vehicle deteriorated progressively on straight road sections but not on curved road sections. Similar results were found during single-task driving following performance of the fatigue-induction technique. Relative to control drivers' lateral performance, fatigued drivers showed a greater impairment of lateral control on straight but not on curved road sections, indicating a transfer of fatigue from dual- to single-task performance. A similar pattern of results were found for drivers' steering wheel reversal frequencies which showed a reduction on straights but not on curves during the late phase of the fatigue drive.

This initial study of task-induced fatigue provided support for adaptive models of performance but not for resource theories of attention. The findings suggest that fatigued drivers have difficulty regulating their effort during low-demand driving conditions (straight road) but are able to adapt their effort successfully to meet the demands of high workload driving conditions (curved road). The decrement in task motivation found in the study suggested that de-motivation might provide a possible explanation for the failure in effort regulation found on straight road sections. In the next section, I will review a series of studies that have addressed the role of the driver's motivational state in fatigue-related performance decrements.

DRIVER FATIGUE AND MOTIVATION

The importance of motivation in individuals' subjective experience of fatigue has been pointed out by Craig and Cooper (1992) who argue that de-motivation and apathy represent primary symptoms of fatigue. Moreover, McDonald (1984) has argued for a motivational explanation in accounting for impairments in driving skills such as decision-making, judgement of risk and attention when driving is prolonged. The research reviewed in the previous section of this chapter has illustrated the importance of the driver's motivational state. As well as experiencing tiredness, boredom and physical and perceptual fatigue, drivers experienced a reduction in task motivation. In a follow-up study of simulated driving performance, Desmond and Matthews (1997b) examined the effectiveness of a motivational manipulation in reducing fatigue-related impairments during low demand, straight-road driving. As in the previous Desmond and Matthews (1996, in press) study, 80 drivers performed both a fatiguing drive, in which fatigue induction was added to the primary task of driving, and a control drive without fatigue induction. The effects of fatigue induction on drivers' subjective state were assessed by the same selection of subjective measures as were used in the authors' previous study. The motivational manipulation used in the study was an instruction that driving skill was under assessment, and was presented to drivers during early and late phases of control and fatigue drives. Drivers' lateral control of the vehicle was assessed before, during and after presentation of the motivational instruction. Since a reduction in effort mobilization was likely during low-demand driving conditions, it was expected that the motivational manipulation would result in an improvement in drivers' lateral control during straight-road driving but would have little impact on driving performance on curved road sections.

The results of the study showed that the motivational manipulation affected driving performance only during the later part of both fatigue and control drives. During the last part of the control drive, the motivational instruction resulted in an improvement in lateral control only when the instruction was actually presented to drivers, after which performance deteriorated. In contrast, fatigued drivers' performance continued to improve even after the instruction had disappeared, converging to the level of performance in the control drive. Thus, the findings appeared to provide support for Desmond and Matthews' (1997b) motivational hypothesis of fatigue-related impairments in lateral control during low demand driving conditions. The findings from the subjective state measures showed a similar pattern of change as was found in Desmond and Matthews (1996, in press) study. Task motivation and active coping were found to decrease following the fatigue drive. Consistent with the previous study, task-induced fatigue affected performance during low demand driving (straight road). Fatigue effects on steering reversal frequencies suggested that a reduction in effort regulation of the vehicle's position and trajectory occurred during the later part of the fatigue drive on straight sections. Thus, fatigued drivers seem to be vulnerable to complacency in that they commit the error of under-estimating the amount of effort which is required to preserve successful driving performance when task demands appear low.

However, a rather different explanation could be offered to account for Desmond and Matthews' (1997b) findings. It might be that curved road driving gives the driver more immediate feedback of impairment in lateral control of the vehicle. When drifting occurs on curved roads, the driver is altered to this event more rapidly than on straight roads where feedback is more delayed. Fatigue may act to impair the driver's processing of feedback signals which is most likely to be susceptible to impairment when task demands are low (straight roads). This possibility led the authors to design a further driving simulator study (Desmond and Matthews, 1998 in press) that would allow them to test the feedback and motivational hypotheses against one another. The study examined these hypotheses by assessing the effects of three types of interventions on drivers' lateral control of the vehicle: the motivational manipulation used in the Desmond and Matthews' (1997b) study; an enhanced feedback manipulation; and a reduced feedback manipulation. The enhanced feedback manipulation provided drivers with visual feedback when impairments in lateral control were detected. Conversely, the reduced feedback manipulation reduced the quality of lateral performance feedback by requiring drivers to control the vehicle during 'foggy' driving conditions. As in Desmond and Matthews' (1997b) study, drivers performed control and fatigue drives. The experimental manipulations appeared to three different groups of drivers during early and late phases of control and fatigue drives. The authors hypothesized that if fatigue-related decrements in lateral control were motivational in nature, then the motivational manipulation should prove to be the most effective manipulation in reducing impairments on straight roads during the later part of the fatigue drive. However, if the fatigue-related impairments were the result of a decrement in the processing of feedback signals, the decrements in lateral control should be reduced by the enhanced feedback manipulation but accentuated by the reduced feedback manipulation.

The findings of the study provided support for the motivational hypothesis but not for the feedback hypothesis. The motivational manipulation led to a decrease in impairments in lateral control during the early and late intervention phases. However, the improvement in lateral control was stronger during the later phase, a finding that is consistent with Desmond and Matthews' (1997b) study. The motivational manipulation resulted in an improvement in lateral control on straight and curved road sections but the improvement was greater on curved sections. The results replicated the improvement in lateral control on straight road sections found in Desmond and Matthews' (1997b) previous study. The findings showed that the enhanced feedback manipulation failed to reduce the impairment in lateral control during straight-road driving. Fatigued drivers appeared to have inappropriately responded to the enhanced feedback intervention, since they were found to exhibit coarse steering wheel movements when the manipulation was presented to them. The reduced feedback manipulation led to an increase in lateral control on both straight and curved road sections during 'foggy' driving conditions. Thus, the findings of Desmond and Matthews (1998, in press) study seem to suggest that impairments of lateral control on straight road sections are dependent on motivational factors rather than on the processing of feedback signals.

The research reviewed so far has provided a motivational explanation for fatigue-related decrements in driving performance. However, a more fine-grained analysis of motivation is necessary to understand what components of the driver's motivational state actually deteriorate under fatiguing conditions. Matthews, Joyner, Gilliland, Campbell, Huggins and Falconer (in press) have made an important distinction between two dimensions of motivation, performance motivation and intrinsic motivation. The first dimension relates to concerns regarding successful performance, and also to achievement motivation and fears of failure, while the second dimension is associated with interest in the task. In Desmond and Matthews' studies, a decrease in the driver's intrinsic interest in the driving task was found but it is unclear whether fatigue leads to a reduction in the driver's motivation to maintain successful driving performance. In a recent unpublished study, Desmond investigated how fatigue affects drivers' performance motivation and intrinsic motivation states. In this study, 60 US drivers performed a 40-minute monotonous drive on the Minnesota Wrap-around environment simulator (WES). The drive consisted of a straight road with no other vehicles present on the road. Drivers' lateral control of the vehicle was assessed throughout the drive. The Dundee Stress State Questionnaire (DSSQ: Matthews et al. in press) was administered before and after the fatiguing drive to assess drivers' intrinsic and performance motivational states, as well as other components of subjective stress. The findings showed a similar pattern of subjective state change as was found in Desmond and Matthews' previous studies.

The monotonous drive led to increases in physical and perceptual fatigue and boredom, anger and cognitive interference. In addition, decreases in energetic arousal, perceived control, and concentration were found. Both intrinsic motivation and performance motivation, were found to decrease following the fatiguing drive. Drivers' lateral control of the vehicle showed a progressive decrease over the duration of the drive. This finding is consistent with Desmond and Matthews' previous studies in which lateral control decreased during low-demand driving conditions. The study has important implications for fatigue countermeasures. The findings suggest that countermeasures should aim to not only raise the driver's level of intrinsic interest in the driving task, but also raise the driver's awareness and concern for impairments in driving performance. Thus, it may be important to monitor the driver's performance during low demand driving episodes and provide warning signals to the driver when performance-impairments occur.

DRIVER FATIGUE AND AUTOMATION

The development of transportation technologies such as Intelligent Vehicle-Highway Systems (IVHS) offers the driver with the possibility of automation of many aspects of the driving task. In such systems, in-vehicle navigation and collision avoidance systems are combined with the control software of the vehicle to automate several features of the driver's task. Implementation of automation may have an adverse impact on the driver's performance and subjective state. We have seen from the research reviewed in previous sections of this chapter that fatigue effects in driving are exacerbated during low-demand driving episodes. Desmond and Hancock (in press) identify a potentially hazardous fatigue state that they term 'passive fatigue'. The authors argue that transportation technologies that remove the driver from the task of driving may act to starve the driver of necessary task engagement that is important to maintaining successful driving performance. A passive fatigue state may develop under such conditions in which an extreme state of boredom may prevail since the driver is required to assume the role of system monitor rather than active controller. Several studies have highlighted the dangers of placing the human in the role of system monitor in automated environments. Studies that have assessed the individual's capability of restoring manual control of a system following a breakdown in automation have shown that performance is impaired when compared with individuals who are continually engaged in the task (e.g. Endsley and Kiris, 1995). Two main explanations have been put forward to account for this finding. Endsley and Kiris argue that such decrements in performance are the consequence of a loss of situation awareness. Endsley defines situation awareness as 'the perception of elements in the environment within a volume of time and space, the comprehension of their meaning, and the projection of their status in the near future' (Endsley, 1998, p.97). A further factor that has been implicated in automation-induced impairments is operator complacency. Singh, Molloy and Parasuraman (1993) have shown that performance may be impaired by complacency resulting from pilots' confidence in automated systems in aircraft cockpits.

An important issue is whether automated systems in driving give rise to negative subjective symptoms, and have an adverse effect on performance. In a recent study, Desmond, Hancock and Monette (in press) have explored the impact of an automated driver system on drivers' vehicle control and subjective states. In this study, 40 US drivers completed a manual and automated drive on the Minnesota Wrap-around environment simulator. Both drives lasted for 40 minutes. In the manual drive, drivers had full control over the vehicle for the duration of the drive. However, in the automated drive, the vehicle's velocity and trajectory were under system control. In both drives, perturbing events occurred at early, intermediate and late phases of the drives. In the manual drive, the perturbation appeared as a 'wind gust' that caused the vehicle to drift towards the edge of the right-hand driving lane. In the automated drive, the perturbing event took the form of a failure in the automated system that caused the vehicle to drift in the same direction and at the same magnitude as the wind gusts. The same subjective measures used in Desmond and Matthews' previous studies were administered before and after both drives to assess the impact of automation on drivers' subjective states. The findings showed that both drives elicited similar subjective reactions in drivers. Both drives led to an increase in physical and perceptual fatigue and boredom, as well as increases in tension, anger, depression, and cognitive interference. Decreases in energy, task motivation, perceived control and active coping were also found following manual and automated drives. These findings are consistent with the pattern of subjective fatigue found in simulated (e.g. Desmond and Matthews, 1997a) and real-life (e.g. Desmond, 1998) studies of driving. Moreover, the findings suggest that monitoring an automated system may be equally as fatiguing as prolonged driving. The study also demonstrated that drivers in the automated drive took longer to recover from the failures in automation than those drivers who were manually controlling the vehicle throughout the experiment. This finding is consistent with the results of previous studies (e.g. Endsley and Kiris, 1995) which have found impairments in performance when individuals are forced to manually control a system following automation failure. These findings have important practical implications for the implementation of automated highway systems. As Hancock, Parasuraman and Byrne (1996) have stressed, we should design transportation systems using human centered design principles: systems should aim to keep the driver within the driving loop as much as possible to prevent performance decrements.

CONCLUSIONS

The research reviewed in this chapter has revealed several important findings of both theoretical and practical significance. The studies have shown that duration of driving performance is only one factor that contributes to the production of driver fatigue. The demands of the driving task have been shown to be a critical variable in the type of fatigue examined here. Adaptive models of performance appear to provide a valid account of fatigued drivers' performance impairments in low demand driving conditions. Further studies must be conducted to discover if similar strategic changes in performance and subjective state changes characterize states such as sleepiness. The research described here has shown that changes in the driver's motivational state may be crucial in underlying the performance changes that accompany fatigue in simulated driving. Future research is necessary to discover if fatigue is associated with similar motivational changes in real-life driving. Finally, automation may have a negative impact on drivers' performance and subjective state but these effects will necessarily vary with factors such as the reliability of the system and the level of control that the driver has over the system. Further work is needed to explore the effects of these factors in detail.

REFERENCES

Craig, A. and R.E. Cooper. (1992). Symptoms of acute and chronic fatigue. In: *Handbook of human performance. Vol. 3: State and trait* (A.P. Smith and D.M. Jones, eds.), pp. 289-339.

Desmond, P.A. (1998). Subjective symptoms of fatigue among commercial drivers. In: *Contemporary ergonomics 1998* (M.A. Hanson, ed.), Taylor and Francis: London.

Desmond, P.A. and P.A. Hancock. (in press). Active and passive fatigue states. In: *Stress, workload and fatigue* (P.A. Hancock and P.A. Desmond, eds.), Lawrence Erlbaum Associates, Inc: Mahwah, NJ.

Desmond, P.A., P.A. Hancock and J.L. Monette. (in press). Fatigue and automation-induced impairments in simulated driving performance. *Transportation Research Record.*

Desmond, P.A. and T. W. Hoyes. (1996). Workload variation, intrinsic risk and utility in a simulated air traffic control task: evidence for compensatory effects. *Safety Science*, **22**, 87-101.

Desmond, P.A. and G. Matthews. (1996, in press). Task-induced fatigue effects on simulated driving performance. In: *Vision in Vehicles VI* (A.G. Gale, ed.), North Holland: Amsterdam.

Desmond, P.A. and G. Matthews. (1997a). Implications of task-induced fatigue effects for in-vehicle countermeasures to driver fatigue. *Accident Analysis and Prevention*, **29**, 515-523.

Desmond, P.A. and G. Matthews. (1997b). The role of motivation in fatigue-related decrements. In: *Traffic and Transport Psychology: Theory and Application* (T. Rothengatter and E.C. Vaya, eds.), Chap. 33, pp. 325-334. Pergamon: Oxford.

Desmond, P.A. and G. Matthews. (1998, in press). The effects of motivational and perceptual-based interventions on fatigue-related decrements in simulated driving performance. In: *Vision in Vehicles VII* (A.G. Gale, ed.), Amsterdam: North Holland.

Endsley, M.R. (1988). Design and evaluation for situation awareness enhancement. In Proceedings of the Human Factors Society 32nd Annual Meeting. Santa Monica, CA: Human Factors Society.

Endsley, M.R. and E.O. Kiris. (1995). The out-of-the-loop performance problem and level of control in automation. *Human Factors*, **37**, 381-394.

Hancock, P.A. and J.S. Warm. (1989). A dynamic model of stress and sustained attention. *Human Factors*, **31**, 519-290.

Hancock, P.A., R. Parasuraman, and E.A. Byrne. (1996). Driver-Centered Issues in Advanced Automation for Motor Vehicles. In: *Automation and human performance: theory and applications*. (R. Parasuraman and M. Mouloua, eds.), Lawrence Erlbaum Associates. New Jersey.

Harms, L. (1991). Variation in drivers' cognitive load: Effects of driving through village areas and rural junctions. *Ergonomics*, **34**, 151-160.

Holding, D.H. (1983). Fatigue. In: *Stress and fatigue in human performance*. (G.R.J. Hockey, ed.), John Wiley.

Kahneman, D. (1973). *Attention and Effort*. Prentice Hall, Englewood Cliffs, NJ.

Matthews, G. (1992). Mood. In: Handbook of human performance. Vol. 3: State and trait (A.P. Smith and D.M. Jones, eds.). London: Academic Press.

Matthews, G., S. Campbell., L. Joyner., J. Huggins., S. Falconer and K. Gililand. (in press). Validation of a comprehensive stress state questionnaire: Towards a state 'Big Three'? In: *Personality Psychology in Europe* (I. Mervielde., I.J. Deary., F. De Fruyt and F. Ostendorf, eds.), Tilburg: Tilburg University Press.

Matthews, G., T.J. Sparkes and H.M. Bygrave. (1996). Attentional overload, stress and simulated driving performance. *Human Performance*, **9**, 77-101.

Maycock, G. (1995). Driver sleepiness as a factor in car and HGV accidents. Report 196, Transport Research Laboratory, Crowthorne, Berks., UK.

McDonald, N. (1984). *Fatigue, safety and the truck driver*. London: Taylor and Francis.

National Highway Traffic Safety Administration. (1992). NHTSA IVHS Plan. Publication No. DOT HS 807 850, NHTSA Office of Crash Avoidance Research.

O'Hanlon, J.F. (1971). Heart Rate Variability: a New Index of Driver Alertness/Fatigue. Report 1712-I, Human Factors Research Incorporated, Santa Barbara Research Park, Goleta, California.

O'Hanlon, J.F. and G.R. Kelley. (1974). A Psychophysiological Evaluation of Devices for Preventing Lane Drift and Run-off-road Accidents. Technical Report 1736-F, Human Factors Research Incorporated, Santa Barbara Research Park, Goleta, California.

Scerbo, M.W. (1998). What's so boring about vigilance? In: *Viewing psychology as a whole: The integrative science of William N. Dember* (R.R. Hoffman, M.F. Sherrick and J.S. Warm, eds.), Washington, DC: American Psychological Association.

Singh, I.L., R. Molloy and R. Parasuraman. (1993). Individual differences in monitoring failures of automation. Journal of General Psychology. **120**, 357-374.

Summala, H., H. Salmi., T. Mikkola and J. Sinkkonen. (1996). Task effects on fatigue symptoms in overnight driving. In Proceedings of the International Conference of Traffic and Transport Psychology. Valencia, Spain.

Wickens, C.D. (1987). Attention. In: *Human factors psychology* (P.A. Hancock, ed.), Amsterdam: North Holland.

Wickens, C.D. (1992). *Engineering Psychology and Human Performance*. Harper Collins, New York.

PART II.

SLEEP LOSS AND OTHER CAUSES OF FATIGUE

5 SUSTAINING PERFORMANCE DURING CONTINUOUS OPERATIONS: THE U.S. ARMY'S SLEEP MANAGEMENT SYSTEM

Gregory Belenky, Thomas J. Balkin, Daniel P. Redmond, Helen C. Sing, Maria L. Thomas, David R. Thorne, and Nancy J. Wesensten, Division of Neuropsychiatry, Walter Reed Army Institute of Research, Washington, DC 20307-5100

INTRODUCTION

Success in any operational setting, including combat operations, depends upon effective performance at all levels of command and control. The operational environment is any environment in which accurate human performance is system-critical and there is a time limit within which the correct response must be made or the system fails. Illustrative examples of operational settings include combat, commercial ground, air and sea transportation, power grid management, and medical emergency rooms. Effective operational performance depends upon complex mental operations (e.g., in combat, maintaining situational awareness and the successful execution of fire and maneuver). Adequate sleep sustains mental performance. With less than adequate sleep, performance degrades over time. Total sleep deprivation degrades human mental performance, in approximately linear fashion, by 25% for each succeeding 24 hours awake (Fig. 1). Speed degrades more than accuracy. Even small amounts of sleep slow this degradation. A 30-minute nap given once a day in otherwise total sleep deprivation reduces the rate of performance degradation to 17% for each succeeding 24 hours awake (Fig. 2). The difference is apparent when the two data sets are co-plotted (Fig. 3). We have conducted studies using Positron Emission Tomography (PET) to investigate the effects of sleep deprivation on brain activation. Volunteers were scanned when rested, and at 24, 48 and 72 hours of sleep deprivation. These studies using 18-fluoro-deoxyglucose as a tracer, and statistical parametric mapping of co-registered PET images for pixel-by-pixel data analysis have shown that sleep deprivation decreases brain activation as measured by brain glucose metabolism. The greatest decreases occurred in those areas involved in complex mental operations and sustained attention. The areas of greatest decrease included the prefrontal cortex, the medial parietal cortex, the anterior and posterior cingulate gyri, and the thalamus as indicated by the labeled/shaded areas in Fig. 4. These basic science findings are consistent with the results of our S.L.A. Marshall-style debriefings of friendly fire incidents during Operation Desert Storm. These debriefings indicated that partly as a result of sleep deprivation, crews of Bradley fighting vehicles and M1 tanks lost their orientation to the battlefield (complex mental task), but still were able to put cross hairs on suspected targets and send rounds accurately down range (simple mental task). Other PET studies conducted by us show that the same brain areas most impaired during prolonged waking are deactivated to a greater degree during sleep suggesting that they have a greater need for sleep-mediated recuperation. Thus, sleep deprivation impairs performance by decreasing brain activation and this impairment maps into failures in performance in operational settings. To aid in managing sleep to sustain performance, we are developing the U.S. Army's sleep management system. Our work is proceeding in collaboration with other U.S. Army and U.S. Department of Defense laboratories, the U.S. Department of Transportation, civilian universities, and private industry. The sleep management system is a performance-sustaining biomedical contribution to the revolution in military affairs.

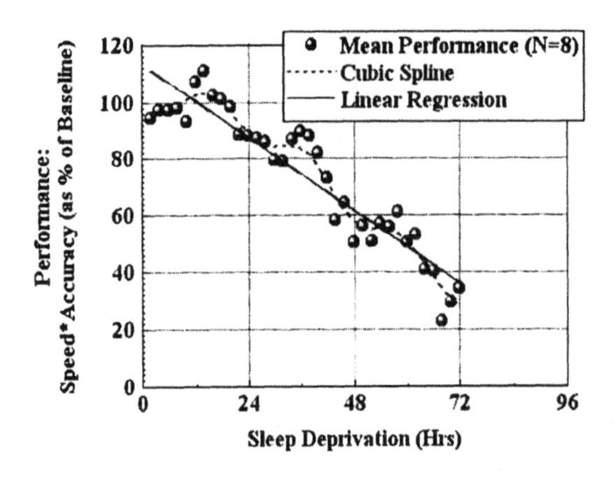

Figure 1. 72 hrs of sleep deprivation: Effect on performance.

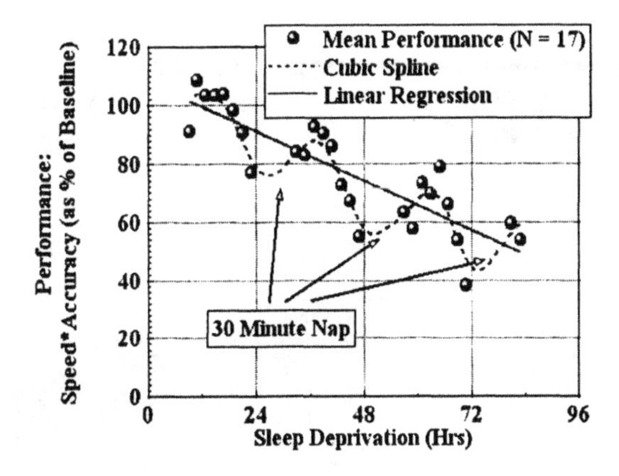

Figure 2. 85 hrs of sleep deprivation: Effect of daily 30-min nap on performance.

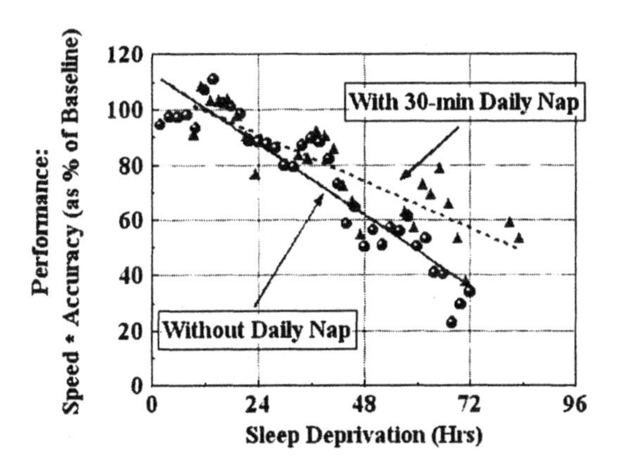

Figure 3. Total sleep deprivation vs. daily 30-min nap: Effect on performance.

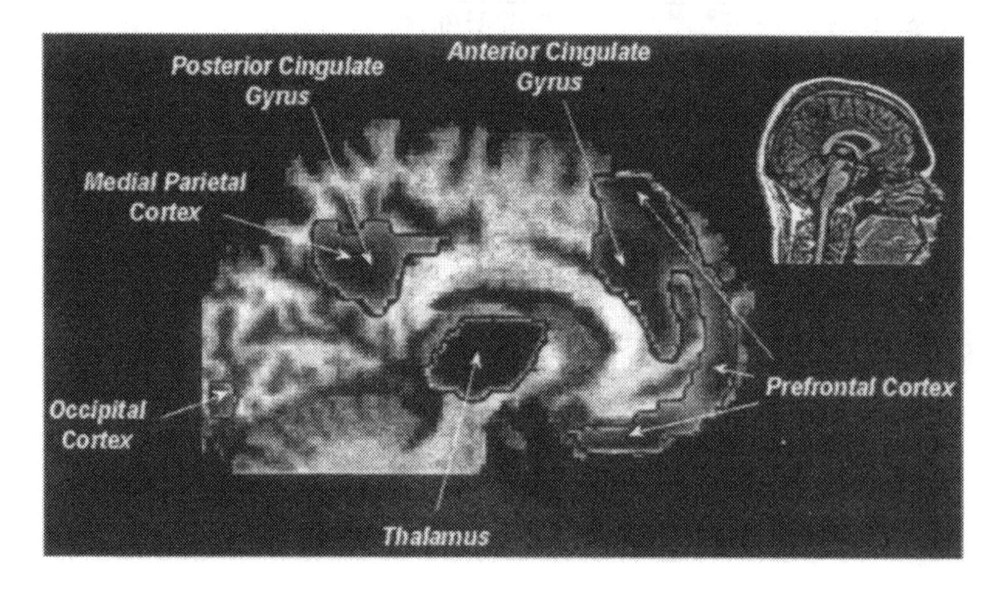

Figure 4. Effect of Sleep Deprivation on Regional Brain Glucose Metabolism

THE SLEEP MANAGEMENT SYSTEM

The sleep management system will quantify sleep and predict performance, and provide specific recommendations for optimizing sleep and performance during combat and operational deployments. The sleep management system will consist of: 1) a wrist-worn sleep/activity monitor to measure and record quantity of sleep under operational conditions; 2) a mathematical model to predict performance on the basis of prior sleep to be integrated into the software of the sleep/activity monitor; 3) a stimulant drug to temporarily sustain performance when no sleep is possible; 4) a sleep-inducing/rapid-reawakening drug combination to induce recuperative sleep with no subsequent performance decrements due to drug hangover; 5) an on-line, real-time monitor of alertness to warn the soldier and the soldier's chain of command of an imminent failure in performance; 6) doctrine for the use of the sleep management system to manage sleep, alertness, and performance; 7) integration of sleep management system into the personnel status monitor/soldier computer. When completed and fielded, the sleep management system will provide data on individual and unit sleep and predict performance over time. The sleep management system will assist soldiers and their commanders in operational planning, enabling them to optimize performance through effective sleep management. It will improve effectiveness at the individual and unit level and reduce casualties by preventing errors and accidents, including friendly fire. In civilian settings, the sleep management system will provide the wearer with information on his or her current state vis-à-vis sleep and predicted performance and give the wearer a tool for effective sleep management. This system could serve as an alternative to current prescriptive hours of service regulations with the possibility of replacing these with informed self-management.

MEASURING SLEEP UNDER OPERATIONAL CONDITIONS

We have developed a sleep/activity monitor and deployed these devices in the U.S. Army Ranger School, the National Training Center, and Operation Desert Storm. This wrist-watch-sized device can accurately measure sleep and wakefulness and will operate for four weeks without the need for a battery change. It contains an accelerometer to detect movement, a central processing unit for data collection and analysis, and random access memory for data storage. Figure 5 shows forty-eight hrs of a sleep/activity monitor record from a soldier going through Ranger School; activity counts below 100 indicate sleep in this record.

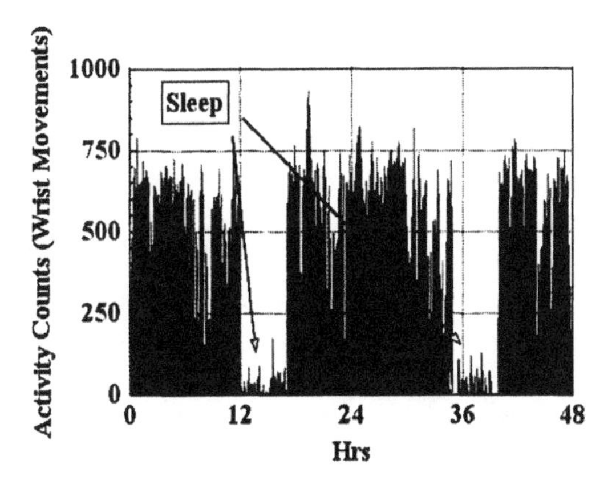

Figure 5. Sample sleep/activity monitor record.

THE SLEEP PERFORMANCE MODEL

We have developed a mathematical model to predict performance on the basis of prior sleep and have integrated this model into the sleep/activity monitor. The model incorporates both the results of experimental studies on the effects of sleep deprivation and circadian (24 hour) rhythmicity on performance and current theory on the characteristics of the processes underlying the degradation of performance by prolonged waking and the restoration of performance by sleep. Figure 6 shows the predictions of the model when applied to artillery battery operations. The sleep/activity monitor with integrated performance prediction model (Fig. 7) is currently being tested under simulated and real operational conditions.

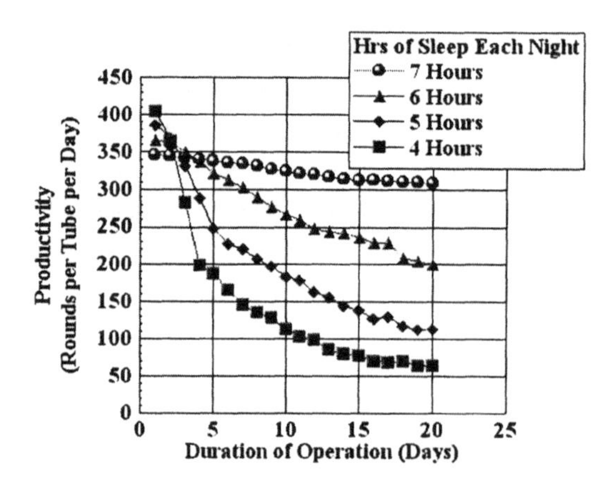

Figure 6. Predictions for artillery battery performance over 20 days of continuous operations with different schedules of sleep.

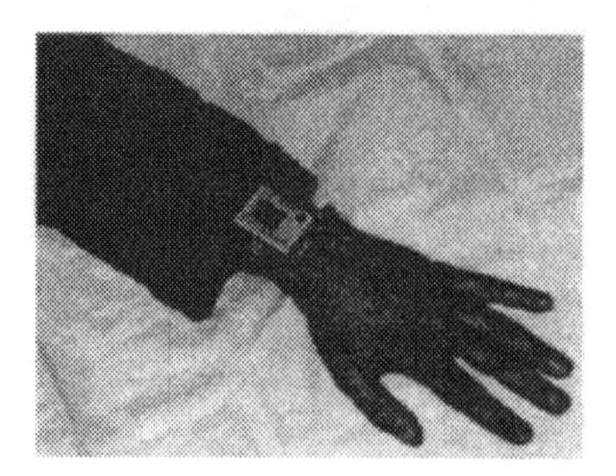

Figure 7. The Sleep Activity Monitor (Actigraph) with Integrated Sleep Performance Prediction Model

STIMULANTS AND SLEEP-INDUCING DRUGS

In laboratory studies comparing the effects of several stimulant drugs on performance during sleep deprivation, we have found that caffeine, given in doses of 300-600 mg, improves performance for 8-10 hours after 48 hours without sleep and does so without undesirable side effects (Fig. 8). In our search for an effective sleep-inducing drug, we have found that both triazolam (Halcion®) and zolpidem (Ambien®) can induce sleep in non-sleep conducive (operational) conditions. However, both have drug-hangover effects that impair performance during subsequent waking. We have found that flumazenil (Mazicon®) given upon awakening reverses these drug-hangover effects and restores full alertness and performance. Flumazenil is not in itself a stimulant. We are pursuing the development of a dual drug system (e.g., triazolam to induce sleep, followed by flumazenil on awakening to restore performance) for use in inducing recuperative sleep during long-range deployments by air.

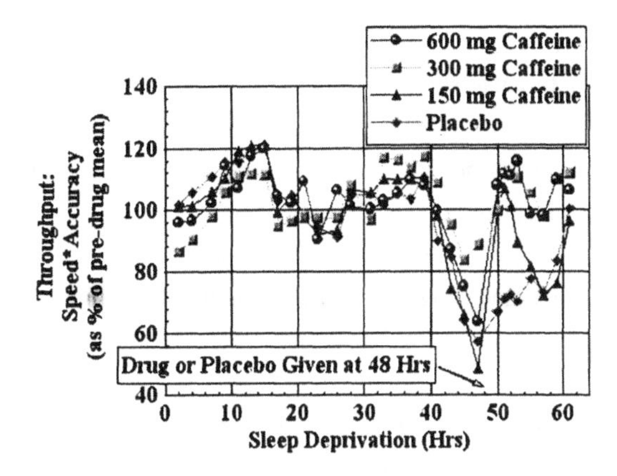

Figure 8. Caffeine reverses the effects of sleep deprivation.

ON-LINE, REAL-TIME MONITORING OF ALERTNESS AND PERFORMANCE

We are evaluating a variety of non-invasive, unobtrusive techniques to monitor alertness in real-time, including embedded reaction time tests, on-line, real-time EEG analysis, eye tracking, and pupillometry. By testing and cross-comparing these technologies, we will develop a system to warn the soldier and his chain of command of imminent performance failures.

THE SLEEP MANAGEMENT SYSTEM IN ACTION: A PROJECTION TO THE YEAR 2004

It is August 2004. An American expeditionary force is deploying to contain aggression by a disciplined, well equipped, technologically sophisticated, and well-led force in Southwest Asia. The Americans are deploying by air after 5 days of preparation during which there was little opportunity for sleep. Sleep management system software located with command and control elements at all echelons periodically interrogates, through a local area radio-frequency network, the sleep/activity monitor of each soldier's personnel status monitor, generating reports on sleep obtained and predicting effects on performance. The integrated sleep management system including hardware, software, modular integration into the personnel status monitor, pharmacological agents to assist in managing the sleep/wake cycle, on-line/real-time alertness monitoring, and appropriate doctrine had been introduced into the American armed forces in the late 1990s. Reports generated by the sleep management system indicate that, on average, most personnel had managed 4.5 hours of broken sleep each night over the last 5 nights while higher echelons of command and control managed only 3.5 hours of sleep each night. The sleep management system predicts that performance by all echelons will be below optimum. On the basis of current intelligence, commanders are anticipating immediate engagement with the enemy upon insertion. Given mission requirements, optimum performance is essential. The sleep management system predicts that 6 continuous hours of sleep for all personnel will improve performance and increase the probability of a successful operation.

Lead elements are now only 2-3 hours from take-off. As called for by doctrine, commanders elect to implement the sleep-induction/rapid-reawakening system for all soldiers once they are airborne. This consists of two pills, orally administered, given sequentially. The first, a sleep-inducer, is administered prior to the sleep period to induce sleep. The second, an antidote to the sleep-inducer, is administered at the end of the sleep interval to restore full alertness and performance. The antidote is a specific blocker of the sleep-inducer; it is not a stimulant and taken by itself it has no effects. Once airborne, soldiers take their sleep-inducer. Light levels, noise, and commotion are kept to a minimum during the sleep period. After 6.5 hours of sleep, the soldiers are awakened. Immediately, they take their antidote. Within 45 minutes all personnel are fully alert. They are refreshed from their sleep. Their thinking is clear and rapid. Their motivation is high. They are ready for combat. A query to the sleep management system indicates that personnel obtained an average of 5.5

hours sleep during the in-flight sleep period. Factoring in this additional sleep, the sleep management system predicts individual and unit performance on arrival to be near 90%, a substantial improvement over pre-sleep, pre-flight estimates.

Enemy resistance to the insertion of the expeditionary force is suppressed. The build-up in-theater continues. The alertness and performance of critical command and control personnel are continually monitored by the on-line, real-time alertness monitoring system. Forty-eight hours into the operation, the expeditionary force comes under pressure as the enemy launches all its forces in a coordinated counterattack. Commanders expect the period of sustained operations to be intense but brief. At this point, again in accordance with doctrine, commanders elect to implement short-term stimulant use to enhance the alertness of personnel at all positions and help ensure adequate performance over the ensuing 10-12 hours. The counterattack is repulsed; the operation proceeds as planned. Commanders continue to use the sleep management system to manage the sleep/wake cycle to optimize performance. Two weeks into the operation, organized enemy action ceases. The first phase of the operation concludes successfully with minimal casualties from enemy action and no losses from accident or friendly fire.

THE SLEEP MANAGEMENT SYSTEM: CURRENT CONFIGURATION

The Sleep Management System is currently implemented as a two-component system. One component is the wrist-worn actigraph with sleep scoring algorithm and the current version of the Walter Reed sleep performance prediction model built in. The sleep performance model predicts performance on the basis of the sleep scored by the algorithm. The actigraph thus gives a continuous, real-time readout of the wearer's current state vis-à-vis sleep and performance. Supplementing the actigraph is a palmtop computer. The wearer can download current state parameters to the palmtop, enter projected mission profiles, and optimize work/sleep schedules.

World Wide Web Site

An in-depth look at the research described above, including text, graphics, and references is available through the Walter Reed Army Institute of Research Home Page on the World Wide Web (http://wrair-www.army.mil).

6

Identification And Control Of Nonwork-Related Contributors To Operator Sleepiness

Dr. John A. Caldwell, U.S. Army Aeromedical Research Laboratory, Fort Rucker, AL, USA and Dr. J. Lynn Caldwell, Sleep Disorders Center of Alabama, Birmingham, AL, USA

Introduction

The safety of transportation operations depends on operators possessing a high degree of both physical and mental well-being. However, factors such as exposure to noise and vibration, day-to-day stress, lengthy work periods, irregular schedules, and inadequate sleep can compromise driver performance by producing dangerous levels of fatigue. Impairments in vigilance, judgement, and coordination can occur after only moderate amounts of fatigue. In fact, research has shown that restricting the sleep period to less than 6 hours per day produces noticeable deteriorations in performance (Belenky et al., 1994). Unfortunately, in situations where timely, critical judgements are essential to safety, these deteriorations may have serious consequences. Aviation research suggests many of the errors made by pilots of aircraft are probably the direct result of fatigue-related inattentiveness and failures to respond to critical information in the cockpit (Dinges and Graeber, 1989).

Traffic safety research suggests that sleepiness on the highway may be as dangerous as sleepiness in the air. Graham (1996) reported that 52 percent of the 1984-1986 crashes on the Pennsylvania Turnpike and 35 percent of the 1995 traffic fatalities on the New York State Thruway were thought to be a result of drivers falling asleep at the wheel. Furthermore, 31 percent of the heavy truck crashes which were fatal to the driver have been attributed to driver fatigue. Such findings are not surprising in light of the fact that fatigue is known to cause degradations in accuracy and timing, unconscious acceptance of lower standards of performance, impairments in the ability to integrate information, and the development of "tunnel vision" which leads to forgetting or ignoring important aspects of operator duties (Perry, 1974). In addition, fatigue leads to decreased physical activity and loss of the ability to effectively divide mental resources among different tasks (Dinges, 1990). When drivers are sleepy, unpredictable and involuntary sleep lapses increase, problem solving and reasoning slow down, and psychomotor skill (tracking accuracy) degrades. Perceptual illusions (such as seeing lights which aren't actually there) may even occur in severely fatigued individuals.

Thus, improperly managed operator fatigue can become a significant problem for transportation workers or any personnel employed in environments where alertness, complex judgements, and quick reactions are necessary. Fortunately, most of the effects of fatigue can be avoided by making sure that suitable periods of restorative sleep are a daily priority. Unfortunately, however, gaining adequate amounts of sleep often is difficult because of work requirements, family demands, short-term insomnias, sleep disorders, and/or poor sleep habits.

The Problem Of Sleepiness In Modern Society

The pace of modern society has become so demanding that daily sleep periods often are sacrificed for the sake of work, recreation, or social obligations. This has resulted in epidemic levels of daytime sleepiness in the adult population. One third of the respondents to a recent survey reported significant daytime sleepiness on the Epworth Sleepiness Scale, and 6 percent indicated they were severely sleepy (Gallup Organization, 1997). Almost a tenth of adults report some type of significant sleep disorder.

It is clear that daytime sleepiness exerts a negative impact on mental and physical well-being as well as workplace productivity. A National Sleep Foundation survey found that much of the U.S. workforce experiences difficulties initiating or maintaining nightly sleep, and as a result, their on-the-job concentration, problem solving, interpersonal relationships, and performance suffer (Lou Harris and Associates, 1997). Most of these workers report sleeping only 6.4 hours per night, and half sleep 6 hours or less. Given these statistics, it is not surprising that many workers are suffering from inadequate alertness on the job. Unfortunately the costs associated with such problems extend beyond the negative impact on personal well being into an estimated 18 billion dollar loss in the annual productivity of U.S. companies. Furthermore, as has already been mentioned, both personal and industrial safety are adversely affected by sleepy workers, and drowsiness on the highways is thought to be responsible for about 200,000 traffic accidents in the U.S. each year.

CAUSES OF EXCESSIVE DAYTIME SLEEPINESS

The causes of daytime sleepiness are numerous, but generally they can be classified into four major categories: 1) long work hours, especially under stressful conditions; 2) rotating work shifts, particularly night work; 3) intentional sleep restriction; and 4) insomnia due to poor sleep hygiene or sleep disorders. The first two causes of sleepiness often cannot be avoided because of the nature of the modern industrial complex. Long work hours have become a fact of life due to economic competition which requires companies to produce the most work with the fewest workers. Likewise, rotating work schedules are a necessity in nonstop societies where 24 hour-per-day operations are essential to meeting customer demands in a timely fashion. Therefore, the best we can hope for is to develop and implement strategies designed to help personnel cope with these less than perfect situations. However, the second two causes of sleepiness can be controlled and eliminated in most circumstances because intentional sleep loss, poor sleep hygiene, and sleep disorders do not necessarily result from external societal or economic pressures. Although workers may choose to spend less time in bed each day, this usually is not a job-driven requirement. Furthermore, as is the case with other medical problems, most sleep disorders are treatable after being properly diagnosed.

Intentional Sleep Restriction

Intentional sleep restriction is a controllable cause of daytime sleepiness which occurs because sleep is considered an option rather than a requirement. Over the past century, we as a society have reduced the average amount of sleep by 20 percent while adding a month to our annual working and commuting time (The National Sleep Foundation, 1995). Recent studies suggest that one-third of young adults restrict their sleep to less than 6.5 hours on week nights and later try to make up this sleep debt on the weekend (Bennett and Rand, 1995). Reasons for intentional sleep restriction include family responsibilities, social pressures, perceived work requirements, and the desire to participate in recreational activities. Night workers especially are tempted to sacrifice daytime sleep for the sake of spending time with family members and friends. Career-minded individuals may feel that it is necessary to lose sleep in order to be successful. In fact, 26 percent of respondents to a recent survey (Lou Harris and Associates, 1997) felt it was not possible to be successful in a career and to obtain sufficient sleep. However, consistent self-imposed sleep deprivation is counterproductive because of the impairments in mood, alertness, and performance that result. Transportation workers should carefully evaluate whether or not they are taking adequate advantage of their opportunities to sleep. Indicators of insufficient sleep include: difficulties awakening without the aid of an alarm clock, repeatedly pressing the snooze button when the alarm sounds, feeling that a nap during the day is necessary to alleviate sleepiness, consistently looking forward to weekends or other times for the purpose of "catching up on sleep," and frequently feeling tempted to go to sleep during meetings, while watching television, or during other sedentary activities. People who are experiencing any of these symptoms probably are not sleeping well enough or long enough at night. These individuals should gradually lengthen their normal sleep schedules (trying each new schedule for at least a week) until they discover the amount that makes them feel alert and rested during the day.

Insomnia

Difficulties falling asleep and/or staying asleep, or problems with nonrestorative sleep are considered symptoms of insomnia (Zorick, 1994) which may be the cause of daytime sleepiness in individuals who are apparently scheduling sufficient daily sleep time. Insomnia is classified into one of three categories depending on the duration of the sleep problem (Roehrs, Zorick, and Roth 1994). Transient insomnia lasts only a few nights, short-term insomnia lasts 2-3 weeks, and chronic insomnia lasts 3 weeks or longer. The causes of both transient and short-term insomnia are relatively easy to diagnose based on sleep histories supplied by the individual with the problem, whereas the causes of chronic insomnia are often more difficult to determine.

Transient or short-term insomnia. Brief periods of insomnia (temporary inability to obtain restful sleep) may result from attempting to sleep in environments that are unfamiliar or those not conducive to sleep because of noise, light, temperature, or an uncomfortable sleep surface. Usually, this type of insomnia disappears after 1-2 adaptation nights if the new environment is conducive to sleep.

Sleep disruptions can occur because of ***environmental factors*** such as noise from traffic or other sources which may degrade the restfulness of sleep by increasing the number of nocturnal awakenings and short term arousals. Sleeping quarters that are too cold or hot also can increase the amount of light sleep and decrease the amount of rapid eye movement (REM), or dream, sleep. There are other factors as well that can be overcome through the use of several strategies. Personnel sleeping in a new place should make the environment comfortable and familiar. Carrying a family picture, personal pillow, or some other small, familiar objects on trips away from home can make the new environment more relaxing.

Spending a brief amount of time to explore novel surroundings before bedtime also may facilitate a good night's sleep by reducing anxieties about the location of facilities for meetings, dining, and exercise which will be needed the next day. Setting more than one alarm clock or requesting a morning wake up call will reduce worrying about oversleeping the next day. Ensuring that curtains are fully closed, blinds are drawn, and the room is completely dark will eliminate the possibility of sleep disruption because of light exposure. Interferences from light also can be minimized by wearing an opaque sleep mask. Using earplugs, a fan, an air conditioner, or a portable stereo can improve sleep by masking disruptive and/or unfamiliar noises. In addition, a short-acting hypnotic during the first night or two in a new situation can be used to promote sleep on a short-term basis (Roth, 1996). An excellent choice would be zolpidem tartrate (Ambien®) since this medication rapidly promotes sleep without significantly altering sleep architecture or impairing next day performance (Sauvanet, Langer, and Morselli, 1988). Zolpidem has an average half life of only 2.5 hours and there are no residual performance effects at 6 hours postdose (Caldwell et al., 1997).

If sleep difficulties are not due to environmental factors, ***psychological stress*** may be the cause. Stressful life circumstances often produce difficulties in going to sleep and maintaining sleep which persist until the stressful event is resolved. If stress is the problem, a program of self-administered relaxation techniques may help to naturally dissipate feelings of tension and apprehension by inducing an incompatible response (relaxation). Progressive muscle relaxation, in which the various muscles of the body are systematically tightened for 5-10 seconds and then gradually relaxed, is easy to self administer. Stress also can be alleviated by exercise. It has been determined that 20-30 minutes of exercise significantly reduces anxiety (Petruzzelo et al., 1991). Aerobic exercise appears to be more effective than anaerobic exercise, particularly for reducing "state" or acute anxiety. In addition, exercise has been shown to promote longer and deeper sleep (Kubitz et al., 1996). The greatest benefits occur after longer-duration aerobic exercise (probably 20 minutes or more) completed several hours prior to bedtime. One other method for minimizing sleep disrupting stress is to set aside some time to "wind down" before retiring for the night. Reading a non-work-related book or listening to music are excellent ways to relax before going to bed.

If temporary sleep problems occur in the absence of environmental factors or psychological stressors, it is possible that ***circadian rhythm disruptions*** are present, especially if shift work or travel across time zones is involved (Krueger, 1991). Shift workers and travelers often suffer from shortened or disrupted sleep because of attempting to sleep at times that are out of phase with internal biological rhythms or external environmental cues. When the body's normal schedule is disrupted, internal desynchronization occurs. This results in reduced energy and motivation, impaired sleep, degraded performance, and gastrointestinal discomfort. Although circadian disruptions often are unavoidable because of work requirements, it is possible to minimize their effects by managing activity levels, meal times, and bright light exposure. Personnel who travel to a new time zone or transition to a new work shift and remain on the new schedule for a week or more, should rapidly adjust meal, activity, sleep times, and wake-up times to normal intervals for the new work/rest periods, and rigidly adhere to this new schedule even on weekends and days off (Comperatore and Krueger, 1990). In addition, sunlight exposure should be carefully managed to promote circadian resynchronization. Generally speaking, persons transitioning from the day shift to the night shift should, if possible, minimize or avoid bright light in the morning between leaving from work and going to bed. This will prevent reinforcing the normal circadian rhythms, allowing sleep to occur later and later each day until the sleepiest time changes from its usual 2200 to as late as 0800 (or even later). While attempting to readjust circadian rhythms, short-acting sleeping medications are helpful for promoting sleep and thus improving on-the-job alertness for 1-5 days while the body is adjusting to the new sleep schedule. Drivers who transition to a new time zone or work schedule for only brief periods should not attempt to readjust their circadian rhythms, but should try to remain on their original meal and activity/rest schedules as much as possible. Individuals can take advantage of naturally-occurring periods of sleepiness (coinciding with their usual bed times or nap times) to gain short periods of restful sleep as needed. For instance, someone who is required to perform a single night shift should adhere to the normal activity schedule prior to the night except that an early afternoon nap should be added. When the work shift is complete, they should remain awake for the rest of the day (except for a short nap if necessary) until the normal evening bedtime. Bedtime can be advanced slightly to compensate for the sleep deprivation, but an effort should be made to remain awake until the usual time. Circadian rhythms adjust rather slowly (about 1.0-1.5 hours per day) so there is little to gain by attempting to completely resynchronize to a new work/rest schedule when it is only temporary.

Chronic insomnia. Chronic sleep problems in situations where environmental discomfort, transient stress, or other obvious factors have been ruled out may be the result of poor sleep habits. However, serious sleep difficulties also can result from physical ailments such as breathing disorders, problems with involuntary muscle activity, chronic pain, or gastroesophageal reflux. It is often difficult to self-diagnose the source(s) of chronic sleep problems, but sleep disorders centers can determine the nature of the insomnia and provide effective treatments.

Poor sleep hygiene, or habits that are not conducive to obtaining restful sleep, accounts for a significant proportion of troubles initiating and maintaining sleep in the 100 million Americans who regularly experience sleep problems. Inconsistent sleep schedules, inadequate exercise, ineffective stress coping, and alcohol, caffeine, and prescription medications are all factors which can interfere with a good night's sleep (Gillin, 1992). Fortunately, individuals who suffer from insomnia due to bad habits can treat themselves by implementing several strategies (American Sleep Disorders Association, 1997). Nightly sleep quality can be improved by adhering to consistent bedtimes and awakening times every day of the week including weekends; using the bedroom only for sleep and sex; setting aside a place and time outside of the bedroom to resolve daily dilemmas at least an hour before lights out; engaging in tension-relieving, daily exercise after work but at least 2 hours prior to bedtime; establishing and following a consistent pre-bedtime routine every night; and creating a comfortable sleep environment. At the same time, people with sleep problems should not consume alcoholic beverages, chocolate, or drinks containing caffeine within 4 hours of bedtime. They also should not take naps during the day, watch the clock once in bed, or remain in bed for longer than 30 minutes if sleep doesn't come readily--instead, get up, do something else for a while (like reading a book), and try again later. Following these recommendations will break counterproductive mental associations that disturb sleep and replace them with mental connections that make sleep easier and more restful. For instance, avoiding activities other than sleeping and sex in the bedroom prevents bedtime from becoming associated with work, T.V., or other activities that are inconsistent with sleep. Setting aside a time and place outside of the bedroom to resolve daily conflicts minimizes associations between bedtime and worrying. Adhering to a consistent bedtime 7 days a week establishes a mental connection between the established bedtime and the act of going to sleep so that eventually sleepiness will begin to occur automatically at a certain time every night. The consistent application of good sleep habits is essential to promoting routine, restful sleep.

Sleep apnea is another cause of excessive daytime sleepiness which, unlike poor sleep habits (that lead to complaints of night time insomnia), may be indicated only by complaints of alertness problems during the day. People with sleep apnea often are unaware of their sleep disruptions since the disorder affects the restorative nature of sleep rather than sleep onset or duration. Sufferers of sleep apnea experience involuntary breathing interruptions at a rate of up to 20-30 times per hour (National Center on Sleep Disorders Research, 1995). These interruptions produce oxygen deprivation resulting in frequent night time awakenings that usually are not remembered in the morning. Apnea is most often caused by airway obstructions which prevent adequate respiration during the sleep period (obstructive apnea), but some cases are the result of the brain's failure to send appropriate breathing signals to the body (central apnea), while others are a combination of the two (Roth and Roehrs, 1996). Regardless of the cause, the net effect of sleep apnea is the same. Sufferers experience chronic fatigue and daytime sleepiness that interferes with their personal lives, psychological well being, and job productivity. The symptoms of sleep apnea include habitual snoring that is loud enough to disturb the sleep partner; breathing, choking, or gasping for breath during sleep; sweating during sleep; significant grogginess upon awakening in the morning; morning headaches; excessive fatigue during the day; falling asleep during daytime sedentary activities (watching T.V., sitting in meetings, etc.); and obesity. It is important to note that individuals who suffer from sleep apnea may not be aware of the nighttime symptoms listed above even though they recognize the headaches, fatigue, and daytime sleepiness. In fact, they often deny snoring despite the fact that their bed partners complain. Thus, it is important for both the individual and his/her partner to schedule a joint appointment at a sleep disorders center for an accurate evaluation. Once sleep apnea has been diagnosed, the problem can be successfully treated with either weight loss, surgery, medication, or a device that promotes proper night time breathing.

Another cause of excessive daytime sleepiness is nocturnal muscle activity or discomfort. ***Restless legs syndrome*** (RLS) and ***periodic limb movement disorder*** (PLMD) are both classified in this category. RLS occurs when leg discomfort (especially in the calfs) interferes with relaxation and sleep (National Center on Sleep Disorders Research, 1995), and as a result, leads to sleep deprivation. The cause of RLS is unknown, but it does tend to run in families, and it may be associated with low iron levels, anemia, kidney disease, diabetes, and arthritis. Sufferers of RLS complain of creeping, tingling, or prickling sensations in the legs that are especially pronounced during periods of inactivity. These sensations tend to be worse in the evenings or at night, and they tend to subside after walking around, stretching, messaging, or taking hot baths. The leg discomfort is severe enough to interfere with sleep onset. People who have RLS can benefit from pharmacological treatments such as benzodiazepines, dopaminergic agents, or opiods (pain killers). The other category of muscular disorder, PLMD, may occur in conjunction with RLS or it may be an independent disorder. PLMD is characterized by repetitive limb movements, most often in the legs, that occur about every 20-40 seconds throughout the sleep period (America Sleep Disorders Association, 1997). Unlike RLS, which occurs when an individual is awake, PLMD occurs at night after going to sleep. PLMD sufferers often are not aware that their sleep is being disrupted by these limb movements, and the problem may instead become an issue because of complaints from the bed partner about being kicked or awakened at night. Individuals who experience excessive daytime sleepiness that seems to be associated with periodic limb movements (based on bed partner complaints) should see a sleep specialist for a proper diagnosis. Mild cases of PLMD can be treated with behavioral interventions designed to improve sleep soundness, and more severe cases can be treated with a variety of medications.

There are other causes of excessive daytime fatigue that are too numerous to be addressed in this brief paper. However, conditions such as gastroesophageal reflux (with heartburn, chest pains, and other symptoms), arthritis, depression, anxiety, and a host of other medical or psychological disorders can significantly interfere with sleep if left untreated. Individuals who are experiencing chronic daytime fatigue should seriously investigate whether or not the root cause is a sleep disorder. In most cases, treatments are available to restore restful sleep once a proper diagnosis is made by a qualified physician or sleep specialist.

CONCLUSIONS

Excessive sleepiness is a serious threat to transportation operations because it jeopardizes both safety and productivity. Research has proven that insufficient sleep is not only costly, but dangerous as well. Although drowsiness on the job can result from unavoidable work-related factors, many of the most common causes are subject to individual control. Impaired alertness attributable to intentional daily sleep restriction; stress, discomfort, and circadian disruptions; sleep apnea, restless legs syndrome, or nocturnal limb movements; or poor personal sleep habits can be minimized once the causes and effects are recognized and understood. Scheduling sufficient nightly sleep, countering transient insomnia, seeking treatment for sleep disorders, and improving personal sleep habits will enhance on-the-job efficiency, increase individual stress tolerance, and facilitate the personal well-being of everyone in the transportation industry.

REFERENCES

American Sleep Disorders Association. (1997a) Sleep Hygiene. Rochester: American Sleep Disorders Association

American Sleep Disorders Association. (1997b). Restless Legs Syndrome and Periodic Limb Movement Disorder. Rochester: American Sleep Disorders Association

Belenky, G., Penetar, D.M., Thorne, D., Popp, K., Leu, J., Thomas, M., Sing, H., Balkin, T., Wesensten, N., and Redmond, D. (1994) The Effects of Sleep Deprivation on Performance During Continuous Combat Operations in Food Components to Enhance Performance. Washington: National Academy Press, 127-135

Bennett, M.H., and Rand, D.L. (1995) We are chronically sleep deprived. Sleep, 18(10), 908-11.

Caldwell, J.A., Jones, R.W., Caldwell, J.L., Colon, J.A., Pegues, A., Iverson, L., Roberts, K.A., Ramspott, S., Sprenger, W.D., and Gardner, S.J. (1997) The Efficacy of Hypnotic-induced Prophylactic Naps for the Maintenance of Alertness and Performance in Sustained Operations. USAARL Technical Report No. 97-10. Fort Rucker: US Army Aeromedical Research Laboratory

Comperatore. C.A., and Krueger, G.P. (1990) Circadian Rhythm Desynchronosis, Jet lag, Shift lag, and Coping Strategies. Occupational Medicine: State of the Art Reviews, 5(2),323-341

Dinges, D.F., and Graeber, R.C. (1989) Crew Fatigue Monitoring. Flight Safety Digest, Suppl, 65-75

Dinges, D.F. (1990) The Nature of Subtle Fatigue Effects in Long-haul Crews. In Proceedings of the Flight Safety Foundation 43rd International Air Safety Seminar; Rome, Italy. Arlington: Flight Safety Foundation

Gallup Organization (1997) (Survey conducted for The National Sleep Foundation). Sleepiness in America. Princeton: The Gallup Organization

Gillin, J.C. (1992) Relief from Situational Insomnia. Postgrad Medicine, 92(2), 157-170

Graham, S. (1996) Don't Drive Drowsy: Fatigue Can Be Just as Lethal as Drunk Driving. Traffic Safety, July/August,:12-15

Krueger, G.P. (1991) Sustained Military Performance in Continuous Operations: Combatant Fatigue, Rest, and Sleep Needs. In Gal, R., and Mangelsdorff, A.D.,(eds) Handbook of Military Psychology. New York: John Wiley and Sons, 255-277

Lou Harris and Associates, Inc. (1997) (Survey conducted for The National Sleep Foundation). Sleepiness, Pain, and the Workplace. New York: Louis Harris and Associates, Inc.

National Center on Sleep Disorders Research. (1995a) Sleep Apnea. NIH publication No. 95-3798. Betheda: U.S. Department of Health and Human Services

National Center on Sleep Disorders Research. (1995b) Restless Legs Syndrome (RLS). Information Retrieved from the Internet. Bethesda: U.S. Department of Health and Human Services

The National Sleep Foundation. (1995) The Nature of Sleep. Washington: The National Sleep Foundation

Perry,I.C., (ed) (1974) Helicopter Aircrew Fatigue. AGARD Advisory Report No. 69. Neuilly sur Seine, France: Advisory Group for Aerospace Research and Development

Petruzzello, S.J., Landers, D.M., Hatfield, B.D., Kubitz, K.A., and Salazar, W. (1991) A Meta-analysis of the Anxiety-reducing Effects of Acute and Chronic Exercise: Outcomes and Mechanisms. Sports Medicine,11(3),143-182

Roehrs, T.A., Zorick, F., and Roth, T. (1994) Transient Insomnias and Insomnias Associated with Circadian Rhythm Disorders. In Kryger, M.H., Roth, T., and Dement, W., (eds) Principles and Practice of Sleep Medicine. Philadelphia: W. B. Saunders Company, 433-441

Roth, T. (1996) Recommended Guidelines for the Clinical Use of Prescription Sleep-promoting Drugs. In Goldberg, J.R.,(ed) The Pharmacological Management of Insomnia: A White Paper of the National Sleep Foundation. Washington, DC: National Sleep Foundation, 15-22

Roth, T., and Roehrs, T.A. (1996) Etiologies and Sequelae of Excessive Daytime Sleepiness. Clinical Therapeurtics, 18(4), 562-577

Sauvanet, J.P., Langer, S.Z., and Morselli, P.L. (eds) (1988) Imidazopyridines in Sleep Disorders: A Novel Experimental and Therapeutic Approach. L.E.R.S. Monograph Series, Vol. 6. New York: Raven Press

Zorick, F. (1994) Insomnia. In Kryger, M.H., Roth, T., and Dement, W., (eds) Principles and Practice of Sleep Medicine. Philadelphia: W. B. Saunders Company, 483-485

7

THE SLEEP OF LONG-HAUL TRUCK DRIVERS

Merrill M. Mitler, Ph.D., James C. Miller, Ph.D., Scripps Clinic and Research Foundation, La Jolla, CA. Jeffrey J. Lipsitz, M.D., Sleep Disorders Centre of Metropolitan Toronto, Canada. James K. Walsh, Ph.D. Sleep Medicine and Research Center, Chesterfield, MO and C. Dennis Wylie, B.A., Wylie & Associates, Goleta, CA

INTRODUCTION

Each year, over 110,000 people in the United States are injured and over 5,500 are killed in motor vehicle accidents involving commercial trucks. (The Center for National Truck Statistics, Truck and Bus Accident Factbook, 1994). Estimates of the percentage of crashes that are partially or completely attributable to fatigue range from 1 to 56 percent, depending on the data base examined and the level of detail available from crash investigations. Knipling et al, 1994 & Treat et al, 1979).

There is increasing public and regulatory interest in the health consequences of fatigue, sleep deprivation, disruption of circadian rhythms, and sleep disorders. (National Institutes of Health. The National Sleep Disorders Research Plan, 1996). Driver fatigue was recently judged to be the number one problem in commercial transportation. (1995 Truck and Bus Safety Summit: Report of Proceedings, 1995). In 1988, Congress directed the Federal Highway Administration to study driver fatigue and its implications with respect to federal 24-hour electrophysiologic and performance monitoring of 80 truck drivers who carried revenue-producing loads (loads carried in the course of their employer's normal business), working day, night, or irregular shifts on common North American routes. We report on the sleep and drowsiness data from that study.

METHODS

Design of the Study

Driving schedules that represented the most demanding operations permissible were selected from the U.S. and Canadian trucking industries. In both countries the longest time on duty (which includes the time spent driving plus all the other time at work) for drivers is 15 hours, the shortest off-duty time is 8 hours, and the longest time on duty during a seven-day period is 60 hours. However, drivers can only drive a total of 10 hours without 8 hours off in the United States and 13 hours without 8 hours off in Canada. We used a parallel-group design to compare four driving schedules, two in the United States and two in Canada. The design and the associated informed-consent form were reviewed and approved by The Federal Highway Administration of the U.S. Department of Transportation and by The Transportation Development Centre of Transport Canada. The trucking companies that contributed trucks and personnel asked not to be identified.

We studied two schedules for the route between St. Louis and Kansas City, Missouri: a "10-hour, steady day" schedule consisted of five trips beginning at 9.00am each day, and a "10-hour, advancing night" schedule consisted of five trips, with the first trip beginning at 9.30am and subsequent trips beginning two to three hours earlier on the following days. We studied two schedules for the route between Toronto and Montreal: a "13-hour, steady night" schedule consisted of four trips beginning at about 11.00pm each evening, and a "13-hour, delaying evening" schedule consisted of four trips, with the first beginning at 11.30am and subsequent trips beginning one hour later each day. All four schedules involved trucks and drivers engaged in revenue-producing runs between these cities.

Subjects

The subjects consisted of 80 male, licensed commercial drivers, 40 from the United States and 40 from Canada. Twenty were assigned to each schedule. The subjects were recruited through participating trucking companies. The drivers were told the purpose of the study and asked to participate. From among those who volunteered, drivers were selected so that the average age of the drivers on each schedule was similar. All drivers read and signed an informed-consent form that detailed study procedures. All drivers followed their appointed routes and successfully completed the study. At any time, drivers were free to stop and rest or to withdraw from the study. The drivers had no financial incentive or disincentive to take naps. There was a uniform pay scheme for all drivers, but individual rates of pay varied depending on such factors as seniority and geographical location. The drivers were also compensated for spending their principal sleep periods in our laboratories and for engaging in other study-related tasks.

Data were collected in the United States between June 14 and August 22, 1993 and in Canada between September 27 and December 3, 1993.

Each driver completed a questionnaire on demographics and sleep habits. (Douglass et al, 1994). The items on the questionnaire were similar to those of population-based surveys by the Gallup Organization (Sleep in America II, 1997) and the American Cancer Society (Kripke et al, 1979). Table 1 summarizes the drivers' schedules, demographic characteristics and perceived need for sleep as reported on the questionnaire. The estimate of perceived need for sleep indicated that our sample was representative. Respondents in a Gallup survey of the general population of the United States were asked, "How many hours of sleep do you feel you need each night in order to remain alert during the daytime?" The mean (±S.D.) was 7.2 hours ± 1.2 hours.

The electroencephalogram and eye movements were recorded continuously, yielding over 7,500 hours of data. (Miller, 1995). Each truck was equipped with an infrared video system that continuously recorded views of the driver's face and the road ahead and a computer that recorded the truck's speed and road position and allowed all data to be synchronized. The video recordings were sampled every half hour, and various judgments were made, including whether the truck was moving and whether the driver appeared drowsy on the basis of drooping eyelids and a bobbing head. (Wylie et al, 1996). The results were entered into a relational data base that permitted assessment of episodes in which a driver appeared drowsy while driving in terms of their frequency during the four schedules and their distribution over day and night.

Each driver provided two urine samples, one immediately before the first principal sleep period and the second before another, randomly selected principal sleep period. No urine samples were obtained during or immediately after duty. No driver was excluded from the study on the basis of findings from the urine screens.

Polysomnography During Principal Sleep Periods

Each driver determined his own bedtimes and awakening times according to his driving schedule. Drivers slept in rooms near their travel routes. During sleep tin electrodes (Oxford Instruments, Inc., Abingdon, Oxon, England) were used for central and occipital electroencephalography of the chin, for polysomnographic scoring (Rechtschaffen and Kales, 1968). Respiratory air flow and effort were also monitored and pulse oximetry was performed. The instruments were hooked up 60 to 90 minutes before the first period of sleep. Bad leads were replaced and respiratory sensors reapplied as needed. Polysomnographic data were gathered on Oxford Medilog 9000-11 recorders (Rosekind et al, 1994) and subsequently stored on optical disks. Oximetric measurements were processed by PROFOX software (Timms et al, 1998) and then transferred to optical disks. The data obtained during principal sleep periods were scored on an Oxford 9200 system by experienced technologists using 30-second segments. (Rechtschaffen & Kales, 1968). For the sleep variables that we measured, the reliability of scoring and rescoring has previously been shown to exceed 0.90 for individual trained technologists and 0.80 between technologists. (Lessnau et al, 1991). We ensured that scoring would be consistent by rescoring randomly selected records, having regular staff meetings, and reviewing the data. Although this study was not designed to measure the prevalence of sleep-disordered breathing in commercial drivers, we could identify sleep apnea by the presence of rapid repeated periods of desaturation (a drop in the oxygen saturation value of more than 4 percent that lasted less than three minutes) on oximetry and scoring polysomnographic records using clinical criteria. (Guilleminault, 1982; Walcak & Chokroverty, 1994).

Electrophysiologic Recording during Driving

After the principal periods of sleep, respiratory sensors were disconnected from the drivers and the remaining leads were checked. Scorers were blinded to drivers' activity or truck speed and, for finer temporal resolution, used 20-second scoring periods during hours in which drivers were supposed to be awake. (Rechtschaffen & Kales, 1968).

Statistical Analysis

We examined the frequency distribution of the duration of sleep across all drivers and found that the homogeneity-of-variance assumption was tenable and that the distribution appeared unimodal and nonskewed. Univariate, repeated-measures analysis of variance (Winer, 1962) was our basic analytic tool. All reported P values are two-tailed.

RESULTS

The 80 drivers had a total of 400 principal sleep periods (5 for each driver) and 200 10-hour trips, and 160 13-hour trips. Over 96 percent of all data points from the principal sleep periods were collected. Our software required all data points, so missing data were replaced with means for the driver in question. The use of sample analyses of variance that replaced missing data with the grand means rather than means for the subject produced similar results.

Work and Rest Schedules

The drivers were given only general guidance about when they could rest during their schedules. However, we knew, from model patterns of work and rest, that eight hours off duty between trips was possible on any of the four driving schedules on most days. We estimated thc lcngths of time off during all 280 intervals between trips from the departure and return times of the trucks recorded by on-board computers or from our technicians' notes. There were 33 intervals (12 percent) during which the drivers had less than eight hours off duty (average time available, 7.4 hours), when they were not involved in job- or study-related matters (10 on the advancing night schedule and 12 on steady night schedule, and 11 on the delaying evening schedule.

Sleep

The amount of sleep a subject could have was defined by the amount of time spent in bed plus the opportunity for napping. The average time spent in bed during principal sleep periods and the number of naps are shown in Table 2. Among all drivers, the average time spent in bed was 5.18 hours. There was a significant difference ($P < 0.001$) between schedules, with the shortest times in bed occurring on the steady night schedule and the longest times on the steady day schedule. The younger drivers (average age 35.5 years) spent more time in bed (5.34 hours vs. 5.03 hours, $P = 0.02$) than the older drivers (average age 50.1 years).

TABLE 1 - Driving schedules and characteristics of the 80 long-haul truck drivers

Driving Schedule	Schedule Description	Number of Drivers	Age (years)	Height (cm)	Weight (kg)	Body Mass Index	Reported ideal Time in Bed (hours)
10-hour, Steady Day	10 hours of day driving beginning at about the same time each day	20	49.0±8	179±5	94±13	29.2±3	7.0±1
10-hour, Advancing Night	10 hours of night driving beginning 2 hours earlier each day	20	43.9±11	181±5	99±22	30.0±6	7.0±1
13-hour, Steady Night	13 hours of late night to morning driving beginning at about the same time each day	20	40.3±11	180±8	92±16	28.6±6	6.9±1
13-hour, Delaying Evening	13 hours of afternoon to night driving beginning 1 hour later each day	20	38.0±7	179±8	89±14	27.9±4	7.8±1
Overall		80	42.8±10	180±5	94±17	28.9±5	7.1±1

TABLE 1. DRIVING SCHEDULES AND CHARACTERISTICS OF 80 LONG-HAUL TRUCK DRIVERS.*

*Plus-minus values are means ±SD. † The body-mass index is calculated as the weight in kilograms divided by the square of the height in meters.

†† Drivers were asked to respond to the following: "My ideal amount of sleep is X hours."

TABLE 2 - TIME IN BED (hours)

Driving Schedule	Younger Drivers	Older Drivers	Average by Schedule
All Schedules	5.34	5.03	
Number of Drivers	40	40	
Average Age ± SD	35.5 ± 7	50.1 ± 7	
Number of Naps	32	31	
10-hour, Steady Day	5.95	5.61	5.78
Number of Drivers	10	10	20
Average Age ± SD	42.7 ± 5	55.2 ± 4	49.0 ± 8
Number of Naps	10	3	13
10-hour, Advancing Night	5.27	4.93	5.10
Number of Drivers	10	10	20
Average Age ± SD	34.5 ± 6	53.2 ± 5	43.9 ± 11
Number of Naps	3	10	13
13-hour, Steady Night	4.58	4.16	4.37
Number of Drivers	10	10	20
Average Age ± SD	31.8 ± 6	48.8 ± 7	40.3 ± 11
Number of Naps	11	11	22
13-hour, Delaying Evening	5.55	5.40	5.47
Number of Drivers	10	10	20
Average Age ± SD	32.8 ± 4	43.1 ± 5	38.0 ± 7
Number of Naps	8	7	15

TABLE 2. THE AVERAGE TIME SPENT IN BED AND THE NUMBER OF NAPS TAKEN, ACCORDING TO THE AGE AND SCHEDULE OF THE DRIVERS.*

*Plus-minus values are means ±SD. A nap was defined as an episode of sleep outside the principal sleep period that was scorable with the use of electrographic criteria. †P=.02 for the difference between age groups. ††<.001 for the difference between the other driving schedules by analysis of variance.

The periods of sleep latency (the length of time between turning off the lights and falling asleep were 19.3 minutes for drivers on the steady day schedule, 12.9 minutes for drivers on the advancing night schedule, 7.4 minutes for drivers on the steady night schedule and 14.8 minutes for drivers on the delaying evening schedule. The difference between the groups was significant ($P < 0.001$). The overall average period of sleep latency was 13.6 minutes.

The drivers slept for an average of 4.78 hours, or about 2 hours less than their reported average ideal sleep (Table 1). Table 3 shows the duration of sleep according to the four driving schedules and the five principal periods of sleep. There was a significant effect of schedule on the duration of sleep ($P < 0.001$), with the longest durations (5.38 hours) on the steady day schedule and the shortest (3.83 hours) on the steady night schedule. The period of sleep also had a significant effect ($P < 0.001$): the longest and the shortest sleep durations occurred in first sleep periods. There also was a significant interaction between schedule and sleep period ($P < 0.001$). Although several interpretations are possible, the simplest is that the durations of sleep in periods 2 to 5 are typical of the various schedules, whereas the durations of sleep in period 1 vary because drivers were coming back to work after being off duty for at least 24 hours. The younger drivers slept for an average of 4.94 hours per principal sleep period, as compared with 4.61 hours for older drivers. Although significant ($P = 0.02$), this difference was small and attributable to the fact that younger drivers spent 0.32 more hours in bed per principal sleep period than older drivers.

TABLE 3 - SLEEP DURATION (hours)

Driving Schedule	Sleep Number 1	2	3	4	5	Average by Condition
10-hour, Steady Day	5.37	5.13	5.64	5.37	5.41	5.38
10-hour, Advancing Night	6.36	4.54	4.73	4.35	3.85	4.76
13-hour, Steady Night	3.22	3.88	4.41	4.28	3.38	3.83
13-hour, Delaying Evening	6.73	4.75	4.71	5.16	4.27	5.12
Average By Sleep Number^	5.42	4.57	4.87	4.79	4.23	
Grand Mean						4.78

TABLE 3. AVERAGE DURATION OF SLEEP DURING THE FIVE PRINCIPAL SLEEP PERIODS ACCORDING TO THE DRIVING SCHEDULE.*

*There was a significant difference between driving schedules ($P<.001$ by analysis of variance). The interaction between driving schedules and period of sleep was also significant ($P<.001$).

Sleep efficiency, (the ratio of time asleep to the time spent in bed) exceeded 0.91 for all schedules. Sleep efficiency as well as other measures sensitive to sleep disturbance such as the number of awakenings during the sleep period indicated that sleep during principal periods was well-consolidated (data not shown).

Naps

A nap was defined as an episode of sleep outside the principal sleep period that was scorable with the use of electrographic criteria. Drivers took 0 to 3 naps per day for a total of 63 naps (Table 2). A chi-square statistic (Siegel, 1956) indicated that neither age nor schedule was predictive of the number of naps.

Thirty-five drivers took at least one nap. Naps increased the total amount of sleep obtained by an average of 0.45 ± 0.31 hour (range, 0 to 1.63 hours), or 11 percent.

Respiration During Sleep

Pulse oximetry disclosed repeated periods of desaturation in two drivers, who were 49 and 55 years of age. Both were on the steady day schedule. Polysomnography revealed that both drivers had sleep apnea, with 10 to 30 respiratory events per hour. However, the sleep data for these drivers were not substantially different from those of other drivers.

Drowsiness of Sleep among On-Duty Drivers

To assess the frequency of drowsiness or sleep among drivers while they were driving, we focused on the times when the trucks were travelling faster than 72 km per hour (45 miles per hour), according to computer records. Using the same sleep scoring criteria applied to the principal sleep periods, (Rechtschaffen & Kales, 1968) we identified 1 trip by a 30-year-old driver and 1 by a 25-year-old driver involving a total of 7 episodes with electrographic features of drowsiness such as slow, rolling eye movements and electroencephalographic alpha activity. During each of these trips, the drivers had an episode of stage 1 sleep, (Rechtschaffen & Kales, 1968). During the remaining five episodes, the electroencephalographic slowing and slow, rolling eye movements were of insufficient duration to be scored as stage 1 sleep. Stage 1 sleep, the lightest of non rapid-eye-movement sleep is characterized by a relatively low voltage, mixed frequency electroencephalogram with prominent activity in the range of 2 to 7 Hz.

For a 30-year-old driver, who was on the delaying evening schedule, had five episodes of stage 1 sleep between 11:12 p.m. and 11:53 p.m. (duration, 20 to 520 seconds) during his first trip. The first episode occurred after he was driving for about 10 hours and 15 minutes. The 25-year-old driver, who was on the steady night schedule, had two stage 1 episodes at 2:24 a.m. (Fig. 1) and 4:38 a.m. (duration, 60 and 80 seconds, respectively) during his fourth trip. Prior to this trip, he had been off duty for 9 hours. The episode at 2:24 a.m. occurred after 2 hours and 3 minutes of driving. Neither driver showed evidence of sleep apnea.

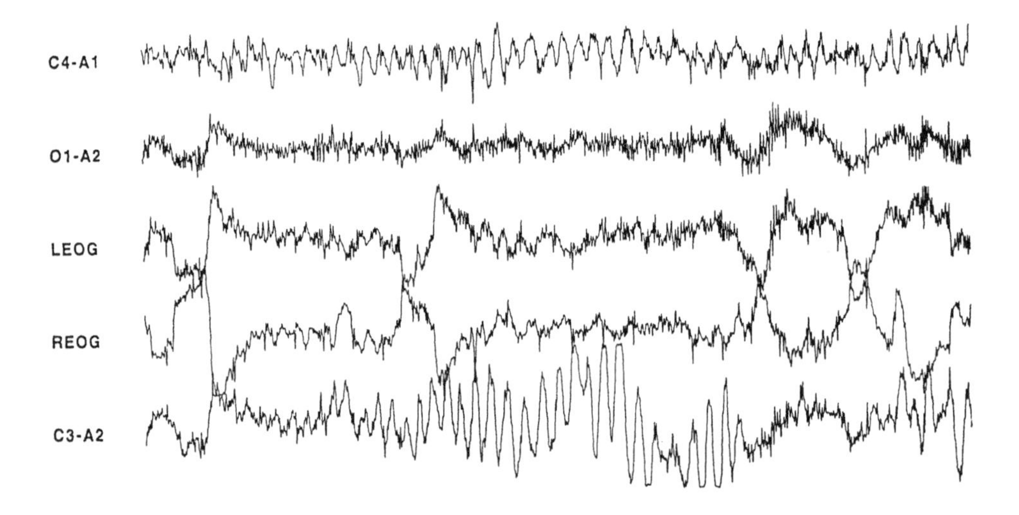

Figure 1. Electrographic Data Showing Sleep-Like Patterns in a 25-Year-Old Driver on the Steady Night Driving Schedule.

The 20-second segment recorded while the subject was driving began at 2:24 a.m. There are slow, rolling eye movements (the large curves on the eye-movement tracings), a high level of alpha activity throughout the occipital tracings, which are known as paroxysmal hypnagogic hypersynchrony. (Walcak & Chokroverty, 1994).

Video Recordings During Driving

When we analyzed the video recordings of the drivers' faces, of the 29,310 six-minute segments (7 percent) were judged to show a drowsy driver. Forty-five of the 80 drivers (56 percent) were judged to be drowsy on at least one segment, but of the segments (54 percent) showing drowsy drivers involved just 8 drivers. Five of these drivers were on the steady night schedule, 2 were on the delaying evening schedule, and one was on the advancing night schedule.

Table 4 shows the number of segments in which drivers were judged to be drowsy according to the driving schedule and time of day. Of the 1,989 segments in which the driver was judged to be drowsy, 1,646 (83 percent) occurred between 7 p.m. and 6.59 a.m. The average number of consecutive segments in which the driver was judged to be drowsy was 6.44 (range, 1 to 37; mode, 1; median, 4). The driver on the delaying evening schedule whose electroencephalograms showed stage 1 sleep while driving had video records in which he was judged to be drowsy at or near the time when stage 1 was recorded. The other driver in whom stage 1 was recorded did not have any corresponding video records in which he was judged to be drowsy.

TABLE 4 - Number and percent of 6-minute video epochs judged to be "drowsy while driving"

Driving Schedule	TIME OF DAY 07:00-18.59 O'Clock Epochs	Percent	19:00-06:59 O'Clock Epochs	Percent	Driving Schedules Overall Epochs	Percent
10-hour, Steady Day						
Judged Drowsy	53		47		100	
Total	4940	1.07%	1785	2.63%	6725	1.49%
10-hour, Advancing Night						
Judged Drowsy	91		222		313	
Total	4385	2.08%	2760	8.04%	7145	4.38%
13-hour, Steady Night						
Judged Drowsy	151		739		890	
Total	3150	4.79%	4515	16.37%	7665	11.61%
13-hour, Delaying Evening						
Judged Drowsy	48		638		686	
Total	3185	1.51%	4590	13.90%	7775	8.82%
Time of Day Overall						
Judged Drowsy	343		1646		1989	
Total	15660	2.19%	13650	12.06%	29310	6.79%

TABLE 4. SIX-MINUTE SEGMENTS OF VIDEO RECORDINGS JUDGED TO SHOW A DROWSY DRIVER, ACCORDING TO THE TIME OF DAY AND DRIVING SCHEDULE. The total number of six-minute segments of driving for each entry in the table was used as a denominator. Since the video recordings were sampled every 30 minutes, we estimated the required number by counting, for each entry in the table, the number of these 30-minute sampling periods that occurred while the truck was moving at a normal speed (as opposed to being parked or stopped) and multiplying this number by 5, since 30 minutes equals five 6-minute segments. According to these calculations, there were 29,310 6 minute segments during which a driver could have been judged

Discussion

Round-the-clock electrographic data were collected for four parallel groups of 20 long-haul truck drivers who were working 10-hour driving schedules in the United States or 13-hour schedules in Canada. Two of the 80 drivers had sleep apnea detectable on the basis of polysomnographic criteria. Drivers averaged 5.18 hours in bed and 4.78 hours of sleep per day. This amount of sleep was about two hours less than their reported ideal.

One limitation of our study was that the drivers' estimates of their ideal amount of sleep were obtained from questionnaires rather than recordings of the men's sleep when they were not at work. However, only 19.3 percent of a representative sample of men reported sleeping less than five hours a night. (Sleep in America II, 1997). The sleep durations we did observe were much shorter (4.78 hours) than most standards. Sleep-restriction experiments show that a person's tendency to fall asleep increases during normal waking hours if he or she has slept less than six hours and also increases across successive days of restricted sleep (Carskadon & Dement, 1981; 1982). Psychomotor performance is impaired if sleep is limited to five hours for two or more consecutive nights (Carskadon & Dement, 1982; Wilkinson et al, 1996). It is also known that fewer hours of sleep leads to inattention and increased error rates and that little sleep and circadian influences act synergistically (Froberg, 1985). Night driving after relatively little sleep is a better predictor of fatigue-related accidents than is night driving alone (Factors that affect fatigue in heavy truck accidents, 1995). An analysis of accidents involving commercial trucks found that drivers in fatigue-related accidents slept an average of 5.5 ours during their last sleep period as compared with 8.0 hours of sleep for drivers in non-fatigue-related accidents. (Factors that affect fatigue in heavy truck accidents, 1995).

Another limitation of our study related to sample size. Since we used a parallel-groups design, each driver could not be studiedon all schedules. Nevertheless, the demographics and ideal sleep times indicated that the four groups were comparable. The short times spent in bed, found in all schedules, are disturbing and are attributable in part to driver choice. Our study required no more than 50 to 60 minutes of the drivers' time per day and over 88 percent of all intervals between trips still allowed an opportunity for eight hours. Possible reasons for such short times in bed include duty-related demands, involvement in social activities, and time-of-day effects, which reduce one's inclination to go to bed during daylight.

The short durations of sleep - inevitable results of short times in bed - probably explain the findings suggesting that te drivers got too little sleep: We found high rates of sleep efficiency (ratio of the time asleep to the time spent in bed), an observation that is consistently reported in sleep-restriction studies (O'Hanlon & Kelley, 1977) and indicates increased tendency to sleep. Thirty-five of the 80 drivers took naps, which averaged 0.45 hour and augmented the amount of sleep obtained by 11 percent. Two drivers had episodes of stage 1 sleep while driving.

O'Hanlon and Kelley also found electroencephalograms with sleep-like patterns in subjects who were driving (Gross & Feldman, 1995). In our study, episodes of stage 1 sleep occurred while driving between 11 p.m. and 5 a.m., suggesting a circadian influence. The observed nighttime increase in drowsiness as assessed by analysis of video recordings of the drivers' faces was consistent with the expected effects of circadian influences, decreased stimuli during night driving, and too little sleep. Based on available rates of accidents severe enough to be reported to the Department of Transportation, (The Center for National Truck Statistics, Truck and Bus Accident Factbook, 1994; Knipling & Wang, 1994; National Institutes of Health. The National Sleep Disorders Research Plan, 1996) and allowing for no differences in the accident rates for roads and schedules, one would expect 1 accident involving a combination truck (a tractor pulling one or more trailers) in about every 1.2 million miles. The trucks in this project drove about 204,000 miles. Thus, we did not expect an accident and in fact, none occurred. Since no episode of sleep-like electroencephalographic data was associated with a crash, the drivers were probably drowsy during these episodes but not actually asleep. During normal sleep, there is a marked unresponsiveness to stimuli (Carskadon & Derment, 1994) that would preclude safe driving. Thus, it may not be correct to use standard sleep-scoring criteria for records obtained when the subject is behaviorally active as opposed to lying in bed, Torsvall & Akerstedt, 1987).

Long-haul truck drivers obtained less sleep than is required for alertness on the job and that the greatest vulnerability to unwanted sleep or sleep-like states is during the late night and early morning, a finding that is consistent with published data on other industries. (Mitler et al, 1988). Other studies have shown a smaller mid-afternoon period of vulnerability to unwanted sleep. (Dinges & Broughton, 1989; Mitler & Miller, 1996). We may have missed detecting such a period because there was considerable irregularity in the times that drivers were on duty and there were only 20 drivers on any one schedule.

Since physicians are a primary source of information about fitness for duty and its relation to sleep, they should be alert to the possibility of sleep deprivation in people who engage in shift work. It is also important to recognize deleterious synergistic effects on alertness of alcohol and other sedatives in the presence of sleep deprivation or medical conditions known to increase the tendency to sleep, such as sleep apnea (Roehrs et al, 1992; Findley et al, 1995).

Our findings underscore the need to educating workers and schedulers about the importance of adequate sleep with respect to public safety.

ACKNOWLEDGEMENTS

Supported by Contracts (DTFH61-89-C-053) with the Federal Highway Administration. The views expressed in this article are solely those of the authors.

We are indebted to Ted Shultz and Joseph Assmus for expert technical assistance, and to J. Christian Gillin, M.D., at the University of California, San Diego, for editorial assistance.

REFERENCES

1995 Truck and Bus Safety Summit: Report of Proceedings. (1995) Kansas City, MO, Department of Transportation 1995.

Carskadon MA, Dement WC. (1981) Cumulative effects of sleep restriction on daytime sleepiness. Psychophysiology, 18:107-13.

Carskadon MA, Dement WC. (1982) Nocturnal determinants of daytime sleepiness. Sleep, 5:S73-S81.

Carskadon MA, Dement WC. (1994) Normal human sleep: an overview. In: Kryger MH, Roth T, Dement WC, eds. Principles and Practice of Sleep Medicine. Philadelphia, PA: W.B. Saunders Company, 16-25.

Dinges DF, Broughton RJ. (1989) Sleep and Alertness: Chronobiological, Behavioral and Medical Aspects of Napping. New York: Raven Press.

Douglass AB, Bornstein R, Nino-Murcia B, et al. (1994) The Sleep Disorders Questionnaire. I: Creation and multivariate structure of SDQ. Sleep 17:160-7

Factors that affect fatigue in heavy truck accidents. (1995) Vol.1. Analysis. Washington, DC: National Transportation Safety Board. (Safety Study NTSB/SS-95/01.)

Findley L, Unverzagt M, Guchu R, Fabrizio M, Buckner J, Suratt P. (1995) Vigilance and automobile accidents in patients with sleep apnea or narcolepsy. Chest, 108:619-24.

Fröberg JE. (1985) Sleep deprivation and prolonged working hours. In: Folkard S, Monk TH, eds. Hours of work: temporal factors in work-scheduling. Chichester, England: John Wiley, 67-75.

Gross M, Feldman RN. (1995) National transportation statistics 1996. Cambridge, Mass: Volpe National Transportation Systems Center. (DOT-BTS-VNTSC-95-4.)

Guilleminault C. (1982) Sleeping and Waking Disorders. Indications and Techniques. Menlo park: Addison-Wesley.

Knipling RR, Wang J. (1994) Crashes and fatalities related to driver drowsiness/fatigue. Research Note. Washington, DC: U.S. Department of Transportation, National Highway Traffic Safety Administration.

Kripke DF, Simons RN, Garfinkel L, Hammond EC. (1979) Short and long sleep and sleeping pills. Is increased mortality associated? Archives of General Psychiatry, 36:103-16.

Lessnau J, Erman M, Mitler M. (1991) Scoring reliability among the sensormedics somnostar system and two experienced polysomnographic technologists. Sleep Res, 20-430.

Miller JC. (1995) Batch processing of 10,000 hours of truck driver EEG data. Biological Psychology. 40:209-222.

Mitler MM, Carskadon MA, Czeisler CA, Dement WC, Dinges DF, Graeber RC. (1988) Catastrophes, sleep, and public policy: consensus report. Sleep, 11:100-9.

Mitler MM, Miller JC. (1996) Methods of testing for sleepiness. Behavioral Medicine, 21:171-83.

National Institutes of Health. (1996) The National Sleep Disorders Research Plan. Bethesda, MD: National Center for Sleep Disorders Research.

O'Hanlon JF, Kelley GR. (1977) Comparison of performance and physiological changes between drivers who perform well and poorly during prolonged vehicular operation.

In: Mackie RR, ed. Vigilance: Theory, operational performance, and physiological correlates. New York: Planum Press, 87-110.

Rechtschaffen A, Kales A. (1968) A manual of standardized terminology, techniques and scoring system for sleep stages of human subjects. Bethesda, MD: Natl. Inst. Neurol. Dis. Blind. (NIH Publ.204).

Roehrs T, Zwyghuizen-Doorenbos A, Knox M, Moskowitz H, Roth T. (1992) Sedating effects of ethanol and time of drinking. Alcoholism, Clinical & Experimental Research, 16:553-7.

Rosekind MR, Graeber RC, Dinges DF, et al. (1994) Crew factors in flight operations IX. Effects of planned cockpit rest on crew performance and alertness in long-haul operations. Moffett Field, CA: NASA Ames Research Center.

Siegel S. (1956) Nonparametric Statistics for the Behavioral Sciences. New York: McGraw-Hill, Inc..

Sleep in America II. (1997) Princeton, NJ: The Gallup Organization.

The Center for National Truck Statistics, Truck and Bus Accident Factbook, (1994) Report No. VMTRI - 96-40, U. of Michigan, Ann Arbor, Ml.

Timms RM, Dawson A, Taft R, Erman M, Mitler MM. (1998) Oxygen saturation by oximetry: analysis by microcomputer. J Polysomnographic Technology. Spring:13-21.

Torsvall L, Akerstedt T. (1987) Sleepiness on the job: Continuously measured EEG changes in train drivers. Electroencephalogr Clin Neurophysiol, 66:502-511.

Treat JR, Tumbas NS, McDonald ST, et al. (1979) Tri-level study of the causes of traffic accidents. Xxx: Indiana University, Bloomington, IN. [Publication no. DOT-HS-034-3-535-77 (TAC)]

Walcak T, Chokroverty S. (1994) Electroencephalography, electromyography and electrooculography: general principles and basic technology. In: Chokroverty S, ed. Sleep Disorders Medicine: Basic Science, Technical Considerations, and Clinical Aspects. Boston: Butterworth-Heinemann, Ch 7.

Wilkinson RT, Edwards RS, Haines E. (1996) Performance following a night of reduced sleep. Psychonomic Science, 5:471-2.

Winer BJ. (1962) Statistical principles in experimental design. New York: McGraw-Hill, 1962.

Wylie D, Miller JC, Schultz T, Mitler MM, Mackie RR. (1996) Commercial driver fatigue, loss of alertness, and countermeasures. Washington, DC: Department of Transportation.

8

STUDY OF COMMERCIAL VEHICLE DRIVER REST PERIODS AND RECOVERY OF PERFORMANCE IN AN OPERATIONAL ENVIRONMENT

Sesto Vespa, Transportation Development Centre, Transport Canada, Montreal, Canada, Dennis Wylie, Wylie and Associates, Goleta, California, Merrill Mitler, The Scripps Research Institute, La Jolla, California, Ted Shultz, Baker & Schultz, Santa Barbara, California

INTRODUCTION

Background

In the fall of 1989, Transport Canada and the FHWA began a cooperative research effort that culminated in the completion of the largest and most comprehensive over-the-road study ever conducted on driver fatigue and alertness in North America – the Commercial Motor Vehicle Driver Fatigue and Alertness Study (DFAS) – the results of which were released in January 1997 (Wylie et al., 1996a, 1996b; Freund and Vespa, 1997a). In addition to its scientific objectives, the study was also intended to provide a scientifically valid basis to determine the potential for revisiting the hours-of-service rules that had remained essentially unchanged for more than 50 years.

Transport Canada was also interested in obtaining objective scientific data on the duration of off-duty time required for driver recovery from cumulative fatigue with a view to examining related aspects in the hours-of-service rules. During 1992, the Canadian trucking industry also proposed to provincial and federal governments the establishment of a "reset" rule as a primary feature of revised hours of service (HOS) regulations. Currently, the Canadian HOS allow a maximum of 13 hours of driving within a 15-hour period of duty before an 8-hour rest period is required. The multi-day cumulative maxima are 60 hours on duty in a 7-day period or 70 hours on duty in an 8-day period. There is no requirement for days off. The proposal of a 36-hour reset provision would institute a rest period of 36 hours to "reset the clock" once drivers reach the cumulative on-duty maximum of 60 hours (Gough and Grey, 1992).

In 1993, Transport Canada initiated a separate but related research initiative to that of the DFAS – referred to as the Recovery Study – which involved collecting data from additional trips made by some of the Canadian drivers who were participating in the two Canadian observational conditions of the DFAS. Transport Canada wished to use the results of this study to help assess whether a 36-hour off-duty period should be allowed to reset the clock on cumulative multi-day on-duty hours, on the basis of whether it allowed sufficient time for drivers to recover from fatigue effects. The results of the Recovery Study (Wylie et al., 1997) are presented here as well as a related assessment that seeks to validate and explain them. Several overall observations are also formulated that emanate from this research and that may be useful when considering rest and recovery periods in an operational context.

Investigative Approach, Constraints and Validation of Results

To minimize issues surrounding the validity of a laboratory-based study and to enhance the face validity of the results, Transport Canada decided that a field study of off-duty periods and associated driver recovery would be advisable. This decision also considered industry's view, expressed during research planning meetings held for the DFAS and Recovery Study, that they would prefer a study based on data obtained in a real-world operational environment – a view also shared by Transport Canada. Further, because operationally useful answers were being sought, a great amount of importance was placed on examining recovery at the end of work periods during which the work regimen as well as daily and cumulative sleep deficits were as close as possible to those which could be expected in an intensive long-haul operation.

The then ongoing DFAS research provided the ideal opportunity to implement the desired study concept and undertake a rigorous data collection program at an affordable cost, while also making possible an immediate start on data collection. However, because of funding and time constraints – since the DFAS data collection program was already under way and some drivers had already completed their participation – only a limited number of drivers could be included in the Recovery Study data collection. Because of the smaller number of drivers who would participate in the Recovery Study by comparison with the DFAS, there was a disadvantage to this approach in that the statistical tests would not have the same power to detect effects and that the statistical reliability of the results were more limited. Nonetheless, there were important counterbalancing arguments to consider. Plans were also made for further study to offset this disadvantage to the greatest extent possible. These issues are discussed briefly in the following paragraph.

In a study of driver rest periods and recovery, sleep data is of crucial importance in the interpretation of outcome. Because of the availability of other sleep data from the same drivers within the larger DFAS group of drivers, statistical uncertainty due to sample size limitations could be examined through comparisons between Recovery Study and DFAS results. Although not ideal, this approach permitted a logical comparative analysis of the results, their consistency, and identification of any underlying trends. Also planned was a comparison of the results with other more general data that might be available from the technical literature, which would allow an appreciation of the reasonableness of the results obtained. Consequently, a broad review of the scientific literature was planned to be concurrently undertaken to assess current knowledge on recovery from fatigue due to extended hours of work during the day and over a period of several days. Because of the difficulties, required lead time, and expense of gathering additional data in the field independently of the DFAS process (Freund and Vespa, 1997b) – which was out of the question – this was a practical approach. Figure 1 graphically presents the conceptual framework for assessment of the Recovery Study results.

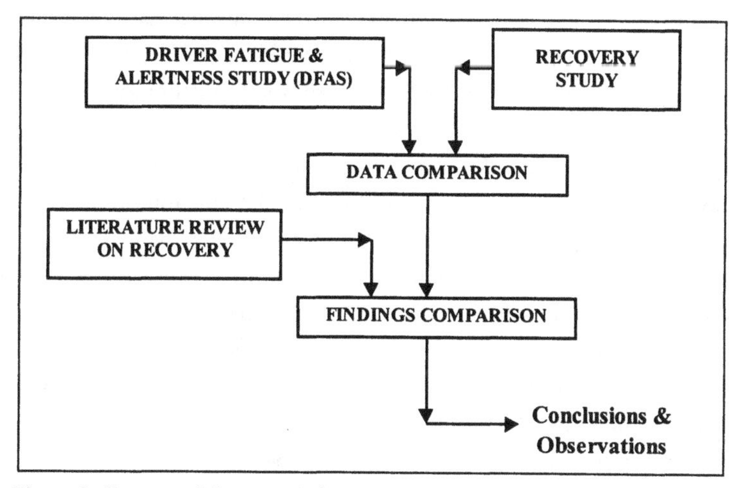

Figure 1. Conceptual framework for assessment of Recovery Study results.

Overview of Driver Fatigue and Alertness Study (DFAS)

A brief overview of the DFAS is presented here to facilitate understanding of the Recovery Study design and various comparisons of results to be presented later in this paper.

The DFAS data collection involved 20 drivers in each of four observational Conditions, two in Canada and two in the U.S. (i.e., a total of 80 drivers), who were monitored over a period of 16 weeks. The overall DFAS data collection program and study design are presented in Figure 2. A number of work-related factors thought to influence the development of fatigue, loss of alertness, and degraded driving performance in commercial motor vehicle drivers, were studied within an operational setting of revenue-generating trips. Factors studied included: the amount of time spent driving during a work period; the number of consecutive days of driving; the time of day when driving took place; and schedule regularity. The main findings of this study, from an operational perspective, are summarized in Table 1.

Recovery Study Methodology

Work/Rest Schedules and Observational Conditions

For the Recovery Study, field data was collected from 25 of the 40 Canadian drivers in the DFAS, who drove a total of 55 additional trips after completing their scheduled DFAS trips. This resulted in five new observational conditions that spanned a maximum of eight workdays. Table 2 shows these observational conditions and the associated work/rest schedules. To facilitate understanding, Table 2 also shows the relationships between DFAS and recovery study observational conditions. The first column, Condition, shows the Conditions (3-5, 4-6, 4-7, 4-8, and 4-9) into which data used for this study is grouped. Each condition is associated with either zero, one, or two workdays off following the fourth trip. One workday off nominally spanned 36 hours of time off between the fourth and fifth trips. Two workdays off was nominally 48 hours. No workday off was nominally 12 hours. For operational reasons, the observed amount of time off varied somewhat from that of the specified nominal. Three of the five conditions included the 36-hour off-duty period, of which two had four more workdays follow the time off while the third had one more workday follow the time off. The remaining two conditions included 12 and 48 hours of time off, respectively, and these both had one more workday follow the time off. The Recovery Study conditions were constructed as follows.

Data was collected from five drivers of DFAS Condition 3 during four additional 24-hour periods on duty following 36 hours off. Condition 3 drivers had been on duty for four days prior to the time off. The follow-up recordings on Condition 3 drivers are coded as Condition 5. We also collected data from five drivers in DFAS Condition 4 during four more 24-hour periods after they had one 36-hour period off duty. These follow-up recordings on Condition 4 drivers are coded as Condition 9. Combining the data collected under Conditions 3 and 4 with that of Conditions 5 and 9, respectively, results in the Recovery Study Condition 3-5 (nightstart) and Condition 4-9 (daystart).

Additional data was collected from other DFAS Condition 4 drivers as follows: for three drivers, during one more workday following the fourth workday – without an intervening

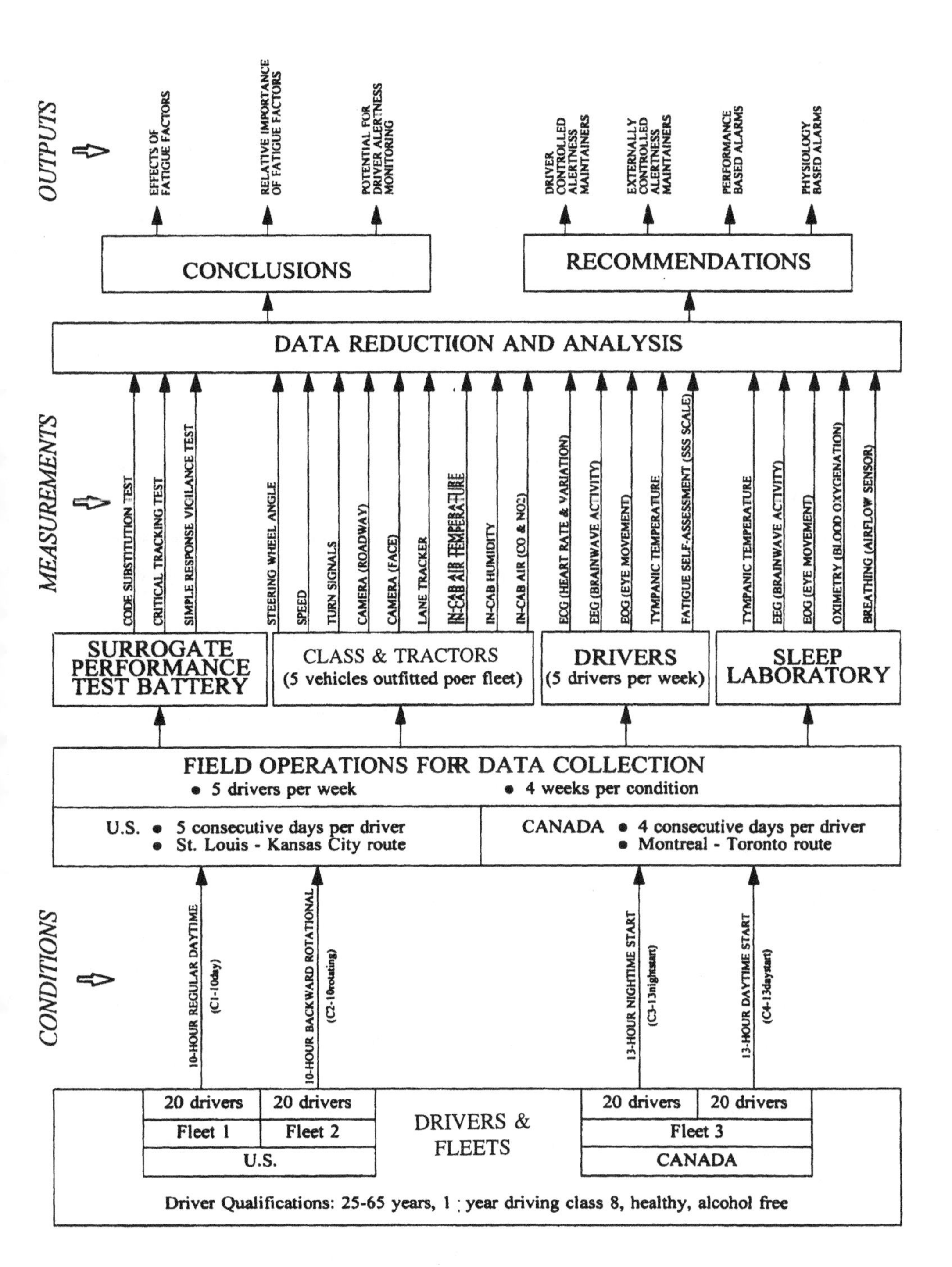

Figure 2. Overall data collection program and design of Commercial Motor Vehicle Driver Fatigue and Alertness Study (DFAS).

Table 1. Main findings of Commercial Motor Vehicle Driver Fatigue And Alertness Study.

1. Drowsiness during driving was much more a result of time of day (circadian rhythm) effects than duration of driving time and cumulative number of days (e.g., given the amount of driver sleep and the four to five days of driving observed in the study).
2. Little or no operationally significant differences in driver drowsiness and performance were identified between the 10-hour and 13-hour schedules during daytime driving for the various measures used. Comparisons for nighttime driving could not be similarly made because the experimental design did not provide for them.
3. Drowsiness was much greater during nighttime driving than daytime driving; an eightfold increase in drowsiness was indicated by judgments of face video recordings.
4. Drivers were judged to be drowsy in approximately 4.9% of the face video recordings reviewed and in 0.008% of the EEG recordings, most of which occurred between about midnight and six a.m.
5. Drivers were poor judges of their own levels of fatigue; they rated themselves more alert than they were and their self-ratings correlated with duration of driving time and cumulative number of trips but not measured performance.
6. There were large differences among drivers in levels of alertness and performance; however, no significant relationships were found between driver age and fatigue. Eleven of the eighty drivers (14%) accounted for 54% of all drowsiness episodes observed in face video recordings.
7. Eight hours off between duty periods might not provide enough time to obtain adequate sleep and take care of personal needs when, as reported by the drivers in the study, 7.2 hours on average is their ideal sleep time.
8. Some drivers in this study did not manage their off-duty time to obtain an adequate amount of sleep even when there was ample opportunity for sleep; average observed time in bed was 5.2 hours and time asleep was 4.8 hours.
9. The sleep shortfall (relative to their reported ideal) was greatest for those drivers who were required to sleep during the daytime, who had an average of 4.4 hours in bed and 3.8 hours asleep.
10. Although they obtained less sleep than they needed, all drivers had efficient and normally structured sleep during their principal sleep periods, as judged by formal clinical criteria.
11. Forty-five percent of drivers stopped and took naps during their trips. They added an average of 27 minutes (11%) to their total daily sleep times.

workday off (i.e., Condition 6); for six drivers, during one more workday following one workday off duty (i.e., Condition 7); for six drivers, during one more workday, following two workdays off duty (i.e., Condition 8). Combining the data collected for these drivers under Condition 4 with that of Conditions 6, 7, and 8 provides the recovery study Conditions 4-6, 4-7, and 4-8, all daystart conditions.

Because a 36-hour off-duty period was of particular interest in this study, sixteen of the 25 drivers were given a 36-hour off-duty period in order to increase the power of the statistical tests to detect effects. The other conditions were used primarily as an aid in interpreting the results of the 36-hour observational conditions by examination of the consistency of the results and any underlying trends in the data.

Data Collection Equipment and Procedures

Data collection equipment and procedures used in the Recovery Study were identical to those of the DFAS. They were described in detail in the final report issued on that project and, for brevity, will not be repeated here. Rather, the Recovery Study observational conditions will be detailed involving drivers and work/rest schedules as well as their relationship to the DFAS observational conditions.

The work/rest schedules implemented for this Recovery Study were designed to take advantage of the fact that each group of five drivers participating in the DFAS would normally reach the on-duty limit specified in the Canadian hours-of-service regulations (60 hours in seven days) by the end of the fourth workday, after which the drivers would have to leave the study and the five instrumented trucks would be idle for the following three workdays until the subsequent set of five drivers arrived. By taking advantage of the 14-day cycle provisions in the Canadian hours-of-service rules (which allow 120 hours of on-duty time during a period of 14 consecutive days, but require a 24-hour off-duty period before completing 75 hours), it was possible to continue the drivers' participation during the remaining three workdays of each week. Additionally, at the completion of data collection at the end of the fourth week of each of the two Canadian conditions, the participation of each set of five drivers was extended by an additional four workdays by delaying the start of data collection for the subsequent DFAS observational condition. Because the decision to undertake this recovery study was taken while the DFAS work was underway and due to the lead time required to institute the study protocol and purchase additional equipment, the first fifteen drivers of observational Condition C3 (C3-13nightstart) could not be monitored.

Table 2. Observational conditions and driving schedules for Recovery Study and their relationships to those of the DFAS.

Condition	Number of drivers	Workdays								
		1	2	3	4	5	6	7	8	9
1	20	X	X	X	X	X	U.S. drivers of DFAS Conditions 1 and 2 drove five turnaround trips each.			
2	20	X	X	X	X	X				
3	20	X	X	X	X		Canadian drivers of DFAS Conditions 3 and 4 drove four turnaround trips each.			
4	20	X	X	X	X					
3-5	5	X	X	X	X		X	X	X	X
4-6	3	X	X	X	X	X				
4-7	6	X	X	X	X		X			
4-8	6	X	X	X	X			X		
4-9	5	X	X	X	X		X	X	X	X

DFAS OBSERVATIONAL CONDITIONS	
Condition 1 (C1-10day)*	10-hrs driving nominally starting at same time each morning.
Condition 2 (C2-10rotating)*	10-hrs driving starting three hrs earlier each day.
Condition 3 (C3-13nightstart)*	13-hrs driving nominally starting at midnight in Toronto.
Condition 4 (C4-13daystart)*	13-hrs driving nominally starting at noon in Montreal.

RECOVERY STUDY OBSERVATIONAL CONDITIONS	
Condition 3-5 (nightstart)	Condition 3 drivers who had one workday (e.g., 36 hours) off after the fourth workday, and then did four added workdays on the same schedule.
Condition 4-6 (daystart)	Condition 4 drivers who drove a fifth workday, without a workday off following the fourth workday.
Condition 4-7 (daystart)	Condition 4 drivers who drove a fifth workday, following one workday (e.g., 36 hours) off.
Condition 4-8 (daystart)	Condition 4 drivers who drove a fifth workday, following two workdays (e.g., 48 hours) off.
Condition 4-9 (daystart)	Condition 4 drivers who had one workday (e.g., 36 hours) off after the fourth workday, and then did four added workdays on the same schedule.

* Expression within brackets refers to name of observational condition used in DFAS final report (Wylie et al., 1996a, 1996b).

The Recovery Study data collection protocol included one significant departure from that of the DFAS and this involved driver monitoring during the one workday (36 hours) and two workdays (48 hours) of off-duty recovery time taken after accumulating the weekly on-duty maximum (i.e., taken after the fourth workday). For both the DFAS and Recovery Study, drivers were continuously monitored - 24 hours per day - by EEG methods during the weekly work/rest cycle, both during sleep and wake cycles. However, for the recovery period of one and two workdays, drivers had their EEG monitoring equipment removed before leaving the sleep centre. They were instructed to return home and conduct themselves normally, but they were to return to the sleep centre at any and whatever times they chose to take their principal sleep periods. Upon arrival at the sleep centre they were then prepared for the usual monitoring by polysomnographic methods during sleep. Drivers were unmonitored during their at-home recovery periods. The reported sleep times over the one and two workdays off are thus for those sleeps taken at the sleep centre and are based on EEG data. There is, however, no information on naps or sleep time that drivers may have taken at home.

Measures and Data Analysis

Although measures and data collection procedures for this recovery study were identical to those of the DFAS, the analysis was substantially different. It focussed principally on identifying changes in dependant measures of driver performance between workdays before and after the various off-duty periods. The analysis also took advantage of the knowledge gained from the DFAS such that emphasis was placed on those measures that were shown to be the most valuable in assessing driver performance. As one consequence of this, the decision was taken not to undertake polysomnography (PSG) analysis of EEG data collected during driving. Funding was constrained and the results of the DFAS had shown that few episodes of "PSG-Drowsy Driving" could be expected to be included in the database, and would probably not have allowed meaningful data analysis.

The resulting set of measures used in the analysis included: sleep time in principal sleep periods based on continuous EEG measurement and done according to Rechtscaffen and Kales methods; prevalence of drowsiness based on face video recordings, lane tracking standard deviation (LTSD); performance on surrogate tests that included Code Substitution (CS), Critical Tracking Test (CTT), and derivative measures from the Simple Response Vigilance Test (SRVT) known as Lapses and Reciprocal Vigilance Score (RVS); and, self-ratings using the Stanford Sleepiness Scale (SSS).

RECOVERY STUDY RESULTS

Sleep Duration

Drivers Working Two Four-Day Cycles with One Intervening Workday (36 Hours) Off

Figure 3 presents average total sleep duration for all principal sleep periods of the two groups of five drivers of Conditions 3-5 (nightstart) and 4-9 (daystart). The first five sleeps are those sleeps these drivers took during their participation in DFAS Condition 3 and Condition 4 respectively. Then, there are five more sleeps for Condition 3-5 drivers, and four more sleeps for Condition 4-9 drivers. These drivers had a nominal 36-hour period off (e.g., one workday with no work-related duties) which included sleeps 5 and 6. Sleep 1 was taken during the off-duty period prior to the first drive. Drivers drove Trip 1 between Sleeps 1 and 2, Trip 2 between Sleeps 2 and 3, and so on, for the duration of their participation. There are only nine principal sleep periods for the Condition 4-9 drivers because they ended their participation in the study after the drive following their ninth sleep.

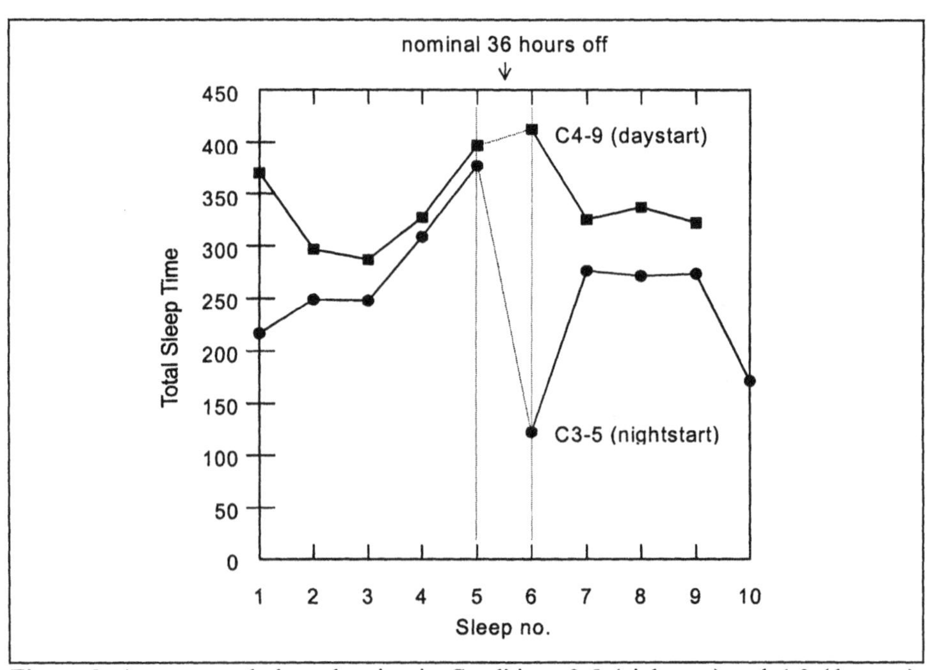

Figure 3. Average total sleep duration in Conditions 3-5 (nightstart) and 4-9 (daystart), both before, during, and after the 36 hours off duty.

Drivers in Condition 4-9 (daystart) obtained significantly more sleep than did drivers in Condition 3-5 (nightstart). This pattern continued throughout the drivers' participation in the study. The marked drop in sleep time for Condition 3-5 (nightstart) drivers at the tail end of the 36-hour off-duty period was statistically significant and reflects a truncation of recovery. A less truncated pattern of recovery sleep is seen for Condition 4-9 (daystart) drivers, who show increased sleep during Sleep 6. Inspection of the sleep parameters relevant to sleep continuity and structure did not reveal systematic differences from the data of DFAS Conditions 1 to 4.

Drivers Working Five Days with Zero, One, or Two Workdays Off after the Fourth Workday Figure 4 presents average total sleep duration during all principal sleep periods of the 15 drivers in Conditions 4-6, 4-7, and 4-8, all daystart Conditions.

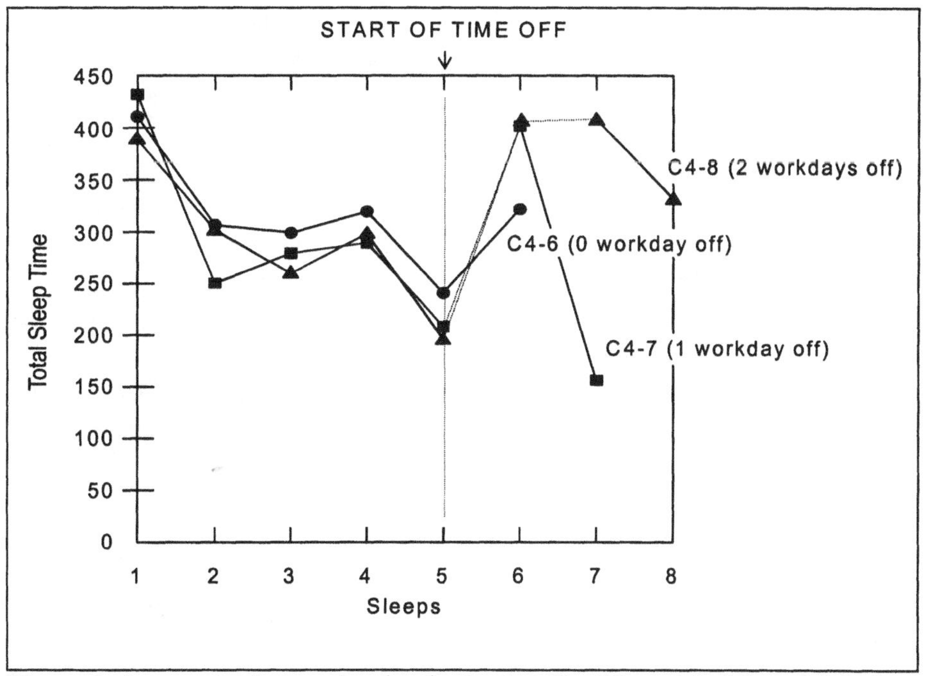

Figure 4. Average total sleep duration in daystart Conditions 4-6, 4-7, and 4-8, (e.g., nominally 12, 36, or 48 hours off), respectively.

There was no overall statistically significant difference between the three groups. However, the extended recovery sleep on sleep 6 relative to sleeps 1 to 5 is obvious for the group with one workday off, Condition 4-7. Likewise, the extended recovery sleep on sleeps 6 and 7 relative to sleeps 1 to 5 is obvious for the group with two workdays off, Condition 4-8. Inspection of the sleep parameters relevant to sleep continuity and structure did not disclose systematic differences from the data of DFAS Conditions 1 to 4.

Drowsiness

The Mann-Whitney U-test was used to compare the average proportion of analysis epochs judged "drowsy" – based on face video recordings – for all the Recovery Study observational conditions (i.e., 3-5, 4-6, 4-7, 4-8, 4-9) aggregated by Condition and half trip. The test failed to reveal statistically significant differences in prevalence of drowsiness between the first four trips completed by each driver and subsequent trips taken after the prescribed recovery periods.

In the symmetrical Conditions 3-5 (nightstart) and 4-9 (daystart), drivers completed four trips, then took 36 hours off, then completed an additional four trips. The repeated-measures analysis of covariance of the arcsine transform of the proportion of analysis epochs judged "drowsy," with between-subjects factors "Condition" (= 3-5, 4-9) and "Age" (covariate) and within-subjects factors "Trip" (= 1st, 2nd, 3rd, 4th) and "Set" (= first set of four trips, second set of four trips) showed no significant effect of Condition or Set, but a significant interaction of Set by Trip by Condition, $F(3,21)=5.0$, $p<.009$. The means are shown in Figures 5a and 5b. It can be seen that there was a relatively high proportion of drowsiness during Trips 2 and 3 of the first four trips of Condition 4-9 (daystart), although this does not seem to bear on the issue of recovery. Contrasting the trip preceding the 36-hour time-off period with the trip following it revealed no difference in both conditions 3-5 and 4-9.

In summary, recovery effects of the 36 hours off duty were not apparent for the prevalence of drowsiness measure.

Lane Tracking

Lane tracking standard deviation (LTSD) by trip for Conditions 3-5, 4-9, 4-6, 4-7, and 4-8 are shown in Figures 6, 7, 8, 9, and 10, respectively.

Each mean in these figures is accompanied by error bars showing standard error of the mean. It can be seen from these figures that on the fifth trip, mean LTSD is at or above its highest preceding value. In the case of the Conditions with the two driving cycles of four days with an intervening workday off, Conditions 3-5 (nightstart) and 4-9 (daytstart), LTSD during the second set of four days, when compared with the first set, shows no evidence of recovery.

In summary, there was no evidence of recovery of lane tracking performance following zero, one, or two workdays off.

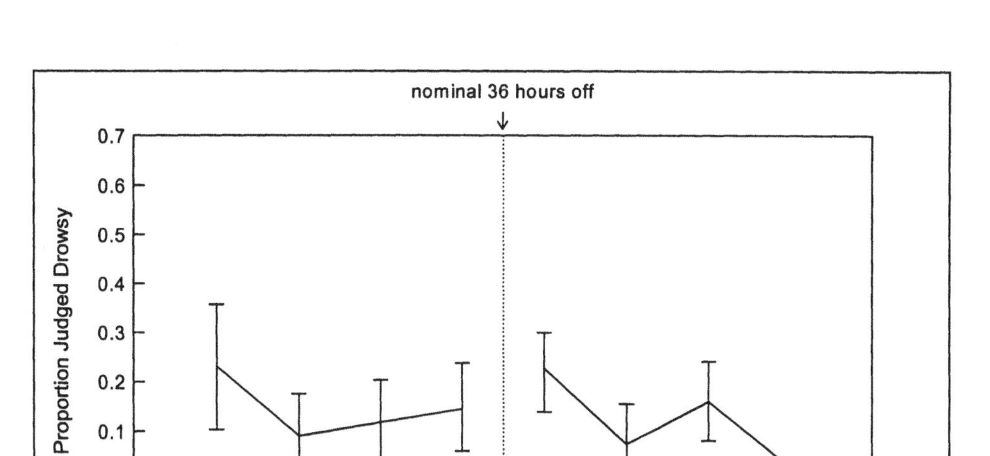

Figure 5a. Proportion of analysis epochs judged "drowsy", Condition 3-5 (nightstart); Set 1 (Trips 1-4), Set 2 (Trips 5-8).

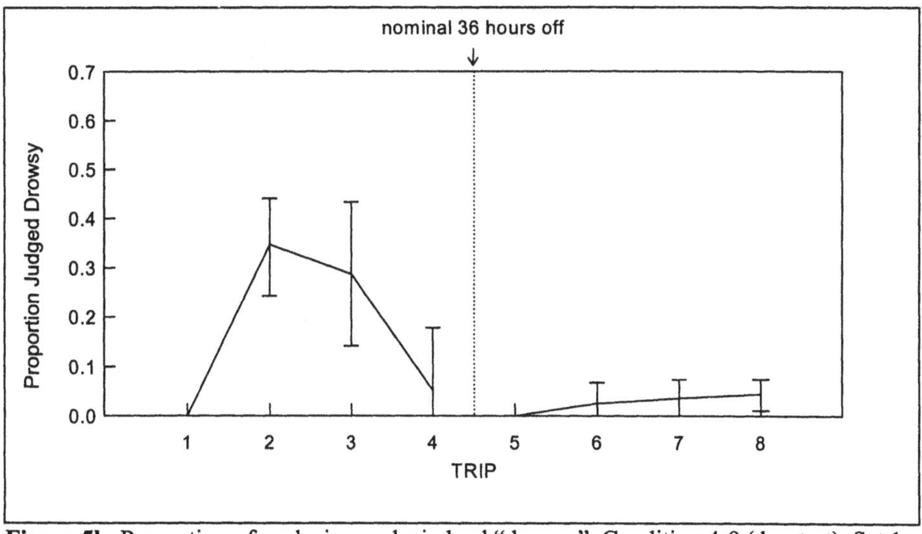

Figure 5b. Proportion of analysis epochs judged "drowsy", Condition 4-9 (daystart); Set 1 (Trips 1-4), Set 2 (Trips 5-8).

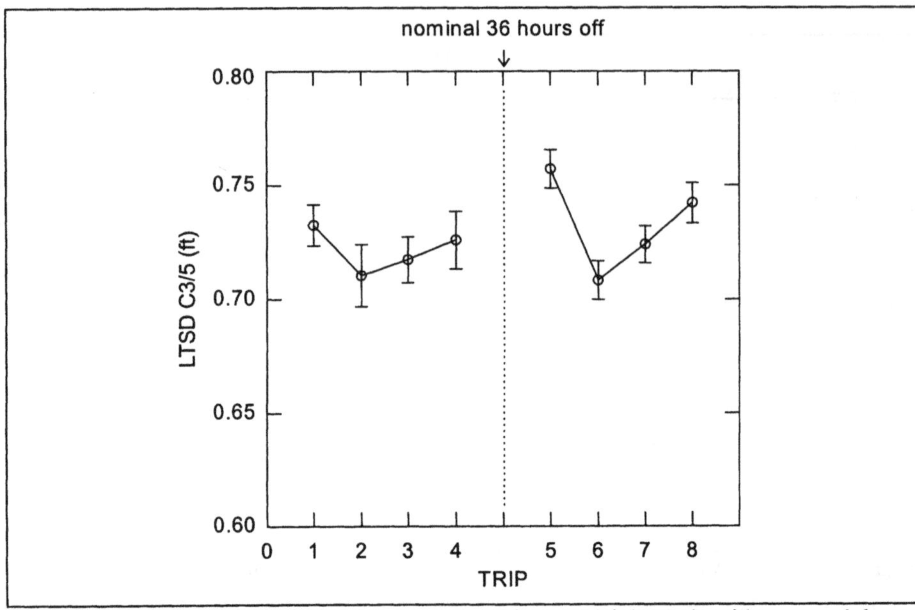

Figure 6. Lane tracking standard deviation, Condition 3-5 (nightstart); with one workday (36 hours) off after Trip 4.

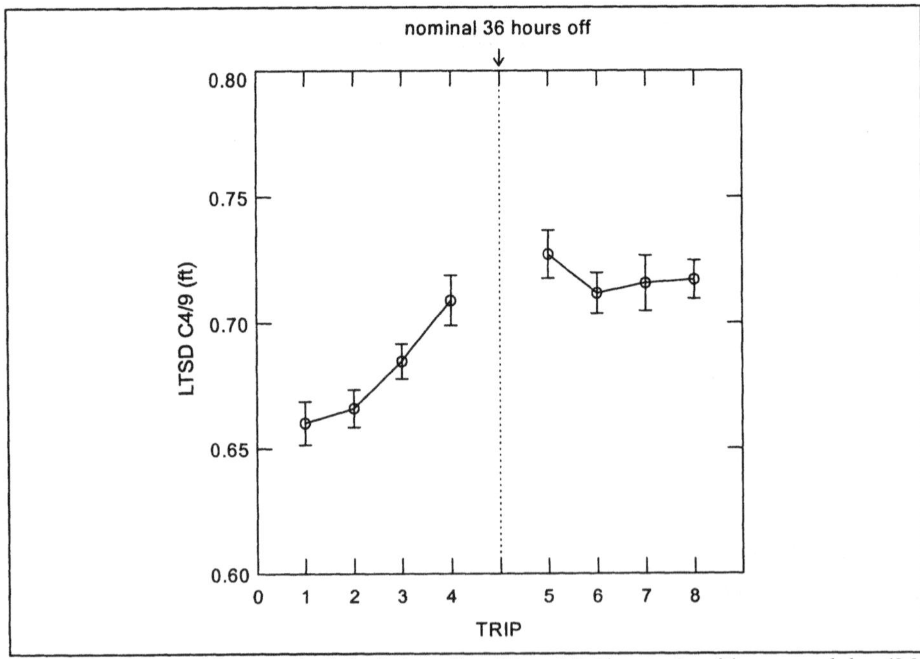

Figure 7. Lane tracking standard deviation, Condition 4-9 (daystart); with one workday (36 hours) off after Trip 4.

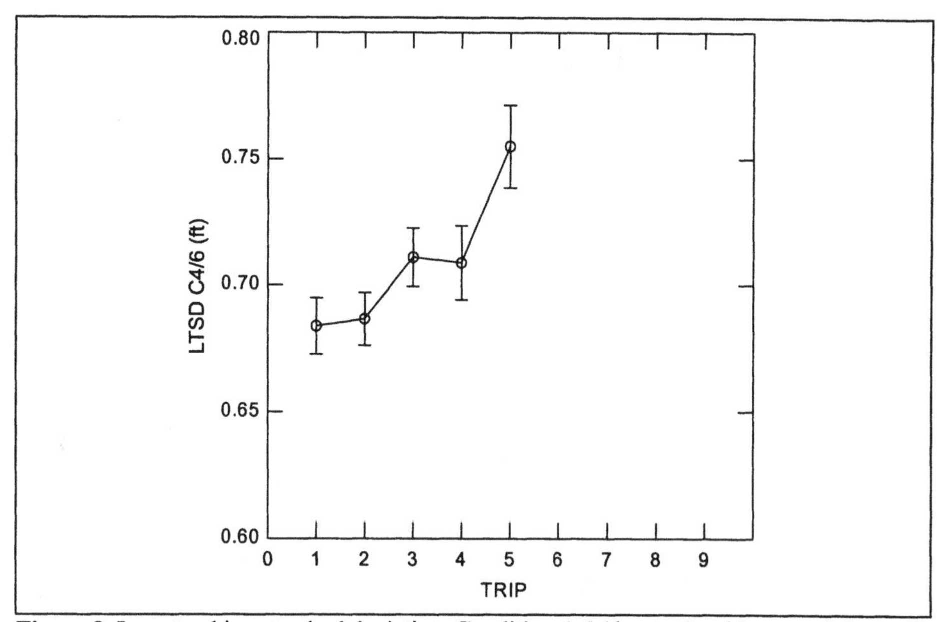

Figure 8. Lane tracking standard deviation, Condition 4-6 (daystart); with no workdays off after Trip 4.

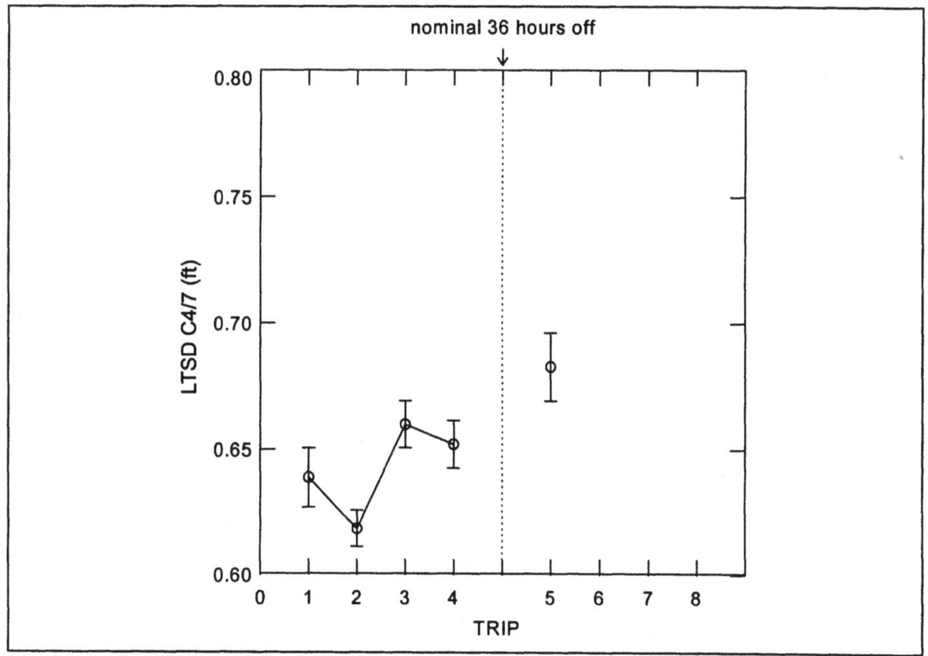

Figure 9. Lane tracking standard deviation, Condition 4-7 (daystart); with one workday (36 hours) off after Trip 4.

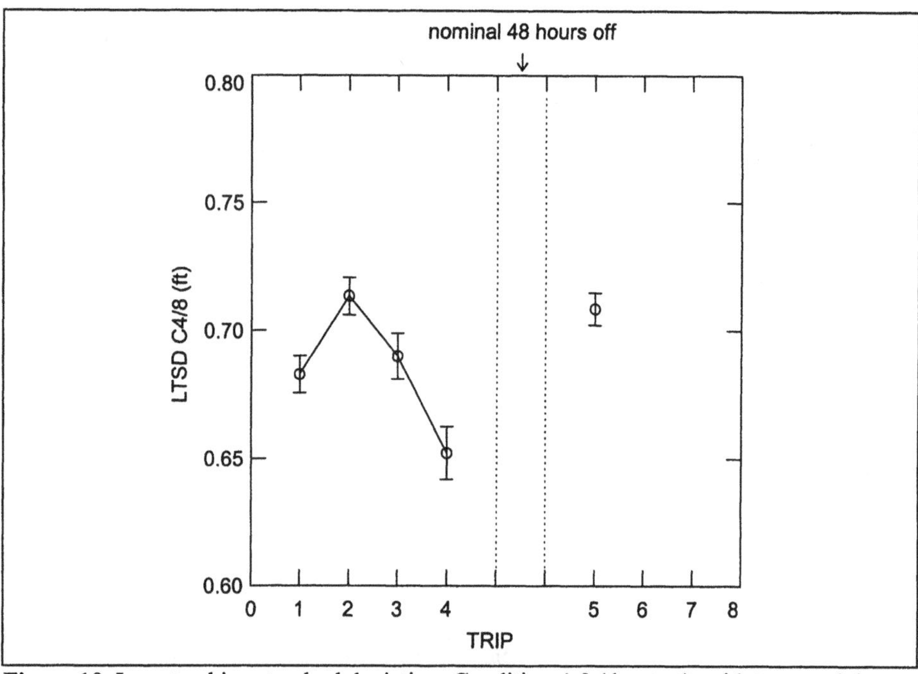

Figure 10. Lane tracking standard deviation, Condition 4-8 (daystart); with two workdays (48 hours) off after Trip 4.

Surrogate Performance Tests

This section describes the results of analyses of surrogate performance tests. Scores on the second set of four trips were compared with scores on the first four trips. The direction of change in scores (degradation versus improvement) was not specified in advance, and therefore two-tailed tests of significance were used throughout. (The definition of each measure, the shape of its distribution, and its correlation with independent variables, other than length of time between the fourth and subsequent trips, are discussed in the DFAS report.) The focus of this section is the change in driver performance between the fourth trip, after which the prescribed off-duty period is taken, and subsequent trips. The ANOVAs (i.e., analysis-of-variance) performed can be divided into two main groups:

- those that compared a second set of four trips with the first set of four trips, with one intervening workday (36 hours) off (Conditions 3-5 and 4-9; N = 10), and
- those that contrast the fifth trip with the fourth trip,
  - with no intervening workday off (Condition 4-6; N = 3),
  - after one workday off (Conditions 3-5,4-7,4-9; N = 16), and
  - after two workdays off (Condition 4-8; N = 6).

The results presented in this subsection are based on three administrations of the performance test battery. Whereas Conditions 1 and 2 of the DFAS had four administrations per trip, Conditions 3 and 4 drivers had only three administrations per trip. A fourth administration (administration number three which was taken just prior to inbound departure at the mid-trip turnaround point in Conditions 1 and 2) had to be eliminated because it would have caused these drivers to exceed allowable on-duty time limits specified by Canadian hours-of-service regulations.

Performance of Drivers Working Two Four-Day Cycles with One Intervening Workday (36 Hours) Off Five nightstart (Condition 3-5) and five daystart (Condition 4-9) drivers took a workday (36 hours) off following their fourth trip, then drove a second set of four trips. Their performance test scores did not differ markedly from those of the other drivers in Conditions 3 and 4. In the discussion that follows, the two-level variable "Set" distinguishes the first set of four trips (Set 1) from the second set of four trips (Set 2). Repeated measures ANOVAs were performed for each of the four performance test measures (e.g., CS, CTT, Lapses, RVS), structured as Condition (two levels), by Set (two levels), by Trip (four levels), and by Administration (three levels). Set, Trip, and Administration are within-subject repeated measures.

Code Substitution (CS) Test: The results of the ANOVA of Code Substitution (CS) test score results are shown in Table 3. There was a statistically significant Administration by Condition interaction, as one would expect from the sizeable impact of time-of-day on performance. Drivers performed more poorly on CS at night than during the day. Figure 11 shows driver CS means on each of the 24 test sessions (three sessions on each of eight trips). T1 through T4 are the first set of four trips, and T5 through T8 are the second set. It would be difficult to describe the improvement between Trips 4 and 5 as recovery of function, since recovery implies a prior degradation of performance, and clearly there was none. The statistically significant Set and Trip effects appear to be the result of ongoing skill acquisition with practice. The nature of the Administration and Administration by Condition interactions is illustrated in Figures 12 and 13. Both figures show better scoring during the day.

The median time of day at the start of each test administration in Conditions 3-5 through 4-9 is shown in Table 4. Median times better characterize the general case, rather than means, since means were strongly influenced by a few drivers who did not keep to schedule.

The nightstart drivers (Condition 3-5) performed better during Administration 4, which occurred in the early afternoon (Figure 12). The daystart drivers performed better during Administration 1, which occurred around noon of each day (Figure 13).

Table 3. Results of ANOVA of CS test for Conditions 3-5 (nightstart) and 4-9 (daystart), contrasting the first and second sets of four trips.

Factor	F-ratio	df1	df2	Probability
Condition	0.968	1	8	0.354
Set	21.641	1	8	0.002
Trip	4.419	3	24	0.013
Administration	4.459	2	16	0.029
Administration * Condition	6.552	2	16	0.008

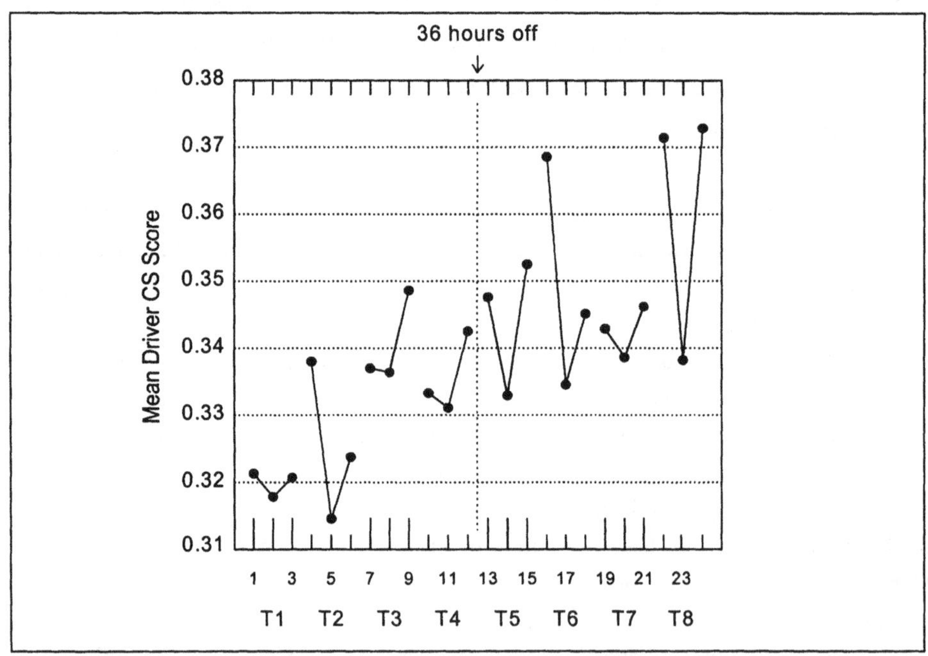

Figure 11. Combined CS scores from Conditions 3-5 (nightstart) and 4-9 (daystart) showing improvement with each successive trip.

Table 4. Median start times of surrogate performance test administrations. (N.B. Administration 3 was not performed in these Conditions.)

Condition	Administration		
	1	2	4
3-5	22:51	6:09	13:27
4-6	10:24	17:48	1:06
4-7	13:00	20:06	4:09
4-8	13:27	21:06	5:03
4-9	11:18	18:27	1:36

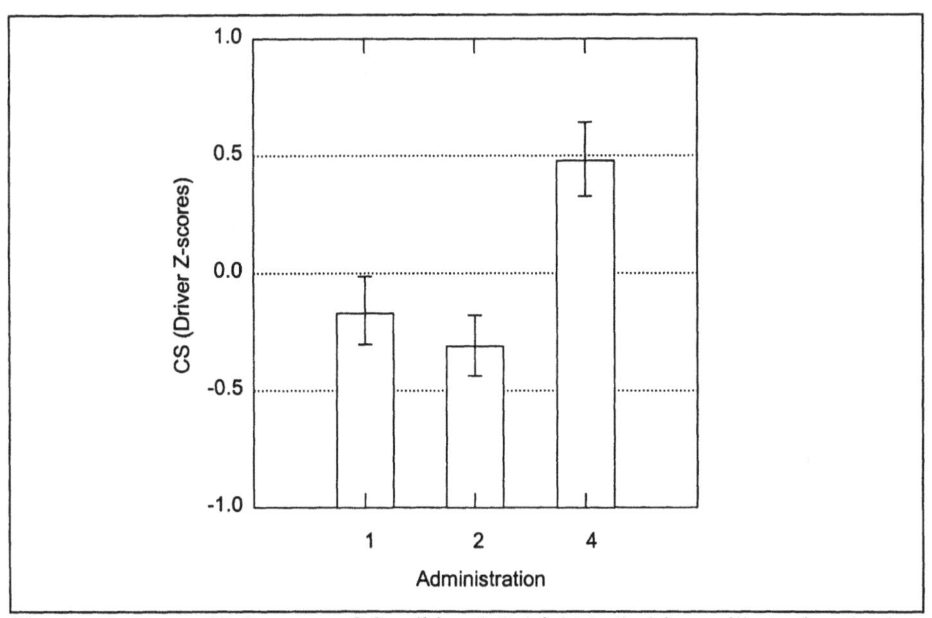

Figure 12. Mean CS Z-scores of Condition 3-5 (nightstart) drivers, illustrating daytime score improvement on Administration 4. (N.B. Administration 3 was not performed in these Conditions.)

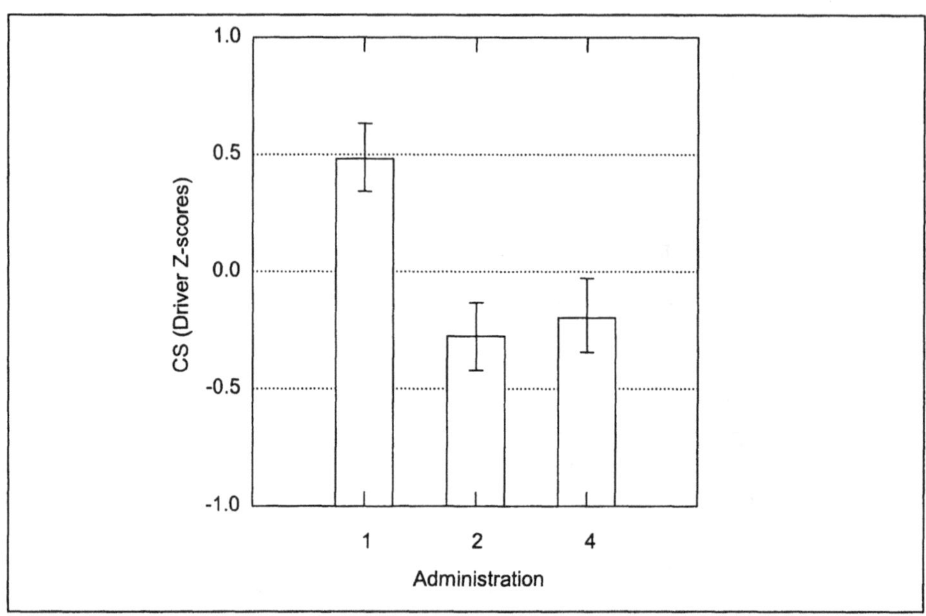

Figure 13. Mean CS Z-scores of Condition 4-9 (daystart) drivers, showing lower scores on Administration 2 and 4, which occurred at night. (N.B. Administration 3 was not performed in these Conditions.)

Critical Tracking Test (CTT): The univariate repeated measures ANOVA of Critical Tracking Test (CTT) scores failed to reveal statistically significant main or interaction effects. Minor differences in CTT between the nightstart and daystart 13-hour trips lack statistical significance. Figure 14 shows the drivers' CTT performance by trip for Conditions 3-5 and 4-9, with 36 hours off between Trips 4 and 5. There was a slight tendency for improvement on the first three trips of both 4-trip sets, and a performance decrement on the last trip of each set. The slight improvement in CTT over the 36 hours between Trip 4 and Trip 5 is not statistically significant.

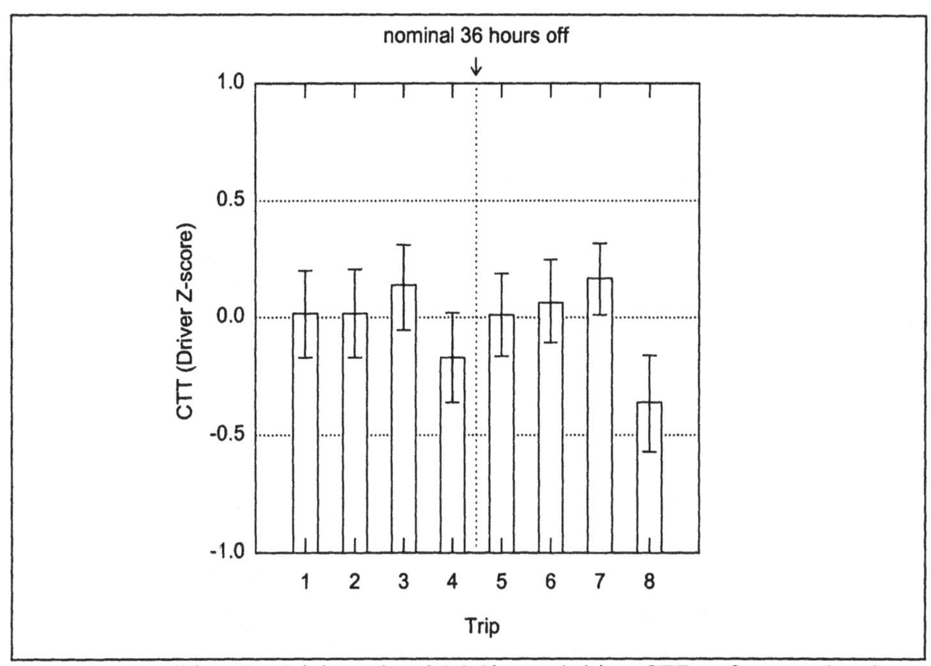

Figure 14. Conditions 3-5 (nightstart) and 4-9 (daystart) driver CTT performance by trip.

Number of Lapses during Simple Response Vigilance Test (SRVT): The ANOVA of Normalized Lapses failed to show any statistically significant differences between Conditions 3-5 and 4-9, the two sets of four trips, or the four trips of each set. The only statistically significant effect revealed was one of decreasing performance (greater number of Lapses) by Administration, $F(2,14) = 9.452$, $p = 0.003$. The effect is visible in both the daystart (Condition 4-9) and nightstart (Condition 3-5) data. Figures 15 and 16 illustrate the trend of increasing numbers of Lapses during each trip. Figure 15 shows performance on the nightstart trips (Condition 3-5), and Figure 16 shows performance on the daytstart trips (Condition 4-9). The reader is reminded that the scores derived from the Simple Response Vigilance Test (Lapses, number of response latencies greater than 500 milliseconds; and RVS, reciprocal of median response latency) appeared to vary with ambient light levels, which would have influenced stimulus intensity on the CRT display used for the SRVT. Ambient light level during test administration was not controlled, but appeared to have less effect on the 13-hour conditions than in the 10-hour conditions. The effect of ambient light levels is discussed more fully in the results section of the main DFAS report.

The structure of the ANOVA obscured a trend of increasing numbers of Lapses with each successive trip after Trip 3. This effect, which was not statistically significant, can be seen in Figure 17, which also shows driver self-ratings of sleepiness on the Stanford Sleepiness Scale. It is particularly interesting that driver self-ratings tended toward recovery after the 36-hour break between Trips 4 and 5, despite no reduction in the numbers of

Lapses experienced. Drivers tended toward greater numbers of Lapses on the first day of the second cycle.

Response Vigilance Score (RVS) from the Simple Response Vigilance Test (SRVT): The ANOVA of RVS data failed to show any statistically significant differences between the two Conditions, the two sets of four trips, or the four trips of each set. Although there was a tendency toward decrement in performance at the end of both daystart and nightstart trips this change fell short of statistical significance. The only statistically significant effect revealed by the ANOVA of RVS data was an Administration by Trip interaction, $F(6,42) = 2.607$, $p = 0.031$ (see Figures 18 and 19). The degradation within each trip became less with each added trip, but this effect was only of marginal statistical reliability given the number of ANOVAs performed on this data.

The structure and results of this ANOVA obscured RVS degradation that occurred across all eight trips of Conditions 3-5 (nightstart) and 4-9 (daystart) – see Figures 18 and 19. Figure 18 shows the combined Conditions 3-5 and 4-9 RVS data for each test session, grouped by trip number. The decrease in RVS (indicating increasing reaction time) across the eight trips of Conditions 3-5/4-9 was statistically significant, $F(7,49) = 3.57$, $p = 0.003$. The 36-hour off-duty period occurs between trips T4 and T5. It is particularly remarkable that drivers had worse RVS scores following their 36 hours off. This is seen in the poor performance of the night start drivers on T5 and T6, visible in Figure 19. The extreme low points were on the second test sessions (Administration 2) of the first two trips following 36 hours off. These test sessions occurred at 05:50 (T5) and 06:43 (T6) after approximately seven hours on duty.

The RVS changes closely mapped driver self-ratings on the Stanford Sleepiness Scale. The correlation of this data with driver self-ratings is -0.29 ($p < .0005$), suggesting that these drivers were to a certain extent aware of impairment of abilities measured by the SRVT.

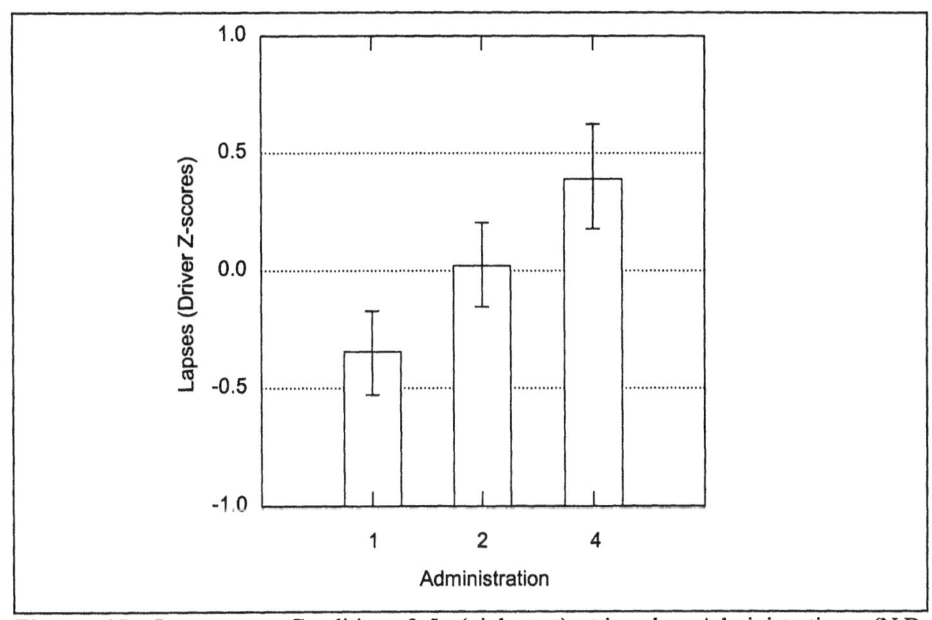

Figure 15. Lapses on Condition 3-5 (nightstart) trips by Administration. (N.B. Administration 3 was not performed in these Conditions.)

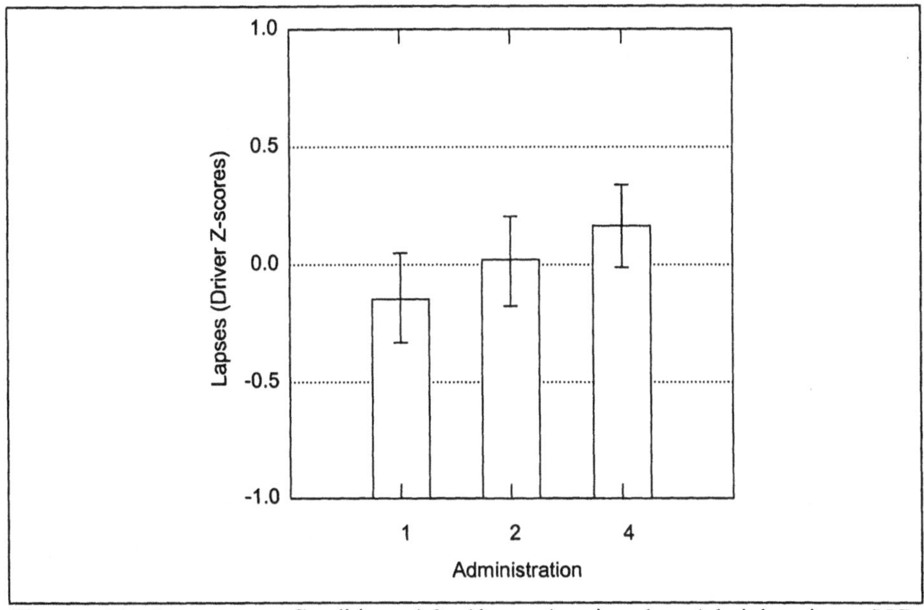

Figure 16. Lapses on Condition 4-9 (daystart) trips by Administration. (N.B. Administration 3 was not performed in these Conditions.)

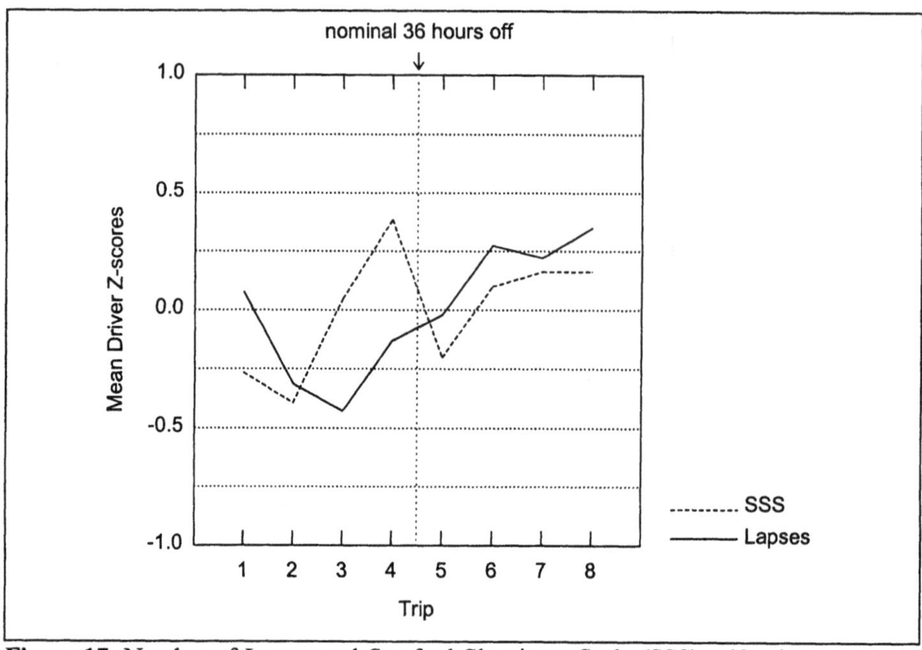

Figure 17. Number of Lapses and Stanford Sleepiness Scale (SSS) self-rating by trip in Conditions 3-5 (nightstart) and 4-9 (daystart).

The Effect of Time Off on Performance the Following Trip

Another series of ANOVAs was performed on the data with the goal of quantifying changes associated with the amount of time off between the fourth and fifth trips. Three drivers drove a fifth trip without taking a workday off between the fourth and fifth trips (Condition 4-6), sixteen drove a fifth trip after one workday (36 hours) off (Conditions 3-5, 4-7, and 4-9), and six drivers drove a fifth trip after two workdays off (Condition 4-8).

No workday off, Condition 4-6: Repeated measures ANOVAs were performed with five levels of Trip and three levels of Administration using the data of the three drivers participating in Condition 4-6 (daystart). Trip 5 data was then contrasted with Trip 4 data. The small subject sample severely limited the ability to make reliable estimates of observed effects. No statistically significant performance test score differences were observed between the fourth and fifth trips. The results of contrasting these two trips on each measure are shown in Table 5.

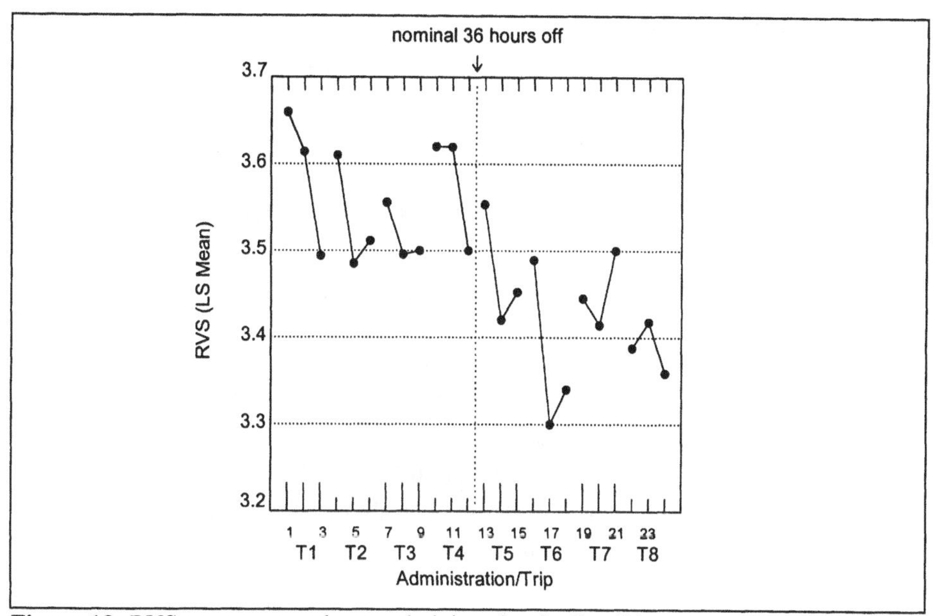

Figure 18. RVS scores on each test administration in Conditions 3-5 (nightstart) and 4-9 (daystart) trips, grouped by trip number.

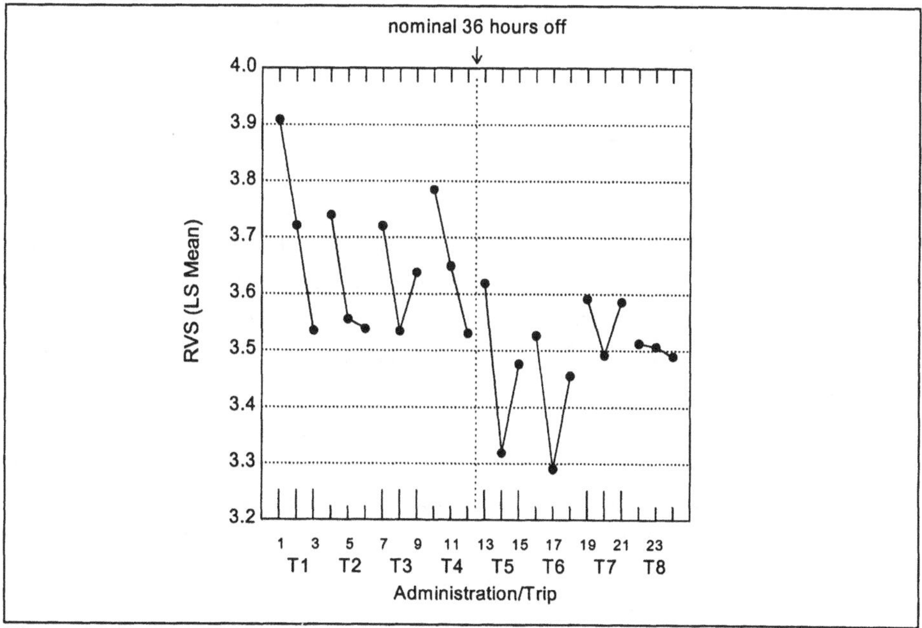

Figure 19. RVS scores on each test administration in Condition 3-5 (nightstart) trips, grouped by trip number.

One workday (36 hours) off, Conditions 3-5, 4-7, and 4-9: Sixteen drivers drove four 13-hour trips, took 36 hours off, then drove a fifth trip. The group included five nightstart drivers (Condition 3-5) and eleven daystart drivers (Conditions 4-7 and 4-9). There were a greater number of observations for this group than in Condition 4-6, permitting estimates with reliability approaching that of the DFAS (which used 20 drivers). The only statistically significant result was a marginal improvement in CS which, in this context, cannot be interpreted as recovery. The results of contrasting the fourth with the fifth trip in this 36-hour off-duty group are shown in Table 6.

Two workdays (48 hours) off, Condition 4-8: Six drivers participated in Condition 4-8 (daystart). Having driven the four 13-hour daystart trips of the base Condition 4, they took 48 hours off, then drove an additional trip. There were no statistically significant improvements in scores between the fourth and fifth trips. The results of contrasting the fourth and fifth trips are shown in Table 7. Only six drivers participated in this observational condition, making reliable estimates of change difficult.

Summary of Surrogate Performance Test Results

Comparisons between two sets of four trips with one intervening workday off: In the comparisons of the first set of four trips with the second set of four trips after one workday off (Conditions 3-5 and 4-9), CS showed ongoing improvement with practice throughout the duration of the study (see Figure 11). Although CTT scores were lower on the last trip of each 4-trip set (see Figure 14), the effect was not statistically significant. The measures derived from the SRVT (Lapses and RVS) showed a trend of ongoing performance decrement across the eight trips (see Figures 17, 18, 19), with no recovery across the 36-hour off-duty period. The decrease in RVS (indicating increasing reaction time) was statistically significant, $F(7,49) = 3.57$, $p = 0.003$. The correlation of the RVS and SSS (driver self-ratings of sleepiness) measures was -0.29 ($p < .005$), which is quite small, but it is possible that drivers were to a certain extent aware of degraded functioning in those abilities measured by the SRVT.

The effect of 0, 1, and 2 workdays off on performance the following trip: There were no statistically significant performance test score changes between Trips 4 and 5 (Condition 4-6) when drivers did not take an intervening workday off. The number of subjects in this Condition (driver N=3) would make it difficult to obtain statistically reliable estimates of even rather large effects.

The larger number of drivers (driver N=16) who took one workday (36 hours) off between Trips 4 and 5 (Conditions 3-5, 4-9, 4-7) afforded better statistical reliability in estimating level of change over the recovery period, with reliability approaching that of the DFAS (which used 20 drivers).

Table 5. Surrogate performance test changes between Trips 4 and 5 of Condition 4-6 (daystart), with no workdays off. (Only three drivers participated in this condition, making reliable estimates of change difficult.)

Measure	F-ratio	df1	df2	Probability	Trip 5 - 4 scores
CS	0.455	1	2	0.570	no significant change
CTT	0.044	1	2	0.854	no significant change
Lapses	0.002	1	2	0.967	no significant change
RVS	0.563	1	2	0.531	no significant change

Table 6. Surrogate performance test changes between Trips 4 and 5 of Condition 3-5 (nightstart), 4-7 (daystart), and 4-9 (daystart) with one intervening workday (36 hours) off. (The data of 16 drivers was used in these analyses.)

Measure	F-ratio	df1	df2	Probability	Trip 5 performance
CS	0.831	3	13	0.018	better than Trip 4
CTT	0.041	1	15	0.842	no significant change
Lapses	0.023	1	13	0.881	no significant change
RVS	4.345	1	13	0.057	no significant change

Table 7. Surrogate performance test changes between Trips 4 and 5 of Condition 4-8 (daystart), with two intervening workdays (48 hours) off. (Six drivers participated in this observational condition).

Measure	F-ratio	df1	df2	Probability	Trip 5 performance
CS	1.411	1	5	0.288	no significant change
CTT	0.514	1	5	0.506	no significant change
Lapses	0.975	1	4	0.379	no significant change
RVS	2.847	1	4	0.167	no significant change

However, the only statistically significant change in scoring was in CS, which showed ongoing improvement with practice in all conditions. CS was better on Trip 5 than on Trip 4 but one could hardly call this performance recovery since, because of practice effects, there was no degradation of performance across the first set of four trips. Recovery implies prior degradation.

In the case of two intervening workdays (48 hours) off (Condition 4-8; driver N = 6), no statistically significant changes were noted between Trip 4 and Trip 5 performance test scores. As was the case with the condition without a workday off (Condition 4-6; driver N = 3), the number of subjects in Condition 4-8 would make it difficult to obtain statistically reliable estimates of even rather large effects.

Recovery Study Conclusions

One Workday (36 Hours) Off

Prevalence of drowsiness: The Mann-Whitney U-test was used to compare the average proportion of analysis epochs judged "drowsy" for all the recovery study Conditions (i.e., 3-5, 4-6, 4-7, 4-8, 4-9) aggregated by Condition and half trip. The test failed to reveal statistically significant differences in prevalence of drowsiness between the first four trips completed by each driver and subsequent trips taken after the prescribed recovery periods.

In the symmetrical Conditions 3-5 (nightstart) and 4-9 (daystart), drivers completed four trips, then took 36 hours off, then completed an additional four trips. The repeated-measures analysis of covariance of the arcsine transform of the proportion of analysis epochs judged "drowsy," with between-subjects factors "Condition" (= 3-5, 4-9) and "Age" (covariate) and within-subjects factors "Trip" (= 1st, 2nd, 3rd, 4th) and "Set" (= first set of four trips, second set of four trips) showed no significant effect of Condition or Set, although it did show a significant interaction of Set by Trip by Condition, $F(3,21)=5.0$, $p<.009$. Contrasting the trip preceding the 36-hour time-off period with the trip following it revealed no difference in both conditions 3-5 and 4-9.

In summary, recovery effects of the 36 hours off duty were not apparent for the prevalence of drowsiness measure.

Lane tracking: In Condition 4-9 (daytstart), the lane tracking standard deviation (LTSD) performance measure trended upward (indicating worse performance) throughout the first four trips, then leveled off and remained at a relatively high level during the second four trips following the 36 hours of time off.

In Condition 3-5 (nightstart), drivers commenced with a relatively high level of LTSD which remained essentially the same throughout the eight trips, with an exceptionally high value on the trip after the one workday (36 hours) off.

In summary, there was no evidence that one workday (36 hours) off brought about any improvement in driving performance as measured by LTSD.

Surrogate Performance Tests: The measures derived from the SRVT (Lapses and RVS) showed a trend of ongoing performance decrement across all eight trips, with no recovery over the 36-hour off-duty period. The decrease in RVS (indicating increasing reaction time) was statistically significant. CS and CTT did not show any statistically significant recovery effects associated with the time off.

Driver self rating: Drivers' self-ratings showed some improvement from the 36-hour off-duty period whereas objective performance measures did not. The drivers may have genuinely felt better, or they may have been reacting to an expectation that the correct response was to show improvement in sleepiness rating.

Sleep duration: For drivers starting their shift by day, some increase was observed in the amount of sleep obtained during the 36 hours of time off. On the other hand, the one workday (36 hours) off appears to have resulted in less sleep for drivers starting their shifts at night. In all likelihood, these drivers resumed day shift sleep-wake patterns on their time off, even though the time off was insufficient for accommodation.

Overall conclusion: For sixteen drivers taking one workday (36 hours) off, there was no objective evidence of driver recovery of performance.

Zero and Two Workdays Off

These conditions had fewer drivers in them and the results are therefore more subject to random variation. For six drivers taking two workdays (48 hours) off, Condition 4-8 (daystart), there was no objective evidence of driver recovery. LTSD went "up-down-up," possibly representing random variation. In any event, LTSD on the day after the break is at a high level relative to the other trips. Surrogate performance test scores did not show recovery effects. For three drivers taking no workday off, Condition 4-6 (daystart), LTSD trended upward across the five workdays. There were no statistically significant surrogate performance test score changes between Trips 4 and 5. Because of the few drivers in these conditions, the prevalence of drowsiness measure was not examined.

Comparative Assessment Of Recovery Study Results

The principal recommendation that came out of the Recovery Study was to repeat the field study using the same methodology but with a larger number of subjects to improve the sensitivity of the statistical tests. It also recommended investigating the effect of longer off-duty periods to establish the duration required for complete driver recovery during day, night, rotating, and irregular schedules.

Although the recommendations were justifiable from a scientific perspective, they did not take into account the funding constraints referred to at the beginning of this document nor the time horizon of the various government and industry stakeholders, all of whom wanted an early answer to the question of whether a 36-hour off-duty period was sufficient for drivers to recover from fatigue effects, and, if it wasn't adequate, what should the duration of the recovery period be. As a result, the Recovery Study findings were examined in more detail and from a different perspective in an attempt to determine whether there was any reason to believe that recovery should have occurred over the 36-hour off-duty period even though it might have gone undetected due to the insensitivity of the statistical tests. The results of this investigation are presented in the following section.

Sleep Duration Comparisons: Recovery Study and DFAS

The sleeps of Recovery Study drivers were compared with those of all DFAS drivers in the same observational condition (i.e., Conditions 3-5 and 4-9 compared with Conditions 3 and 4, respectively) to determine whether sleep behavior of Recovery Study drivers was similar to that of the overall group to which they belonged or whether there were significant differences. Further, because DFAS drivers worked only one cycle while Recovery Study drivers worked two cycles, the DFAS cycle 1 sleeps were also compared with the Recovery Study cycle 2 sleeps. This was done to examine potential end- and start-of-cycle effects. The Recovery Study results for sleep time are provided in Table 8 and are plotted in Figures 20 and 21 for daystart and nightstart drivers, respectively.

Comparison of Daystart Conditions. During work cycle 1 (i.e., sleeps 1 to 5) of Condition 4-9 (daystart), it is clear that, except for the last sleep, Recovery Study and DFAS drivers had similar sleep patterns and sleep times. On the night prior to the first drive, average total sleep time for Recovery Study drivers was 369.7 minutes versus 403.8 minutes for DFAS drivers. The sleeps following the first three drives averaged 302.4 minutes for Recovery Study drivers and 292.3 minutes for DFAS drivers. After the fourth drive, however, there were substantially different end-of-cycle effects between the two groups of drivers. On sleep 5, Recovery Study drivers – who were beginning their 36-hour off-duty period prior to the start of another work cycle of four days – slept an average of 395.2 minutes. This was an increase of 92.8 minutes by comparison with the average obtained on the previous three nights. DFAS drivers, on the other hand, slept an average of 256.3 minutes on sleep 5 – which was their last sleep of the work cycle. This was 36 minutes less than their average on the previous three sleeps and 138.9 minutes less than what the Recovery Study drivers obtained on sleep 5. Examining this end-of-cycle effect more closely and comparing what all drivers did subsequent to sleep 5 is revealing in terms of driver behavior. Figure 22 shows average daily sleep for drivers in all daystart conditions, which includes Conditions 4-6, 4-7, 4-8, and 4-9. For the drivers with one and two workdays off duty, including the Condition 4-9 drivers working two full cycles, driver sleep duration during the principal night sleep periods was very similar, averaging about 405.7 minutes.

During work cycle 2 (i.e., sleeps 6 to 9) of Condition 4-9 (daystart), the sleep pattern was similar to that of cycle 1 although sleep times were somewhat higher. On the first sleep prior to the start of the first drive of the second cycle (i.e., sleep 6), total sleep time averaged 410.8 minutes (compared with 369.7 minutes on sleep 1 of cycle 1), dropping to an average of 326.6 minutes during sleeps 7 to 9 (compared with 302.4 minutes for sleeps 2 to 4). Unlike work cycle 1, however, the drivers chose to leave for home rather than stay for a tenth sleep after their last drive.

The mid-week sleep times (i.e., for sleeps 2, 3, and 4 of cycle 1 and 7, 8, and 9 of cycle 2) are consistent, being an average of 302.4 and 326.6 minutes for the Recovery Study drivers during cycles 1 and 2, respectively, and 292.3 minutes for the DFAS drivers. These are within a range of 34.3 minutes of each other. The usefulness of the mid-week average sleep time value is that it appears to be relatively free of start- and end-of-cycle effects.

The sleep patterns of daystart Recovery Study drivers were also compared with those of DFAS drivers using a multiple discriminant analysis, in which each driver was represented by a vector comprised of the sleep times of his first four sleeps in the first cycle. There was no statistically significant difference between the groups.

Comparison of Nighstart Conditions The sleep pattern and sleep times of Recovery Study drivers in Condition 3-5 (nightstart) began in a fashion similar to those of DFAS drivers in Condition 3 (nightstart). On the night prior to the first drive, average total sleep time for Recovery Study drivers was 214.7 minutes versus 193.1 minutes for DFAS drivers. The sleeps following the first two drives averaged 246.6 minutes for Recovery Study drivers and 248.5 minutes for DFAS drivers. After the third drive, however, the sleep trends between the two groups of drivers diverged in a systematic fashion. On sleeps 4 and 5, Recovery Study drivers increased their sleep to an average of 308 and 375.8 minutes, respectively, while DFAS drivers reduced their sleep to 256.7 and 202.7 minutes. Thus, while the sleep time of DFAS drivers peaked at 264.4 minutes at the mid-point of their work cycle (i.e., on sleep 3 of 5), the sleep of Recovery Study drivers continued to increase from day to day and reached a peak of 375.8 minutes during the last sleep of their first work cycle (i.e., the first sleep of their 36-hour off-duty period). This difference in behavior is probably explained in part by "fatigue anticipation" since the Recovery Study drivers, unlike those of the DFAS, were scheduled to work a second cycle as intensive as the first. It could be argued that this was instead due to increasing fatigue. However, this view is not supported by the data which shows that the DFAS nightstart drivers on a similar, but single, work cycle did not demonstrate the effect. For further confirmation, this behavior was looked for in the daystart Condition 4-9 of Figure 20. The effect was not as strongly apparent there and a strong conclusion could not be reached. However, it should be noted that, after the final drive of the second cycle, which was completed at about 0130, these drivers chose not to take their sleep but rather left the study to go home – delaying recovery to a later time – presumably because, based on their own sense of fatigue level, they judged it preferable to go home first. The daystart conditions were then examined more closely as a group in Figure 22. The "fatigue anticipation" effect was more apparent. However, this was again observed in a somewhat different context since, because of the early morning start time for sleep 5 in Conditions 4-6, 4-7 and 4-8, drivers postponed getting their full recovery sleep until the first available full night period.

During work cycle 2, the sleep behavior of Recovery Study drivers did not mirror their own of the first cycle but rather that of the DFAS drivers. On the first sleep prior to the start of the first trip (i.e., their last sleep during the 36-hour off-duty period), they obtained an average of 118.7 minutes. This is less than their own 214.7 minutes obtained during the first sleep of cycle 1 and the 193.1 minutes obtained by the DFAS drivers during sleep 1. Their sleep then peaks at about 271 minutes averaged over sleeps 2, 3, and 4 of cycle 2 (i.e., sleeps 7, 8, and 9) – compared with a peak of 251.2 minutes for the DFAS drivers averaged over sleeps 2, 3, and 4 of cycle 1). Interestingly, the "fatigue anticipation" explanation cited earlier for the increasing cycle 1 sleeps is supported by the fact that this behavior is absent during cycle 2, just as it is absent for the DFAS drivers.

The mid-week sleep times (i.e., for sleeps 2, 3, and 4 of cycle 1 and 7, 8, and 9 of cycle 2) are consistent, being an average of 267 and 271 minutes for the Recovery Study drivers during cycles 1 and 2, respectively, and 251.2 minutes for the DFAS drivers. These are within a range of 19.8 minutes of each other.

The sleep patterns of nightstart Recovery Study drivers were also compared with those of DFAS drivers using a multiple discriminant analysis, in which each driver was represented by a vector comprised of the sleep times of his first four sleeps in the first cycle. There was no statistically significant difference between the groups.

Comparison of Daystart with Nightstart Conditions The mid-week sleep times of the Recovery Study daystart and nightstart drivers averaged about 302.4 and 267 minutes for cycle 1, and 326.6 and 271 minutes for cycle 2, respectively. These compare with mid-week DFAS daystart and nightstart average driver sleep times of about 292.3 and 251.2 minutes, respectively. Mid-week sleep time differences between daystart and nightstart drivers thus averaged about 35.3 and 55.6 for Recovery Study drivers during work cycles 1 and 2, respectively (i.e., an average difference of about 45.5 minutes), and 41 minutes for DFAS drivers.

During the 36-hour off-duty period, the daily sleep time difference between daystart and nightstart drivers amounted to an average over the two sleeps of about 155.8 minutes because of the short second sleep which was taken by the nightstart drivers prior to the first drive of the second cycle.

The daily sleep gap between daystart and nightstart drivers is plotted in Figure 23 for both Recovery Study and DFAS drivers. The short sleep associated with the nightstart drivers during shift changeover prior to the first drive is clearly evident on sleeps 1 and 6.

Table 8. Average total sleep duration (in minutes) during principal sleeps of daystart and nighstart drivers in Recovery Study Conditions 3-5 and 4-9 compared with all drivers in DFAS Conditions 3 and 4, respectively.

Sleep No.	Sleep Duration; C4-9 (daystart) drivers	Sleep Duration; C4 (daystart) Drivers	Sleep Duration; C3-5 (nightstart) Drivers	Sleep Duration; C3 (nightstart) Drivers
1	369.7	403.8	214.7	193.1
2	295.5	285.05	247.3	232.6
3	285.9	282.4	245.9	264.4
4	325.9	309.5	308	256.7
5	395.2	256.3	375.8	202.7
6	410.8	*403.8**	118.7	*193.1**
7	323.3	*285.0**	273	*232.6**
8	335.2	*282.4**	268.9	*264.4**
9	321.4	*309.5**	271.1	*256.7**
10		*256.3**	168.3	*202.7**

*times shown in italics for sleeps 6 to 10 of DFAS Conditions 3 and 4 are sleeps 1 to 5 repeated for the reader's convenience.

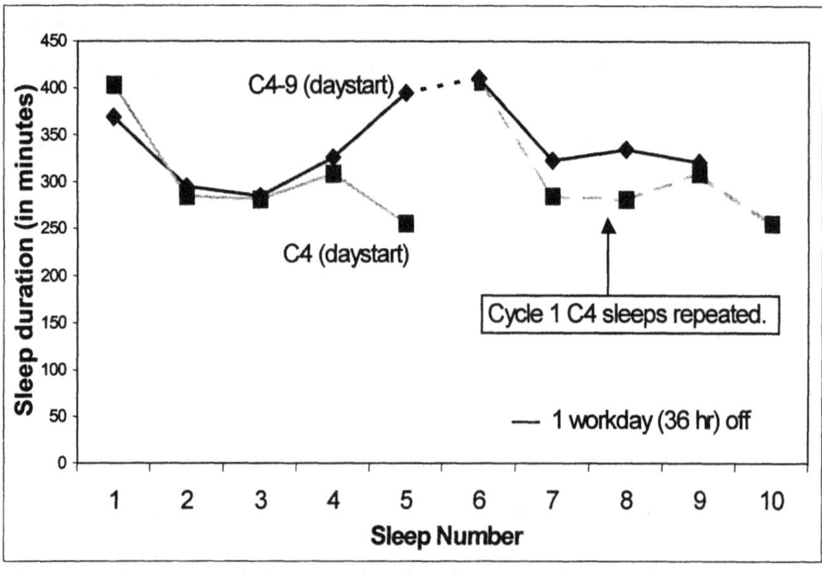

Figure 20. Average total sleep duration (in minutes) during principal sleeps of daystart drivers in Recovery Study Condition 4-9 compared with all drivers in DFAS Condition 4.

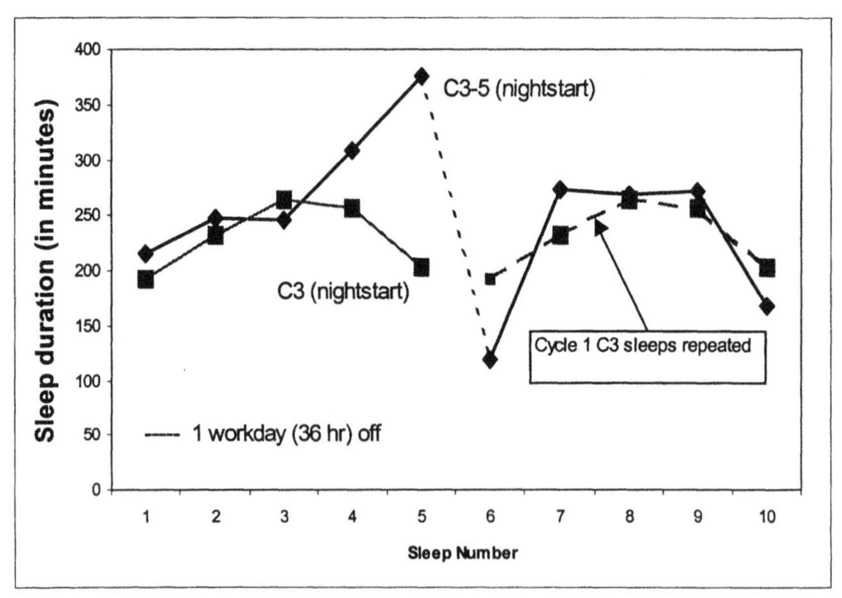

Figure 21. Average total sleep duration (in minutes) during principal sleeps of nighstart drivers in Recovery Study Condition 3-5 compared with all drivers in DFAS Condition 3.

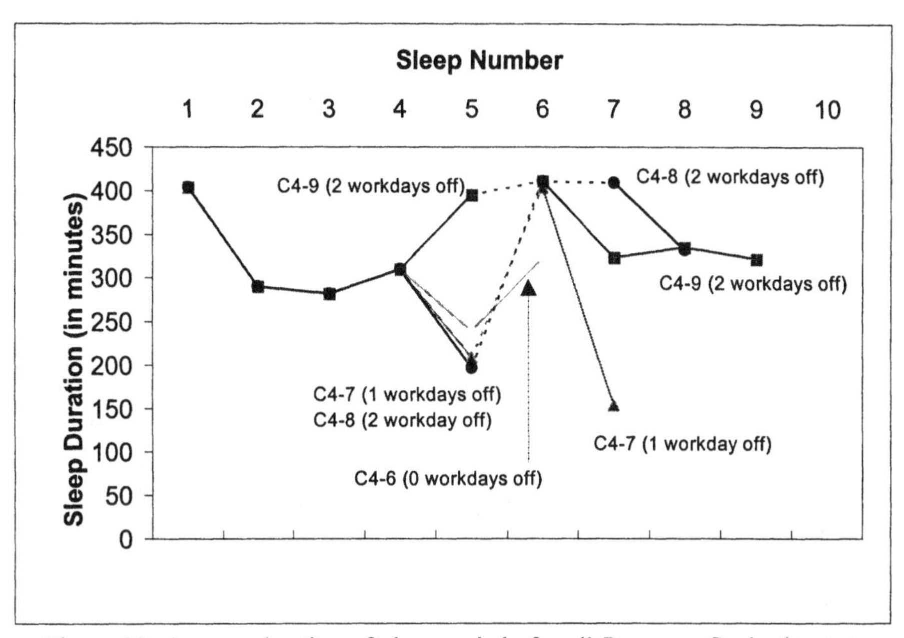

Figure 22. Average duration of sleep periods for all Recovery Study daystart conditions (i.e., first four sleeps have been averaged to facilitate reading since similar trends prevailed).

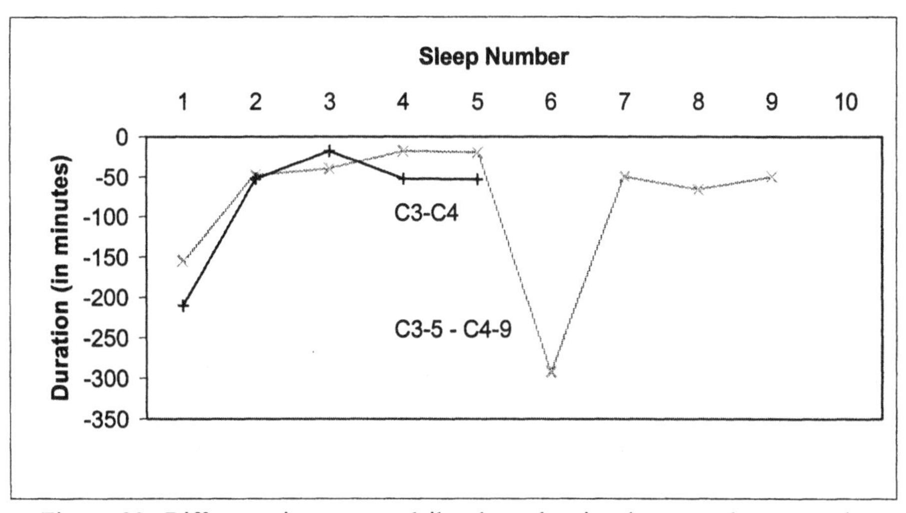

Figure 23. Difference in average daily sleep duration between daystart and nightstart drivers for Recovery Study Conditions 3-5 and 4-9, and respective DFAS Conditions 3 and 4.

Comparison of Sleep Shortfall The DFAS drivers, as a group, reported that 7.2 hours on average is their ideal sleep time. The DFAS study also reported that, overall, drivers obtained about 2 hours less time in bed and 2.5 hours less actual sleep than their reported ideal daily sleep requirement. For the daystart drivers of Condition 4 and nightstart drivers of Condition 3, the sleep shortfall (i.e., relative to their reported ideal) was, respectively, about 2.1 and 3.4 hours overall and 2.3 and 3.01 hours for the mid-week sleep periods.

The sleep shortfall for the Recovery Study drivers – based on average mid-week sleep times and the same 7.2 hours of ideal sleep reported by the overall DFAS group of drivers – was 1.96 and 2.72 hours for daystart and nightstart drivers, respectively. For the two principal sleeps recorded during the 36-hour off-duty period, the sleep shortfall was an average of 0.48 and 3.08 hours for the daystart and nightstart drivers, respectively. Thus, both daystart and nightstart drivers continued to experience a sleep shortfall relative to their reported ideal requirement during the 36-hour off-duty period. However, while daystart drivers substantially reduced the shortfall relative to that at mid-week, the nightstart drivers marginally increased theirs.

Although the reported ideal sleep duration of the daystart and nightstart drivers is 7.8 and 6.9 hours, respectively, the 7.2 hours associated with the overall group of 80 DFAS drivers is used here. This is primarily because it is a common reference value that is useful for comparison purposes and is consistent with the perceived nightly need for sleep of the general (i.e., U.S.) population of about 7.2 ± 1.2 hours (Mitler, 1997).

Recovery Periods and Observations from the Literature Review

A literature review (Smiley and Heslegrave, 1997) was conducted primarily to establish whether the Recovery Study results were consistent with other research. It was also useful in interpreting the results obtained.

Many articles were found which dealt with the impact of length of shift, time of day, and sleep deprivation effects on driver performance and accident rates. Very few studies were found that looked at performance in terms of number of days worked in sequence or number of hours worked since the last period of days off. Even fewer studies were found that looked at the recuperative value of rest periods and time off. No study was found that specifically dealt with an operational schedule that would be permitted under a 36-hour reset scenario.

Relevant observations from the literature review, in point form for brevity, are presented in the following section. These involve duration of recovery periods, and sleep behavior often attributed to commercial drivers, nightworkers, and shiftworkers.

Concerning Duration of Recovery Periods

- Increasing the number of very long shifts in a week can increase the chances for inadequate recovery from acquired sleep debt due to two related effects: cumulative sleep debt increases with the number of shifts, and duration of off-duty period available as opportunity for sleep decreases;
- Circadian rhythm and environmental factors both work against the night driver in sustaining on-the-job performance, by at-work hours during which alertness is naturally reduced and by off-duty hours that impede attainment of sufficient recovery sleep;
- One day is usually not sufficient for full recovery from a sleep debt accumulated over a protracted period of reduced daily sleep (Smiley and Heslegrave, 1997; Dinges et al., 1997);
- Two nights of recovery sleep may be required to allow near full recovery following protracted periods of sleep loss or sleep restriction (Johnson and Naitoh, 1974; Smiley and Heslegrave, 1997; Dinges et al., 1997);
- Most shift workers reported needing at least two days (with normal sleep episodes) to recover after a period of night work involving three shifts in a row and that the need for recovery increased by one day when the period of night work increased to seven shifts in a row (Kecklund et al., 1994; Akerstedt, 1997);
- Longer recovery sleep may be obtained on the second of two rest days (Lille, 1967);
- There are indications that a rest period of three days is superior to two days (Malette, 1994).

Concerning Sleep Behavior

- A very high correlation (0.95) has been found between the length of free time between shifts and sleep duration. Most shift schedules have 16 hours of free time between consecutive shifts, and at least this amount is needed for a sleep duration of 7 or 8 hours (Kurumatani et al., 1994; Akerstedt, 1997);
- Quick changeovers between shifts, usually allowing about 8 hours of free time, lead to reduced sleep. When a morning shift or a day shift is followed by a night shift (with less than 6 hours of free time in between), sleep duration can be less than 3 hours. (Knauth et al., 1983; Totterdell and Folkard, 1990; Kurumatami et al., 1994; Akerstedt, 1997);
- Sleepiness is most pronounced during the first night shift, presumably due to increased prior wake time. Most shift workers have been awake for at least 13 hours before the start of the first night shift, whereas the time awake decreases by at least 4 or 5 hours before the start of the remaining night shifts (Akerstedt, 1997);
- Sleep after a night shift is usually not longer than 5 or 6 hours, and may be further reduced when followed by an evening shift (Akerstedt et al., 1991, 1997);

- Many drivers, nightworkers, and shiftworkers may be (Rutenfranz et al., 1972; Hertz, 1988; Rhodes et al., 1995; Donderi, Smiley, and Kawaja, 1995; Wylie et al., 1996; Wylie et al., 1997):
  - obtaining only moderate amounts of sleep during duty days;
  - building up a large sleep debt over the shift cycle due to reduced sleep; and,
  - not taking advantage of sleep opportunities sufficiently to promote recovery during off-duty days.
- In a study of Coast Guard watchkeepers working 12 hours in two shifts (Donderi et al., 1995), self-reported sleep times were 5.5 and 5 hours for day and night workers, respectively;
- In a study of air traffic controllers on 12-hour shifts, night workers self-estimated an average of 5 hours of sleep (Rhodes et al., 1994) but, when monitored, were found to actually obtain about 4.3 hours (Rhodes et al., 1996);
- In a study of railroad locomotive engineers (Comstock, 1997), 76% reported getting less than 5 hours of sleep on duty days, and the quantity and quality of sleep were generally reported to be poor.

Overall Conclusions And Discussion

Conclusions

Recovery Study sleep and performance data indicated that there was no objective evidence of driver recovery over the 36-hour off-duty period. In an attempt to assess the validity of this result as well as to explain it, the driver sleep data was examined and the following overall conclusions were reached.

Consistency of Recovery Study Sleep Data

- Time series examination of sleep times of drivers participating in the Recovery Study compared with the overall group of drivers in the DFAS showed that, for the respective study conditions, sleep times were comparable and consistent over the study duration both for daystart and nightstart groups, and did not indicate random effects.
- Sleep times observed in both the Recovery Study and the DFAS are comparable with figures available in the technical literature for comparable workers (i.e., about 5 to 6 hours), with night and shiftworkers getting less than day workers.
- The sleep patterns of Recovery Study drivers were also compared with those of DFAS drivers using a multiple discriminant analysis, in which each driver was represented by a vector comprised of the sleep times of his first four sleeps in the first cycle. There was no statistically significant difference between the groups.

Driver Sleep during Work Cycle

Driver sleep patterns during the work cycle indicated that:

- Both daystart and nightstart drivers accumulated a sleep debt due to obtaining less sleep than they needed on a daily basis, with that of the nightstart drivers being significantly worse.
- Mid-week sleep times of daystart and nightstart drivers, which appear to show less significant start- and end-of-cycle effects, respectively averaged about 5.2 and 4.4 hours during principal sleep periods compared with 4.8 and 4.1 hours for the respective DFAS drivers. This was a daily shortfall of about 2 and 2.8 hours for the Recovery Study drivers compared with the 7.2 hours that the overall group of 80 DFAS drivers reported as the duration of their ideal daily sleep.

Driver Sleep during 36-Hour Off-Duty Period

Driver sleep patterns during the 36-hour off-duty period (i.e., for the two principal sleeps recorded) indicated that:

- Both daystart and nightstart drivers had full night sleep of similar duration during the available opportunities, averaging about 6.5 hours. (i.e., this appeared to be the case irrespective of schedule, with an average of 6.6 hours per night when all Recovery Study drivers are included).
- Both daystart and nightstart drivers continued to experience a daily sleep shortfall, with that of the nightstart drivers being significantly worse. Sleep duration of daystart and nightstart drivers was respectively about 6.7 and 4.1 hours averaged over the two sleep periods, a shortfall of about 0.5 and 3.1 hours compared with the 7.2 hours of reported ideal sleep.
- While daystart drivers substantially reduced their average daily sleep shortfall during the 36-hour off-duty period compared with that mid-way through the work cycle (i.e., from 2 to 0.5 hours), nightstart drivers marginally increased theirs (i.e., from 2.8 to 3.1 hours) because of the short sleep on the second night.
- For the sleep prior to the start of the next work cycle (i.e., the second sleep of the 36-hour off-duty period), the daystart drivers had their longest sleep (6.8 hr) while the nightstart drivers had their shortest (2 hr). The short sleep of the nightstart drivers may be associated with quick changeover of wake-sleep pattern from day to night.
- The average daily sleep shortfall over the two sleep periods taken during the off-duty time, amounts to, for daystart and nightstart drivers respectively, about 7% and 43% of the reported ideal daily sleep.
- The average daily sleep gap between daystart and nightstart drivers increased during the 36-hour off-duty period relative to that at mid-week, from about 45.5 to 155.8 minutes (i.e., which respectively represent about 10% and 36% of the reported ideal daily sleep).

Recovery Over 36-Hour Off-Duty Period

- Although it could be argued that the statistical tests may not have had sufficient power to detect recovery effects over the 36-hour off-duty period because of the number of drivers involved in the Recovery Study, driver sleep data provides little support for expecting recovery to have taken place. There is a sleep shortfall that exists from day to day throughout the work cycle and that continues into the off-duty days. Although the daystart drivers substantially reduced their sleep shortfall over the 36-hour off-duty period, the nightstart drivers increased theirs primarily because of the short sleep prior to the first drive of the second work cycle.

DISCUSSION

A number of observations from the results of this study can be used to provide some guidance on the duration of recovery periods that should be provided to minimize the accumulation of driver fatigue over extended periods of work. These are discussed in the following points together with a number of ideas concerning potential reasons for the driver behavior observed in this study. It would be very helpful if some future research could be oriented toward providing more definitive answers in view of the limited amount of data upon which the results of this study are based.

36-Hour Recovery Period

- The sleep shortfall averaged over the two sleeps within the 36-hour off-duty period amounted to about 7% and 43% of the reported ideal daily sleep required by daystart and nightstart drivers respectively. This is additional to the sleep debt accumulated over the preceding work cycle. Accepting that two nights of recovery sleep may be required to allow near full recovery following protracted periods of sleep loss (Johnson and Naitoh, 1974; Kecklund et al., 1994; Dinges et al., 1997), then these results may indicate that, while the 36-hour off-duty period may be marginal for the daystart drivers, it was inadequate for the nightstart drivers. However, important issues to be resolved from an operations perspective concern the level of performance recovery achieved with two off-duty sleeps of the average duration observed in this study (i.e., 6.5 hours) and the relative performance improvement that would be obtained from an additional day off duty. Performance in this context should be considered from the perspective of those tasks that are important for safe driving.
- To offset the sleep insufficiency arising from the short sleep associated with shift changeover, the nightstart drivers appear to require, as a minimum, the opportunity for an

additional full night sleep over and above that available under the 36 hours off-duty period. Under this scenario, the duration of the off-duty period would be about 60 hours for a driver resuming a night shift cycle on a schedule similar to that on the previous cycle. In addition to the opportunity for two full night sleeps, these drivers would also automatically benefit from the availability of additional free time (i.e., between morning wake time after the second night and the start of the first night shift of the next cycle) to accommodate the short shift changeover sleep.

- It can be expected that, for drivers on back-to-back night shift cycles, the 60-hour off-duty period will not eliminate the short changeover sleep and the long wake time prior to the first drive. The expectation is that the two full night sleeps will provide for driver recovery from the fatigue effects of the previous work cycle so that they will not propagate into the next work cycle. However, the following questions need to be answered in order to more fully assess the value of this 60-hour off-duty period. To what extent do the short shift changeover sleep and long wake time affect driving performance on the next day, and what is the relative importance of their effects? To what extent are the effects of the short changeover sleep and long wake time mitigated by subsequent sleeps of duration similar to those observed in this study?

Short Shift Changeover Sleep

- To help compensate for the short shift changeover sleep of the night drivers during the time off prior to the first drive (i.e., which appears to be the worst case sleep under normal conditions), consideration should be given to putting special measures in place to help these drivers deal with their first trip. These might include among others: limiting the number of hours worked during the early morning between midnight and 6 a.m. by, for example, requiring a 2-hour sleep period to be included; providing good sleeping facilities at the appropriate locations and nominal financial incentives for their use at appropriate times; educating drivers and their families about the importance of getting a good first sleep and the dangers potentially associated with not doing so.

Sleep Gap Between Drivers on Day and Night Shifts

- The sleep gap between daystart and nightstart drivers during the work cycle appeared to be on the order of about 45.5 minutes. Using a sleep efficiency figure of 0.93 calculated for the Recovery Study nightstart drivers, the sleep gap translates into about 49 minutes of time in bed. This may indicate a need to provide nightstart drivers with about an additional hour of time off daily by comparison with the daystart drivers in order to approach roughly similar sleep times.

Driver Fatigue Anticipation

- During the first work cycle, the nightstart drivers working two cycles with an intervening 36-hour off-duty period appeared to display "fatigue anticipation" behavior by comparison with those on a single work cycle such that they systematically obtained more sleep toward the latter part of the first cycle. On the last two sleeps of the first cycle, they were less than 20 minutes short of sleep times recorded by the daystart drivers. However, irrespective of the more rigorous work schedule, during the second cycle these drivers conducted themselves similarly to other drivers on a single cycle, both with respect to sleep times during the work cycle and the short sleep associated with the shift change just prior to the first drive of the second cycle. By inference, the sleep pattern of these drivers appears to be based more on the fatigue expected to be associated with the future work period rather than the fatigue accumulated over the past cycle. The daystart drivers appeared to display similar behavior although it was not as visible for a variety of possible reasons discussed in a previous section.
- If sleep behavior is at least in part explained by driver "fatigue anticipation" then unexpected perturbations to driver schedules can be particularly detrimental because driver sleep debt and fatigue could rise significantly higher than expected and reserves could be depleted. This would indicate that driver scheduling should be planned and made known to the driver in advance, while perturbations to the plan should be minimized as much as possible. A flexible strategy, which might involve providing additional sleep opportunities and day/night shift cycle switching to compensate for perturbations to schedules, should be developed in advance with the driver.

Sleep-Performance Trade-Off

- It is clear from this and other studies that drivers and other shiftworkers do not get the amount of sleep that they themselves say they need. In this study, it was found to be the case during the relatively intensive work cycle as well as during the off-duty recovery period. It is also known that, even when given substantially more sleep opportunity, drivers do not necessarily use the bulk of it to obtain their reported ideal sleep. In the DFAS, drivers who worked a relatively less rigorous schedule incorporating substantially more time as available opportunity for sleep, with stable day shift cycles and regular night periods, got only marginally more sleep than did other drivers in more difficult conditions.

These "day" drivers (Condition 1), had available 10.7 hours of continuous off-duty time (excluding the time required for the study protocol), spent 5.8 hours in bed and got 5.4 hours of clinically measured sleep time. This is comparable with the results from the nightstart drivers of DFAS Condition 4 (i.e., which includes the drivers in Condition 4-9 of this study during work cycle 1) who had 8.9 hours off-duty, spent 5.5 hours in bed and got 5.1 hours of sleep. This behaviour has been attributed to the impacts of sleep-related factors that affect sleep quality and quantity (i.e., environmental and circadian) as well as personal, family, and work-related reasons.

- While the above reasons for getting less sleep than reportedly needed are well accepted, it is not certain that they fully explain the phenomena. The drivers in the DFAS Conditions 1 and 4, for example, well understood the objectives of the study and the importance of behaving as naturally as possible. Their enthusiasm for the study and the importance with which they viewed it were obvious. Yet, it was evident that these drivers often did not organise their off-duty time to obtain the maximum possible sleep. It appears that these drivers, within the bounds of available time, may also have made a conscious or subconscious decision to obtain more or less sleep based on their own self-assessment of need "in the field". It would be of substantial operational benefit to determine if there is a valid psychophysiological basis for this self-assessment and to objectively establish the sleep-wake trade-off that may be imbedded within it. The trade-off (i.e., a diminishing sleep return criteria) should be in measurable terms, such as in the form of sleep-performance curves, with performance being related to those elements that are important to safe driving.
- Irrespective of the amount of sleep that may be considered ideal or that may be found to be psychophysiologically required for recovery from acute and cumulative fatigue under controlled conditions, there is little doubt that drivers will not sufficiently avail themselves of sleep opportunities unless these take into account normal personal and family-related needs. For night shift drivers it must also be taken into account that most of society is on a day shift wake-sleep pattern, and drivers, as much as possible, seek to live that way too (Vespa, 1997).

Minimum Recovery Period Options

- Developing a scientific basis for a minimum recovery period is important from the perspectives of driver quality of life and operational efficiency as well as safety. It might serve as an important tool to help drivers better deal with the frequent schedule delays that are a common fact of life in many sectors of trucking. It might help drivers get home earlier to spend more time with their families, rather than being required to spend unnecessary days in distant locations and absorbing additional on-the-road costs. It might lead to greater compliance with the hours of service.

- From the scientific and operational perspectives, establishing a minimum fatigue recovery period need not necessarily mean providing identical off-duty times for all drivers and irrespective of schedule. It does not necessarily have to mean providing the same at-home and away-from-home recovery time. This study has shown, for example, that drivers operating on a "night" schedule have substantially different and more adverse sleep behaviour compared with those on "day" schedules. These "night" drivers must additionally cope with more psychophysiologically related adversities both while working and while driving. It might be considered, for example, that "night" drivers going away from home on the first work cycle could, for the return home journey, be given a recovery strategy that includes shifting to a "day" cycle and, potentially, be allowed a shorter recovery period than would otherwise be allowed. This might permit a safer journey and a longer off-duty recovery period to be taken at the home terminal.
- Recovery periods might be considered from the perspective of the number of full night sleep opportunities rather than number of hours since the last period worked. This strategy would help to maintain regularity of schedules and, by comparison with the more benign "day" shifts, would automatically provide more recovery time for drivers working two consecutive "night" cycles.
- Short-term, shorter duration recovery period options of 36, 48 and 60 hours should be considered in which the choice is made on the basis of the back-to-back schedules that are to be worked. This study and others have demonstrated that day driving is much less fatiguing than night driving and that night sleep is more restorative than day sleep. Furthermore, drivers do not appear to obtain the sleep they reportedly require on days off; e.g., in this study, drivers averaged about 6.5 hours of full night sleep on days off rather than their reported ideal of 7.2 hours. Potential day sleep opportunities were sometimes partially deferred to the night. Regardless of the off-duty recovery period duration, drivers working night shifts appear to obtain a short shift changeover sleep on the day before the first drive; this lack of sufficient rest and long wake time is normally considered to affect performance on the next day. From the fatigue perspective, there is justification for encouraging drivers to get two full nights of sleep, work the more benign day schedules, and minimize the occurrence of short changeover sleeps.
- It might be useful to consider a schedule that provides a 36-hour off-duty recovery period at away-from-home terminals for drivers working back-to-back day-shift cycles, or for drivers shifting from a night- to a day-shift cycle. Drivers changing from day to night shifts could be provided with 48 hours off duty, while drivers on two consecutive night-shift cycles could receive 60 hours off. These recovery periods would have to ensure the opportunity for two full nights of normal sleep. They should be considered short-term measures to allow drivers to go home earlier to gain more time for a full restorative rest.

This research was supported by funding provided by the Transportation Development Centre, Safety and Security, Transport Canada. The Canadian Trucking Research Institute of the Canadian Trucking Association is acknowledged for its financial contribution to the Recovery Study, as is the Office of Motor Carriers of the U.S. Federal Highway Administration for making available much of the equipment used for field data collection. The views expressed in this article are solely those of the authors.

REFERENCES

Akerstedt, T. (1997). *Readily Available Countermeasures Against Operator Fatigue.* American Trucking Associations Foundation, International Conference Proceedings, Managing Fatigue in Transportation, April 29-30, Tampa, Florida, 105-122.

Akerstedt, T., Kecklund, G., and Knutsson, A. (1991). Spectral Analysis of Sleep Electroencephalography in Rotating Three-Shift Work. Scandinavian Journal of Work and Environmental Health, 17. 330-336.

Comstock, M.L. (1997). *Alertness Assurance in the Railroad Industry.* American Trucking Associations Foundation, International Conference Proceedings, Managing Fatigue in Transportation, April 29-30, Tampa, Florida, 29-38.

Dinges, D.F., Pack, F., Williams, K., Gillen, K.A., Powell, J.W., Ott, G.E., Aptowicz, C., and Pack, A.I. (1997). *Cumulative Sleepiness, Mood Disturbance, and Psychomotor Vigilance Performance Decrements During a Week of Sleep Restricted to 4-5 Hours per Night.* Sleep, 20(4), 267-277.

Donderi, D.C., Smiley, A., and Kawaja, K. (1995). Shift Schedule Comparison for the Canadian Coast Guard. Transportation Development Centre, Safety and Security, Transport Canada, Montreal, Quebec. Report no. TP 12438E.

Freund, D.M., and Vespa, S. (1997a). *U.S./Canada Study of Commercial Motor Vehicle Driver Fatigue and Alertness.* Proceedings of Traffic Safety on Two Continents Conference, September 22-24, 1997, Lisbon, Portugal.

Freund D.M., and Vespa, S. (1997b). *The Driver Fatigue and Alertness Study: From Research Concept to Safety Practice.* International Large Truck Safety Symposium, October 27-29, Knoxville, Tennessee.

Gough, B., and Gray, R. (1992). *OTA Hours of Work Initiative: A Study of Industry and Government Concerns and Issues and Analysis of Potential Legislative*

Amendments. October 1992, report by The Gough and Grey Group, Ontario Trucking Association, Ontario, Canada.

Hertz, R.P. (1988) *Tractor-trailer driver fatality: The role of non-consecutive rest in a sleeper berth.* Accident Analysis & Prevention, 20(6), 431-439.

Johnson, I.C., and Naitoh, P. (1974). *The Operational Consequences of Sleep Deprivation and Sleep Deficit.* In Advisory Group for Aerospace Research & Development – AGARDograph No. 193. (pp 1-43 + A1-A3). North Atlantic Treaty Organization (NATO). France: 7 Rue Accelle 9220 Neuilly Sur Seine. Revue Internationale des Services de Santé. 1975. TOME XLVII, p. 675.

Kecklund, G., Akerstedt, T., Lowden, A., and von Heidenberg, C. (1994). *Sleep and Early Morning Work.* Journal of Sleep Research, 3, Suppl 1. 124.

Knaught, P., Kiesswetter, E., Ottman, W., Karvonen, M.J., and Rutenfranz, J. (1983). *Time-Budget Studies of Policemen in weekly or Swiftly Rotating Shift Systems.* Applied Ergonomics, 14.247-252.

Kurumatani, N., Koda, S., Nakagiri, A., et al. (1994). *The Effects of Frequently Rotating Shiftwork on Sleep and the Family Life of Hospital Nurses.* Ergonomics, 37. 995-1007.

Lille, F. (1967). *Le sommeil de jour d'un groupe de travailleurs de nuit.* Le Travail Humain, 30, 85-97,.

Malette, R. (1994). *Shift Study and Assessment of 48 and 72-Hour Rest Breaks.* Ontario Hydro HRP and Development.

Mitler, M.M., Miller, J.C., Lipsitz, J.J., Walsh, J.K., and Wylie, C.D. (1997). *The Sleep of Long-Haul Truck Drivers.* NEJM, 337(11): 755.

Rhodes, W., Heslegrave, R., and Ujimoto, K.V. (1994). *A Study of the Impact of Shiftwork and Overtime on Air Traffic Controllers.* Transportation Development Centre, Safety and Security, Transport Canada, Montreal, Quebec. Report no. TP 12257E.

Rhodes, W., Heslegrave, R., and Ujimoto, K.V. (1996). *Impact of Shiftwork and Overtime on Air Traffic Controllers – Phase 2: Analysis of Shift Schedule Effects on Sleep, Performance, Physiology and Social Activities.* Transportation Development Centre, Safety and Security, Transport Canada, Montreal, Quebec. Report no. TP 12816E.

Rutenfranz, J., Aschoff, J., and Mann, H. (1972). *The Effects of Cumulative Sleep Deficit, Duration of Preceding Sleep Period and Body Temperature on Multiple-Choice Reaction Time.* W.P. Colquhoun, ed. Aspects of Human Efficiency, Diurnal Rhythm and Loss of Sleep . The English Universities Press Limited, 217-228.

Smiley, A., and Heslegrave, R. (1997). *A 36-Hour Recovery Period for Truck Drivers: Synopsis of Current Scientific Knowledge.* Transportation Development Centre, Safety and Security, Transport Canada, Montréal, Québec. Report no. TP 13035E.

Totterdell, P., and Folkard, S. (1990). *The Effects of Changing from a Weekly Rotating to a Rapidly Rotating Shift Schedule.* Shiftwork: Health, Sleep and Performance. Frankfurt am Main: Verlag Peter Lang. 646-650.

Vespa, S. (1997) *Human Conditions.* Recovery, 8(2), 18-19.

Wylie, C.D., Shultz, T., Miller, J.C., Mitler, M.M., and Mackie, R.R. (1996a). *Commercial Motor Vehicle Driver Fatigue and Alertness Study.* Transportation Development Centre, Safety and Security, Transport Canada, Montréal, Québec. Report no. TP 12875E (also available as U.S. FHWA report no. FHWA-MC-97-002).

Wylie, C.D., Shultz, T., Miller, J.C., Mitler, M.M., and Mackie, R.R. (1996b). *Commercial Motor Vehicle Driver Fatigue and Alertness Study: Technical Summary.* Transportation Development Centre, Safety and Security, Transport Canada, Montréal, Québec. Report no. TP 12876E (also available as U.S. FHWA report no. FHWA-MC-97-001).

Wylie, C.D., Shultz, T., Miller, J.C., and Mitler, M.M. (1997). *Commercial Motor Vehicle Driver Rest Periods and Recovery of Performance.* Transportation Development Centre, Safety and Security, Transport Canada, Montréal, Québec. Report no. TP 12850E.

9

FATIGUE: PERFORMANCE IMPAIRMENT, SLEEP AND AGEING IN SHIFTWORK OPERATIONS

Ronald J. Heslegrave, Ph.D. Department of Psychiatry, St. Michael's and Wellesley Central Hospitals, University of Toronto

INTRODUCTION

Shiftwork has been part of the working world since the world has been working. When shepherds tended their flocks by night being vigilant to threats from the natural environment, they undoubtedly experienced the same difficulties as today's shiftworkers who must remain vigilant at night. Similarly, as threats from rival groups prompted the need for sentries and then soldiers, today's police forces need to be available and vigilant on a 24-hour basis. As civilization progressed, so did the need for some segments of society to engage in shiftwork that spanned the 24-hour day. Among the earliest shiftworkers were those involved in security services (guards and soldiers), health services (shamans and infirmaries), hospitality and service industries (food services and lodging), and transportation services (sailors and ground transportation). However, once Thomas Edison invented the light bulb in 1879 (after $40,000 of investment and 1200 experiments), the possibility existed for virtually all sectors of the economy to engage in continuous 24-hour operations.

Today, it is estimated that at least 25% of the workforce in most industrialized countries works some form of shiftwork and this percentage is growing as more and more types of industries need to be available on a 24-hour basis. Shiftworkers are no longer confined to service industries, transportation, power production, and security and health services but are increasingly in manufacturing, communication, and financial sectors. Nevertheless, the transportation sector remains one of the largest sectors of the economy involved in shiftwork. Obviously, since the earliest time, mariners had little choice but to work shiftwork. As noted above, the need for shiftwork in ground transportation developed early and with the "just-in-time" delivery systems in place in most economies, shiftwork and long hours have become not only a reality but is increasing the need for "around-the-clock" operations. More recently, air transportation has become a 24-hour operation with the need to keep 80-85% of fleets in the air around the clock to function competitively. This paper will review some recent work that targeted one aspect of air transportation.

The effects of shiftwork are numerous but among the most acute effects are increased fatigue and sleepiness, impaired job performance, and reduced and poor sleep; all of these effects become more prominent with age. However, when shiftwork is being discussed, the greatest concern is with shiftwork that involves night work. There is perhaps no better recent example for the transportation industry as that from the trucking industry in Canada and the US where a complex study was designed to evaluate the different hours of service regulations in Canada and the US (Mitler, Miller, Lipsitz, Walsh, & Wylie, 1997). In that study, male truck drivers worked four different driving schedules; 2 10-hour, 5-day schedules that included primarily day or night driving in the US and 4 13-hour, 4-day schedules that included primarily day or night driving in Canada (though the actual schedules were not so symmetrical in the experimental design). Interestingly, in terms of performance impairment ("drowsy driving") and sleep, the hours of service regulations contributed little to the differences between conditions whereas night work versus day work accounted for the important differences in performance and sleep. These data provide recent field evidence that are in support of previously collected experimental data on sleep deprivation (e.g., Angus & Heslegrave, 1985; Heslegrave & Angus, 1985) that have identified night work to be of paramount importance when considering shiftwork operations in any area of transportation.

As automation in many industries increases, the shiftworker is increasingly becoming a highly-skilled, vigilant monitor of automatic processes. This changing role of shiftworkers toward one of passive vigilance rather than active involvement is one that makes the shiftworker more vulnerable to the effects of fatigue than ever before especially when shiftwork involves night work where one's circadian rhythms are the least likely to support alertness on the job. Given the increasing prevalence of shiftwork and the changing role of shiftworkers toward passive vigilance and monitoring, the problems associated with shiftwork will become increasingly important.

With the increasing use of shiftwork and night work and the increasingly recognition that fatigue in the workplace has significant implications for worker's health and safety both on the job as well as off the job (with respect to their relationships with families and others), some safety sensitive industries have begun to investigate the implications of shiftwork on their employees and begun to introduce fatigue countermeasure strategies. However, with the recent corporate restructuring exercise in the industrialized world to develop more efficient, productive and competitive workplaces, many industries streamlined their operations by reducing the management layers of bureaucracy and limiting their workforces. The result has been that the workers available to fill highly-skilled, technical occupations are becoming increasingly older due to reduced opportunities to eliminate shiftwork through reassignment or promotion to non-shiftworking jobs and because their are fewer skilled workers competent to take on their responsibilities. As a result, many industries are facing an ageing workforce who are likely more susceptible to the demands of shiftworking regimens.

This paper will present some the recent findings from an ongoing project to study fatigue in the aviation sector of the transportation industry. With respect to the issues outlined above, the air traffic control sector of the aviation industry typifies many of the concerns regarding shiftwork. The job of the air traffic controller is becoming increasingly more automated and the tasks are primarily one of vigilance and increasingly passive vigilance making controllers more susceptible to the effects of fatigue. In addition, with the ability of new systems to handle increased volumes of traffic and the need to increase efficiencies, the air traffic controller workforce has been reduced in size and the average age of the air traffic controller population has become older and it continuing to become older. In Canada, the population of air traffic controllers are in their mid-forties and the average age is likely to increase due to the slow acquisition of new, younger, qualified controllers. Therefore, for air traffic controllers, like many occupations in transportation, fatigue is becoming an increasing concern since their jobs are becoming increasingly automated making the vigilance component of their job increasingly greater and they are a population that is becoming increasingly older making them more susceptible to fatigue effects.

Canadian Air Traffic Control Fatigue Management Project

The Canadian Air Traffic Control project is a multi-year, multi-phase project that was begun in 1994 and is ongoing. Conceptually this project has three phases. The first phase consisted of a review of the current literature and a comprehensive national survey of air traffic controllers to examine the impact of shiftwork on controllers and relate this impact to a number of parameters including job performance impairment, sleep, risk of automobile accidents, somatic complaints, and job satisfaction. Moreover, this survey attempted to examine how the impact of shiftwork might be modified by a number of work parameters, demographic characteristics, and chronobiological typologies. Finally, the survey attempted to determine the prevalence of various coping mechanisms and strategies. For the purposes of this paper, the impact of shiftwork on performance impairment and sleep will be considered and the data relevant to the question of ageing will be reported.

The second phase of this project sought to objectively quantify the degree of performance impairment and sleep reduction that occurred as a result of working a variety of different shift types. In this study, performance was assessed using generic performance assessment techniques at the beginning, middle and end of each shift. In addition, sleep data were collected during a baseline night's sleep and twice during the 5-day shift schedule. In all, 5 different shift schedules, typical of those worked by air traffic controllers in Canada, were assessed. However, for the purposes of this paper only the more extreme shift schedules that have included midnight shifts will be reported. (The government reports on the first two phases of this work can be obtained from the Transportation Development Center in Montreal, Canada.)

The third phase of the project is currently underway. The third phase has two purposes: first, to develop an educational program about shiftwork for air traffic controllers and second, to empirically examine the impact of a variety of fatigue countermeasures on the performance and sleep in air traffic controllers. Specifically, the four types of countermeasures that will be explored independently include strategic napping, use of bright light therapy in the workplace, sleep training (including autogenic training, relaxation training, sleep hygiene training, and sleep hardiness training), and a prescribed exercise program.

Phase 1 - Survey Of The Impact Of Shiftwork On Air Traffic Controllers

As mentioned above, the purpose of this phase of the project was to examine the impact of shiftwork on air traffic controllers through the use of a comprehensive national survey designed to assess how shiftwork affects job performance impairment, sleep, risk of automobile accidents, somatic complaints, and job satisfaction. In addition, the survey attempted to examine how the impact of shiftwork might be modified by a number of work parameters, demographic characteristics (such as age) and chronobiological typologies.

To implement this phase of the work, a survey was developed in collaboration with the controllers, their union representatives, occupational health and safety staff, and the administration and management at the Department of Transport (Canada). The survey instrument contained 114 structured questions, largely requiring Likert-type categorical judgments, designed to gather information on the following topic areas: demographics, work environment, performance and somatic complaints, chronobiological type, family demographics and perceived support, shift characteristics, overtime characteristics, coping strategies, sleep and well-being.

Method

The survey was sent to all 1836 controllers in the system. These results are based on the response to the first mailing where there were 921 anonymous responses for a response rate of over 50%. The demographic data were analyzed with respect to the known distribution of controllers on a regional, gender and age basis to determine the existence of response bias. Based on this analysis, it was found that there were no inherent biases in the sample in terms of regional, gender or age compared to the population. (All effects reported below are highly statistically significant (p<01).)

Results

Sample The surveys returned constitute a representative sample with respect to geographic distribution, years of experience, gender, age, and primary duties. In general, 60% of the controllers work in Eastern Canada, in part because of the heavier volume of trans-Atlantic flights. In terms of experience, 70% of the sample had over 10 years of experience as an air traffic controller and 60% had over 10 years of experience in the specific type of job in which they were currently engaged. Males accounted for 94% of the respondents (though more than 50% of the female population participated) and 82% were married. On average the sample was 41 years of age though females were generally younger averaging only 33.4 years old.

Performance Impairment To assess performance impairment, controllers were asked to rate their job performance (in terms of alertness, memory, productivity) near the end of specific shifts lasting various lengths. Specifically, they provided ratings for the day, evening, and midnight shifts that lasted either 8, 12, or more than 12 hours in duration. They were asked to rate their performance impairment on a 6-point scale from 1 ("no impairment") to 6 ("severe impairment"). Twelve-hour shifts were not uncommon and with shortages in the air traffic control population in the few years preceding the survey, controllers occasionally worked longer shifts and could work double (8-hr) shifts under exceptional circumstances (though many respondents would be speculating about shifts lasting longer than 12 hours based on their 12-hour experience).

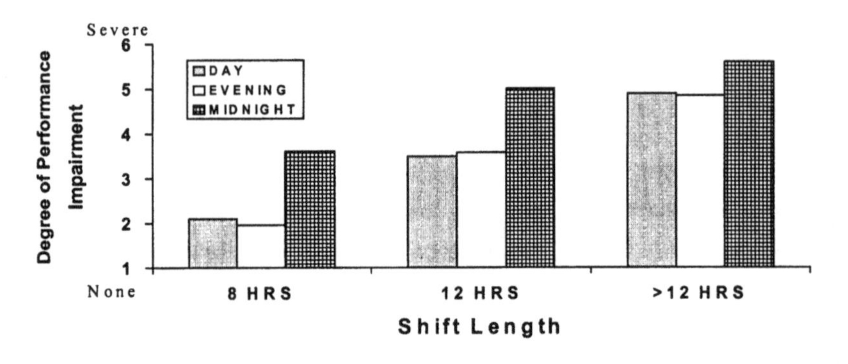

Figure 1

Figure 1 shows the degree of perceived job performance impairment as a function of different shift types and different shift lengths. In general there are several key points to be made with respect to Figure 1 (based on the highly significant shift type by shift duration interaction ($F(4,1828)=97.27$, $p<.001$) and the highly significant shift type and shift duration main effects). First, it can be seen in the figure that the degree of perceived performance impairment on day and evening shifts is about the same on across all shift durations. Second, it is clear that as the shift duration increases, the degree of performance impairment experienced by controllers significantly increases (except when the shifts are extremely long and then a ceiling effect occurs with respect to the performance impairment). These data also show the effects of different shift lengths (8 vs 12 hours) on different levels of job performance impairment. From the data gathered from these 921 workers, it is clear that working overnight leads to significant job impairment compared to working either day or evening shifts regardless of the length. It is interesting to note that Figure 1 shows that 8-hour midnights shifts show about the same degree of performance impairment as 12-hour day or evening shifts. Twelve-hour shifts show convincing impairment on all shift schedules with the same pattern as 8-hour shifts. For 12-hour shifts, moderate to high performance impairment can be expected on day and evening shifts while midnight shifts appear to produce more extreme levels of performance impairment.

The data in Figure 1 indicate that when shifts are of longer duration (namely 12 hours) or when shift begin in the late evening and require working overnight, shiftworkers report being particularly vulnerable to performance impairment effects. Even though these ratings of performance impairment are subjective in nature, such perceived effects should be sufficient to raise concern over job performance effects associated with working overnight particularly. It is noteworthy that when controllers responded to this anonymous survey that under certain circumstances, they experienced "severe" performance impairment. Phase 2 of this project attempted to quantify the actual level of performance impairment under field conditions.

Another issue of particular relevance to shiftworkers concerning their ability to perform relates to their ability to remain alert during their commute to and from work. In this survey, controllers were asked whether they had fallen asleep during their commute to and from work in the context of their air traffic control occupation. These data essentially constitute a lifetime experience for these shiftworkers. The results showed that 21.8% of controllers have fallen asleep while driving during their commuting. This is another expression of the inability of shiftworkers to maintain alertness under vigilance conditions.

Sleep Figure 2 show the amount of sleep that controllers report on the different shift schedules including day, evening, midnight and during their days off. These data clearly show that shiftworkers obtain the least sleep during midnight shifts where they report only just over 5 hours of sleep each night. Given that this is self-reported sleep, their actual sleep, if measured polysomnographically, would likely be about 30 minutes less. Interestingly, these shiftworkers obtain only 6.5 hours of sleep on the day shift which illustrates the problems associated with the early morning start times for these day workers who typically begin about 0600-0630. On evening shifts and on their days-off, shiftworkers make up for the sleep debt accumulated over other shift rotations sleeping 7.5 hours on the evening shift and 8 hours on their days off on average. From these data, it can be concluded that shiftworkers lose the most sleep on midnight shifts yet the second most difficult shift is their day shift in terms of their degree of sleep loss associated with these types of 8-hour shifts. Clearly, evening shifts provide the greatest amount of sleep while the shiftworkers' days off allow shiftworkers to acquire additional sleep. The acute sleep debt across different shift schedules can be calculated by multiplying the daily sleep loss shown in the figure by the number of days working the shift. For instance, if controllers work 5 midnight shifts consecutively then the cumulative sleep debt associated with that shift could be as great as 10-15 hours across that shift schedule depending on their normal baseline night's sleep (which may be difficult to determine for shiftworkers who appear to be chronically sleep-deprived).

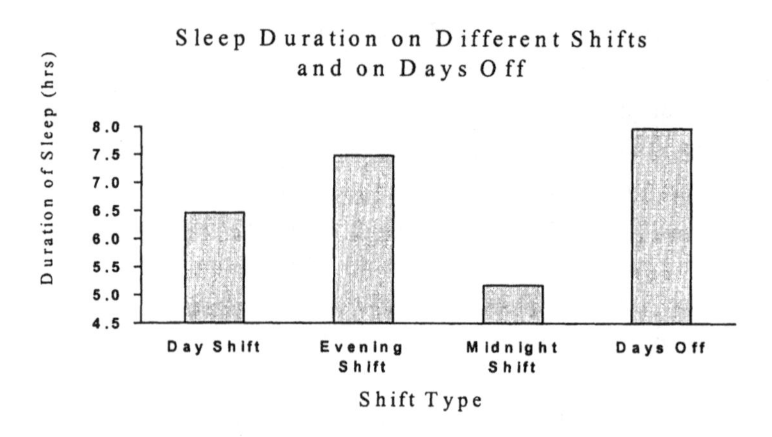

Figure 2

The Ageing Shiftworker Given that the impact of shiftwork is thought to increase as one ages and the fact that the population of shiftworkers in highly skilled occupations, such as air traffic controllers, is an ageing population, it is important to examine whether shiftwork differentially impacts on job performance and sleep across age groups. In addition, since the goal of the present fatigue management project is to introduce interventions to the shiftworker as soon as they begin to be significantly impacted by their shiftwork routine, the question arises "at what age do the performance and sleep disruption effects begin to impact on the shiftworker?

The usual answer to this question of age effects on shiftworkers has been that shiftworkers over 50 are at greater risk for shiftwork effects than younger shiftworkers. Some reports have suggested that shiftworkers begin to show deterioration in job function on night shifts in the 45-50 year age range (Akerstedt, Patkai, & Dahlgren, 1977). In a recent report on this topic, Harma (1996) has recommended that perhaps when workers are over 40 years of age, continuous night work should become voluntary which suggests that under the most extreme circumstances, shiftwork may begin to affect workers below 45 years of age. In this phase of the project, shiftwork effects were examined cross-sectionally across age groups by examining whether respondents thought that shiftwork has become more difficult as they aged and whether their job performance and sleep were affected more as they aged.

In the survey, shiftworkers were asked whether coping with shiftwork had become (slightly or much) more difficult as they had grown older. The results are shown in Figure 3. It can be seen in the figure that increasing numbers of shiftworkers found that shiftwork became slightly more difficult to cope with until their mid to late thirties (35-39 years of age). However, when shiftworkers reached 35-39 years old, there was a shift in the view respondents with a sharp increase in the curve indicating that shiftwork had become much more difficult. By 35-39 years of age, 30% of shiftworkers are reporting that shiftwork had become much more difficult compared with only 10% of those between 30-34 years of age. By 35-39 years of age, over 80% of shiftworkers reported some increase in the difficulty coping with shiftwork. By 40-44 years of age, 80% of shiftworkers also reported an increased in difficulty coping with shiftwork but half of the respondents reported that the increase difficulty was more severe ("much more difficult"). The pattern established by 40 years of age remained stable for older age groups. These data suggest that by the increasing difficulty coping with shiftwork is fairly stable in the population by 40 years of age with about half of the population finding coping much more difficult than when they were younger. In addition, these data suggest that shiftworkers begin to detect significant impacts related to shiftwork between 35-39 years of age. Clearly, shiftworkers are impacted much earlier than previously suggested.

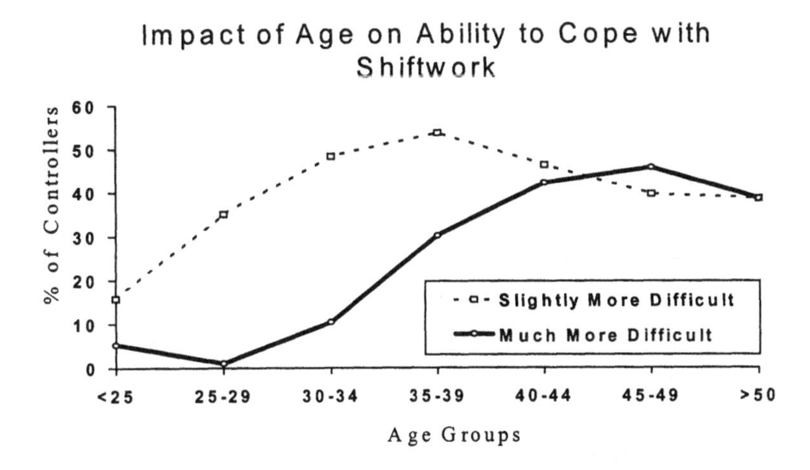

Figure 3

If coping with shiftwork becomes more difficult generally by 40 years of age, does this also mean that job performance is affected as well since shiftworkers could perceive themselves as coping successfully with shiftwork demands. To examine this question, performance impairment was examined across the different types of shifts and across age groups for both a less impaired 8-hour shift and a more impaired 12-hour shift (as shown in Figure 1).

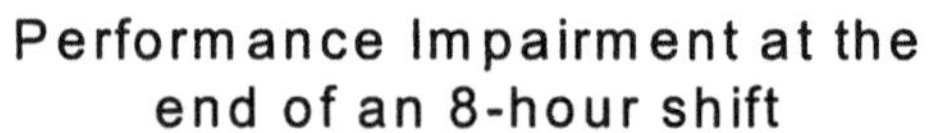

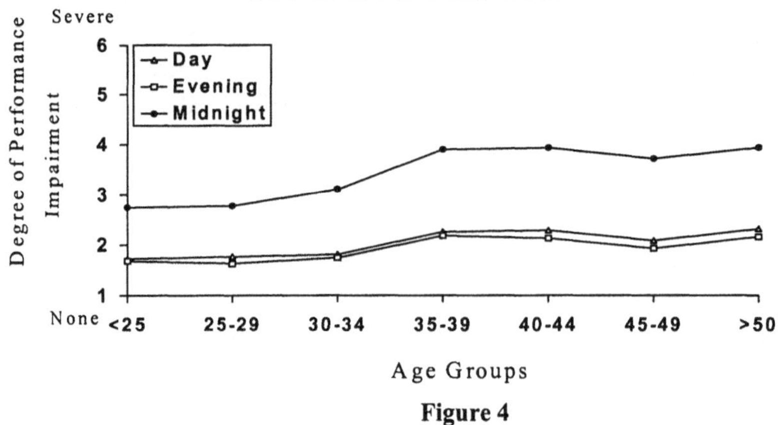

Figure 4

Figures 4 and 5 shows the degree of performance impairment that shiftworkers experienced at the end of either an 8 or 12-hour shift. Clearly 12-hour shifts lead to more impairment than 8-hour shifts and midnight shifts lead to much greater impairment than either day or evening shift. However, for the aging shiftworker, the performance impairment seems to become more evident when workers reach 35-39 years of age (particularly on midnight shifts). Figure 4 shows that even under the less severe conditions of an 8-hour shift, performance impairment shows a marked increase between 35-39 years of age which is particularly noticeable under the more severe midnight shifts.

Figure 5 shows the performance impairment associated with the more difficult 12-hour shift. For the day and evening shift, there is a marked increase in performance impairment between 35-39 years of age. For the more difficult midnight shift, increases in performance impairment may even begin at an earlier age but stabilize by the 35-39 year age group.

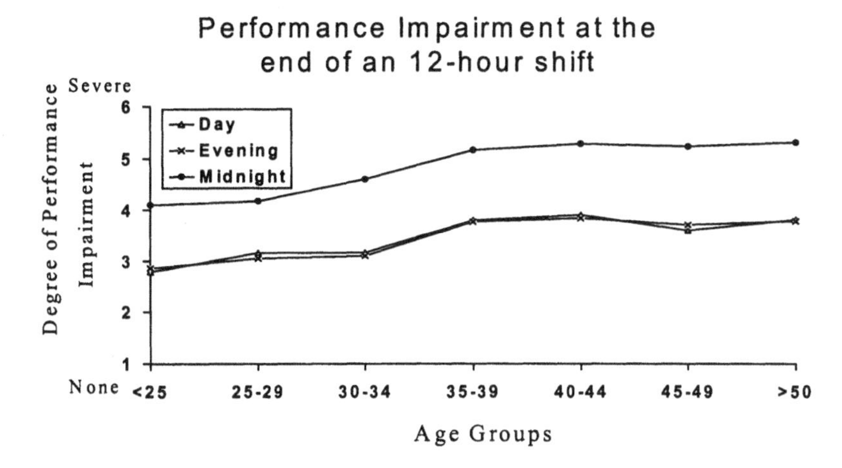

Figure 5

Together these data show that performance impairment as a function of shiftwork becomes evident during the mid to late thirties of shiftworkers' lives and stabilizes thereafter. Thus, interventions must be targeted at the young shiftworker in order to reduce the impact of shiftwork on job performance or at least delay the onset of shiftwork effects. Interestingly, other somatic data from this study are confirmatory with respect to ageing effects. When questioned about appetite disturbance and other somatic complaints, there was an increase in these somatic complaints by 35-39 years of age as well as a marked increase in the use of sedatives for sleep by 40 years of age. These results add further support to the conclusion that shiftwork begins to impact significantly on shiftworkers by their mid-thirties.

Figure 6 shows the amount of sleep obtained on different shifts as a function of age. In terms of age, these data show the expected general effect of age on sleep in that this population obtains less sleep as they age regardless of whether they are sleeping during a midnight shift schedule or during their days off. However, on closer examination, the amount of sleep obtained on the midnight shift begins to decline during the 35-39 year age group and seems to stabilize by the 40-44 year age group.

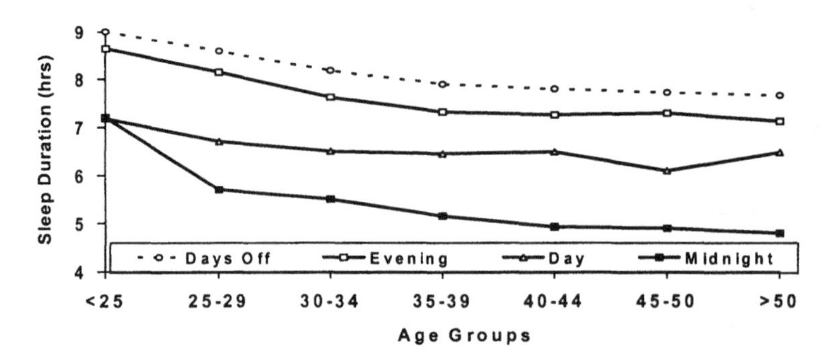

Figure 6

Phase 1 Summary

The data collected in Phase 1 of this project provided support for the suspicion that shiftwork significantly impacts this population in terms of fatigue-related impaired job performance which occurs as a result of working at chronobiologically-inappropriate times and reduced sleep. It was found that by the end of either an 8 or 12 hour shift, performance was significantly more impaired after a midnight shift than after day or evening shifts. In addition, the fatigue-related performance deficit associated with an 8-hour midnight shift was equivalent to that which occurred after a 12-hour day or evening shift. Finally, the levels of perceived performance impairment were marked under many circumstances suggesting that these performance deficits should be of concern.

In terms of sleep, sleeping while on a midnight shift schedule led to the greatest amount of sleep restriction with midnight shiftworkers reporting just over 5 hours of sleep during a 24-hour period. Given that these data are self-report data, it is likely that these estimates are overestimates of the amount of actual sleep obtained during their schedules. For those workers on consecutive midnights, it would be expected that there may be 10-15 hours of sleep debt accumulate over a 5-day shift. Interestingly, the day shiftworkers also reported significant sleep loss of about one hour per day.

With respect to the ageing shiftworker, the data from this phase of the project indicated that shiftworkers find it much more difficult to cope with shiftwork as they age. Moreover, this population of shiftworkers also report increases in performance impairment, reduced sleep, and increases in somatic complaints as they age. Perhaps the most startling result from an ageing perspective was that these impact associated with shiftwork begins to occur and be detected by shiftworkers as young as 35 years of age, perhaps younger under the most difficult circumstances. As a consequence, the interventions aimed at reducing the impact of shiftwork must begin early in the shiftworkers career to be the most effective.

Phase 2 - Survey Of The Impact Of Shiftwork On Air Traffic Controllers

The second phase of this project was an attempt to more objectively quantify the degree of performance impairment and sleep reduction that occurred as a result of working a variety of different shift types. While the data reported in Phase 1 of this study suggests that substantial performance impairment occurs with midnight shifts and that marked sleep debt occurs during schedules that involve midnight shifts, the data collected in Phase 1 were subjective in nature. Before proceeding to Phase 3 interventions studies, it was important to attempt to quantify the degree of performance impairment and sleep loss in more objective manner. Toward this end, 5 different 5-day shift schedules routinely employed by the air traffic controller population were studied. For the purposes of this paper, the results of the most difficult shift schedules will be reported. Unfortunately, the volunteers for this study were from a narrow age range so age effects related to shiftwork could not be adequately analyzed. (The full report is available through the Transportation Development Center in Montreal, Canada or from Transport Canada.)

Method

Volunteers were recruited from five different air traffic control centers to participate in the study. Each center used a different shift schedule. The five shift schedules were as follows: 1) 5 consecutive midnights (MMMMM), 2) a highly varying EDDMM schedule (1500-0000, 1000-1800, 0800-1600, 0000-0800, 2000-0400), 3) double quick change schedule (EEDDM), 4) a backward rotation with no midnights (EEEDD), and a forward rotation schedule (DDDEE). Twelve volunteers participated in the first 3 schedules with 10 and 6 volunteers in the last two schedules. For all conditions, performance was assessed using a test battery and sleep was collected polysomnographically for the first 3 schedules. This report will concentrate on the first schedule of 5 consecutive midnights in terms of performance but report the sleep data collected during the first 3 shift schedules.

To assess performance, data were collected using a 10-min generic test battery which was completed at the beginning, middle and end of each shift. (While every effort was made to train all subjects on the test battery before the actual data collection, complete training did not occur and learning effects remained evident during the course of data collection.) The test battery consisted of the Profile of Mood States, Stanford Sleepiness Scale and 2 minutes each of the following 4 tests that have been well documented in sleep deprivation paradigms: 4-choice reaction time, grammatical reasoning, spatial orientation (Manikin test), and a pattern recognition test which reflected many of the aspects of the air traffic controller's job.

To assess sleep, polysomnographs were collected during the first 3 shift schedules only. A baseline sleep was collected on the night prior to starting their shift schedule and then twice during their schedule. For the MMMMM shift, sleep data were collected during the day sleep following the first and fourth midnight shifts. For the EDDMM shift, data were collected during the evening sleep following the third shift and during the day sleep following the fourth shift. For the EEDDM shift, sleep data were collected during the night sleep following the second evening shift (prior to the day shift 8 hours later) and following the second day shift (prior to the midnight shift 8 hours later).

Results

Arousal, Fatigue and Sleepiness Effects Figure 7 shows the changes in arousal and fatigue levels based on the Profile of Mood States and sleepiness based on the Stanford Sleepiness Scale. It is clear in the figure that across the night, fatigue and sleepiness increase and arousal decreases. However, the patterns of subjective experience remain about the same across the 5 consecutive midnight shifts.

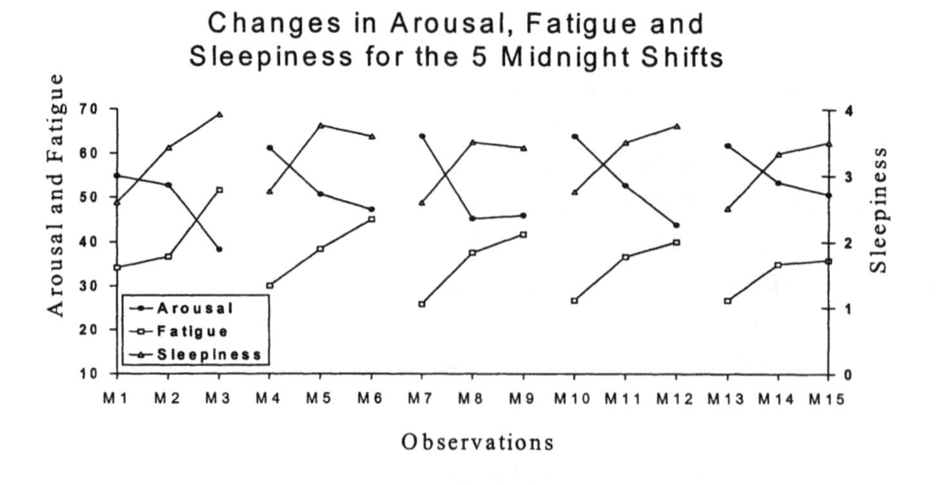

Figure 7

Performance Impairment Figure 8 shows the quantified levels of performance impairment across the night for the five consecutive midnight shifts. Since learning effects were still evident in the raw data, the performance on the 4 2-min tasks each night are plotted as a percentage of the initial score each night which occurred at about 22:30 each night. The data are reported in terms of throughput which is the product of the total responses and the accuracy of responding per minute. This measure integrates speed and accuracy components of performance.

Figure 8 shows that learning remained an issue throughout the first midnight shift where performance improved over the first midnight shift. However, after that first shift, performance becomes impaired during the night on each successive shift with performance showing a 10-17% drop in the middle of the night on most tasks. Performance deterioration appears to be greatest during the third and fourth midnight shifts. In addition, performance impairment appears to be greatest on more difficult tasks with reasoning, spatial orientation, and pattern recognition being more impaired on the third and fourth nights than the reaction time task. These data suggest that performance impairment is about 10% on most tasks. This degree of performance is less than what has been found in previous laboratory-based sleep deprivation studies (e.g., Heslegrave & Angus, 1985), however, in this field study subjects were able to self-select their test opportunity and this may have minimized the observed performance deficit.

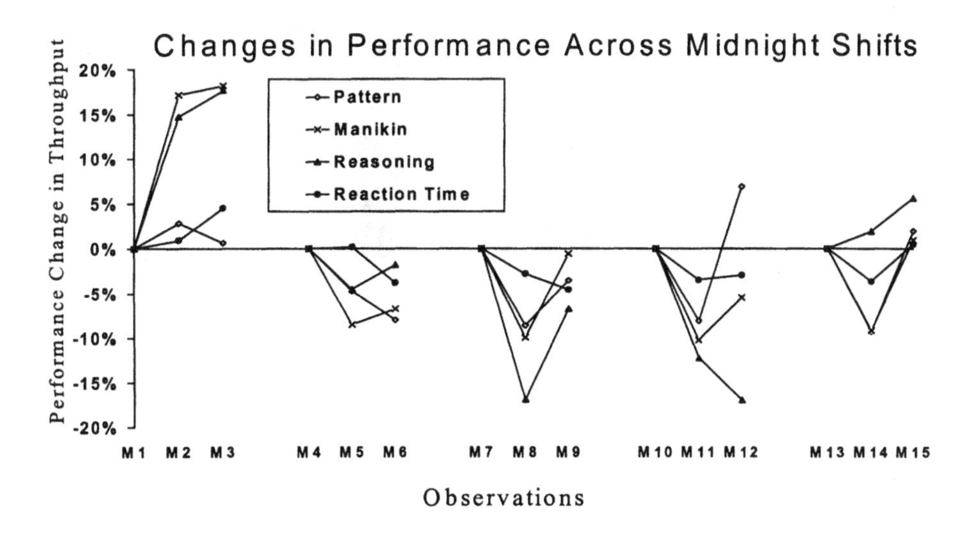

Figure 8

Sleep In Phase 1 it was suggested that those who work midnight shifts only get just over 5 hours of sleep each night resulting in 2-3 hours of sleep loss each night. Figure 9 shows the amount of sleep loss that occurred relative to a normal baseline night sleep prior to the shift schedule on various shift schedules when sleep was measured polysomnographically. For the 5 midnight shifts, the first day sleep (labeled "D") after one midnight shift was about 125 minutes less than the baseline sleep whereas the day sleep following 4 midnights shifts was about 85 minutes less than the baseline sleep. Perhaps an increased level of fatigue associated with more consecutive midnight shifts led to the reduction in sleep loss. The day sleep for the EDDMM schedule following the first midnight shift showed a similar degree of sleep loss to that following the first midnight shift for the MMMMM schedule (about 120 minutes). These data suggest that working midnight shifts may result in a sleep debt of about 2 hours but as the sleep debt accumulates, fatigue may reduce the cumulative sleep debt.

The other finding of interest in Figure 9 is that sleeping in the evening appears to result in the greatest sleep loss. In both evening sleep periods measured in the EDDMM and EEDDM shifts, evening sleep resulted in more than 2.5 hours of sleep loss.

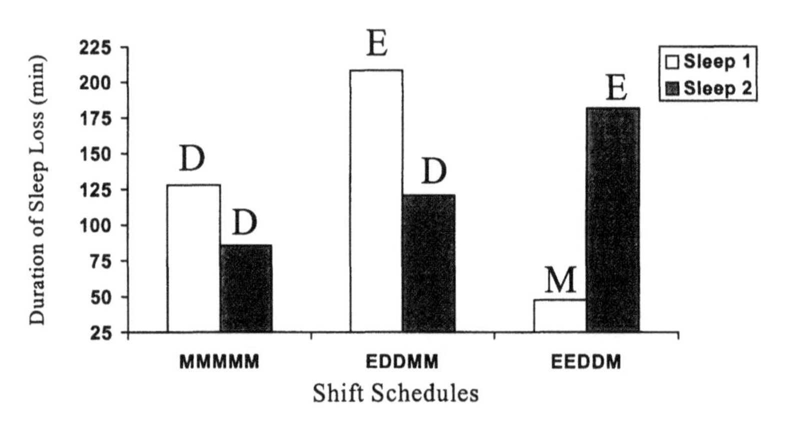

Figure 9

Phase 2 Summary

The results of Phase 2 extend the findings of the Phase 1 survey in a number of ways. In terms of performance impairment, Phase 2 data suggest that there is indeed performance impairment as a result of working shifts that include working overnight. Phase 2 data demonstrate that the magnitude of performance impairment can be on the order of 10-15% of initial performance at the start of a midnight shift and may be greatest on more complex tasks. This quantitative estimate is likely conservative due to the methodology necessarily employed in this field trial. Since the tasks used here were only 2 minutes long, longer tasks requiring greater vigilance would likely have resulted in greater performance decrements. In addition, since the controller had the freedom to choose when the test battery was completed, it is likely that they would chose a time when they felt more confident in doing the test battery.

With respect to sleep, the data from Phase 2 demonstrated that sleeping during the day following midnight shifts resulted in 90-120 minutes of sleep loss on a daily basis though day sleep later in the shift schedule resulted in less sleep loss. However, if controllers slept in the evening rather than during the day, their sleep was further reduced. Evening sleep resulted in sleep loss on the order of 180-200 minutes.

Conclusions

Fatigue in transportation is usually a product of working and sleeping at chronobiologically-inappropriate times: working at chronobiologically-inappropriate times, such as working a midnight shift, leads to acute performance impairment due to increased fatigue associated with circadian rhythm changes; sleeping at chronobiologically-inappropriate times results in reductions in sleep duration and the chronic accumulation of a sleep debt. The degree of performance impairment that can result from the fatigue associated with shiftwork and sleep debt can have devastating consequences.

This report focuses on the work being carried out in the **Canadian Air Traffic Control Fatigue Management Project** and reports the primary results of the first two phases of this multi-phase project. In Phase 1, air traffic controllers were surveyed concerning the degree to which they perceived differential performance impairment and sleep restriction as a result of shiftwork. The results showed that controllers perceived significant performance impairment during midnight shifts and during longer 12-hour shifts. In fact, controllers equated a 12-hour day shift with an 8-hour midnight shift in terms of performance impairment. More importantly, under certain circumstances, controllers perceived their degree of performance impairment to be at least moderate to marked and close to severe in some circumstances. In terms of sleep restriction, Phase 1 clearly pointed out that the greatest reduction in sleep occurred when controllers were working midnight shifts, however, working day shifts which typically start between 0600 and 0630 also leads to sleep restriction.

The other important aspect of Phase 1 concerned the issue of the ageing shiftworker. The increased difficulty coping with shiftwork as well as the changes in performance, sleep, and supplementary health data collected in Phase 1 with respect to age clearly show that shiftworkers begin to sense the impact of shiftwork on performance, sleep and health by their mid to late thirties. The data suggest that by 40 years of age there is a stabilization of performance impairment and reduced sleep. These results suggest that for interventions to be as successful as possible, such interventions should be implemented as early in the shiftworkers' life as possible. These data suggest that the commonly-held notion that shiftwork does not impact on the shiftworker until they reach at least 50 years of age should be replaced with a notion that shiftwork begins to affect the lives of shiftworkers in their thirties and progressively influences them as they age.

Phase 2 sought to quantify the degree of performance impairment and sleep restriction experienced by shiftworkers working their typical shifts in a field study. While there are numerous caveats on the interpretation of these data due to the nature of it being a field study, the results are nevertheless enlightening. In terms of performance impairment, the results demonstrated that working an 8-hour midnight shift resulted in a 10-15% reduction in performance which was greater for more complex tasks. It was argued that these results may be conservative for a number of reasons. Nevertheless, if one considers the results of Phase 1 and 2 together, a rating of performance impairment on an 8-hour midnight shift in Phase 1 resulted in rating of about 3 on a 6-point scale. The same rating was obtained for a 12-hour day and evening shift. Therefore, it could be expected that longer 12-hour shifts might also result in a 10-15% drop in performance. Given that under laboratory conditions (e.g., Heslegrave & Angus, 1985) when such tasks are well learned (perhaps overlearned), performing these tasks overnight results in a 20-25% drop in performance, the results from this study may be somewhat conservative.

Phase 2 also considered the impact of shiftwork on sleep loss. Based on these polysomnographic recordings, it appears that sleeping during the day when working overnight results in 1.5-2 hours of sleep loss though working consecutive midnight shifts may actually reduce the degree of sleep loss due to cumulative increases in fatigue. The more important data from this phase appears to be that sleeping during the evening hours results in the greatest degree of sleep loss. Since it is unclear whether shiftworkers working midnight shifts choose to sleep when the get home during the day or choose to sleep in the evening before going to work, the data in Phase 1 on the sleep restriction associated with midnight shifts may be a combination of both types of sleep routines.

Phase 3 of this project is currently underway. This phase will develop an educational program for controllers modelled in part after a new book written to provide practical strategies to shiftworkers (Shapiro, Heslegrave, Beyers & Picard, 1997). In addition, Phase 3 will experimentally investigate a variety of fatigue countermeasure strategies including strategic napping, use of bright light therapy in the workplace, sleep training (including autogenic training, relaxation training, sleep hygiene training, and sleep hardiness training), and a prescribed exercise program. This project will provide much needed empirical data for the effectiveness of these interventions strategies.

ACKNOWLEDGEMENTS

I would like to gratefully acknowledge the many people who have contributed to these projects, in particular Wayne Rhodes of Rhodes and Associates, Inc., with whom I have worked closely on this project. I would also like to acknowledge the support of Transport Canada (with special recognition to Annette Dunlop) and the Transportation Development Center (in particular Remi Joli) for their support of this work.

REFERENCES

Angus, R.G., and Heslegrave, R.J. (1985). The effects of sleep loss on sustained cognitive performance during a command and control simulation. *Behavior Research Methods, Instruments, & Computers, 17*, 55-67.

Akerstedt, T., Patkai, P., & Dahlgren, K. (1977). Field studies of shift work: II. Temporal patterns in psychophysiological activation in workers alternating between night and day work. *Ergonomics, 20*, 621-631.

Harma, M. (1996). Ageing, physical fitness and shiftwork tolerance. *Applied Ergonomics, 27*, 25-29.

Heslegrave, R.J., and Angus, R.G. (1985). The effects of task duration and work-session location on performance degradation induced by sleep loss and sustained cognitive work. *Behavior Research Methods, Instruments, & Computers, 17*, 592-603.

Mitler, M.M., Miller, J.C., Lipsitz, J.J., Walsh, J.K., and Wylie, C.D. The sleep of long-haul truck drivers. (1997). *New England Journal of Medicine, 337*, 755-761.

Shapiro, C.M., Heslegrave, R.J., Beyers, J., & Picard, L. (1997). *Working the Shift: A Self-Health Guide*. Toronto: Joli Joco.

PART III.

ASSESSMENT OF FATIGUE

10

A Work-Related Fatigue Model Based On Hours-Of-Work

Adam Fletcher and Drew Dawson, Centre for Sleep Research, The Queen Elizabeth Hospital & University of South Australia, Woodville, Australia

Introduction

Sleepiness and fatigue contribute to many of the occupational health risks associated shiftwork (Torsvall and Åkerstedt, 1987; Lauber and Kayton, 1988; Mitler *et al.*, 1994; Rosekind *et al.*, in press). Previous research suggests that shiftwork, and particularly night work, significantly contribute to reductions in sleep durations (Tilley *et al.*, 1981; Torsvall *et al.*, 1981; Torsvall *et al.*, 1989; Åkerstedt *et al.*, 1990; Åketstedt *et al.*, 1991) and increased sleepiness during time awake (Gander and Graeber, 1987; Kecklund *et al.*, 1994). The reduction in sleep duration leads to reduced alertness (Borbely, 1982; Åkerstedt and Folkard, 1990; Dijk *et al.*, 1992; Czeisler *et al.*, 1994), decreased neuro-behavioural performance (Folkard and Monk, 1979; Tepas *et al.*, 1981; Tilley *et al.*, 1982) and potentially larger risks of fatigue-related accident and injury (Lauridsen and Tonnesen, 1990; Laundry and Lees, 1991; Gold *et al.*, 1992; Åkerstedt, 1994). While the scientific data supports this association between hours-of-work, sleepiness and occupational health and safety, few government or organisational policy makers manage work-related fatigue in any methodical or measurable manner.

In part, the scientific data has not filtered in to operational policy because it is difficult to generalise findings of laboratory studies into workplaces. For example, laboratory studies generally oversimplify the psycho-social strata of shiftwork. Lab studies typically do not allow subjects to perform many activities that compete with sleep and recuperation. In addition, very few rosters have been extensively tested in experimental settings. Given the inconsistent findings for specific rosters, difficulties arise in extrapolating from laboratory results to specific shift schedules in a specific workplace. Furthermore, economic or service rationalism often plays the most important role in the determination of a roster for many organisations.

Not surprisingly, it can be extremely difficult for policy makers who are trying to improve employee health and safety to generalise laboratory results to a workplace or to prove overall cost savings. A practical modelling approach which allows organisations to predict work-related fatigue 'on site' would provide an invaluable method to improve shiftwork management. Such a model could therefore evaluate potential rosters prior to their implementation. In addition, organisations could attempt to further clarify the human and monetary cost consequences and relationships between hours-of-work, fatigue and health and safety outcomes.

This chapter attempts to develop the framework of a work-related fatigue modelling approach. We acknowledge that no model can ever predict work-related fatigue entirely, however we believe that current scientific knowledge is sufficient to initiate systematic quantitative modelling. Our opinion is that the method by which policy-makers have generalised laboratory findings, that is by extrapolating from laboratory assessments of specific rosters to an operational workplace, is flawed. We contend that a more appropriate method is to use the determinants of fatigue observed from experimental research and to develop a generalised model suitable for any workplace.

We also believe that there is little incentive for organisations to implement scientifically-based policies until the exact relationship between hours-of-work, fatigue and organisational costs can be shown. Without simple, effective tools which can illustrate the human and economic costs of fatigue, organisations will continue to use policies which reflect day-to-day organisational demands and ignore scientific information.

THEORETICAL CONSIDERATIONS

As with any general predictive model, the current approach relies on conceptual determinants of work-related fatigue. Essentially, the approach to hours-of-work and fatigue outlined in this chapter is an attempt to operationalise our knowledge from shiftwork, sleep and fatigue research literature. The methodology outlined is not theoretically novel, but rather an attempt to develop a useful tool based on our current theoretical understanding of fatigue.

In theory, the impact of hours-of-work and fatigue on any psychologically meaningful construct could be modelled using this approach. That is, the current approach could be used to more clearly understand the effect of shiftwork on absenteeism, sick leave or workload equity. However, in this chapter we will restrict discussion to the relationship between hours-of-work and work-related fatigue.

The level of fatigue or alertness in an individual reflects the balance between two distinct forces. That is, forces which produce 'fatigue' and forces which reverse the effects of fatigue, that is 'recovery'. For simplicity, the time of day that a shift occurs, the length of a shift and how recently it occurred will be considered as 'fatiguing' forces. In addition, the time of day that a break occurs, the length of a break and how recently it occurred will be considered as 'recovery' forces.

This view is, to some extent, overly simplistic since it represents the forces which vary as a function of time. From the moment an individual commences work, fatigue generally continues to increase until the point at which the individual finishes work. The fatigue level of the individual will then decline across the non-work period since a certain amount of sleep will presumably (although not always) occur. The competition between two opposing forces represents a continuous and to some extent periodic process.

In practice, the rate at which fatigue levels increase during a work period, or decline during a non-work period, will vary according to several factors. Firstly, fatigue or recovery are likely to increase as the duration of the wake or sleep period lengthens. The longer an individual is awake, the more fatigued one is likely to become. Similarly, the longer an individual is away from work the more one recovers from fatigue.

Secondly, it is important to remember that the recovery value of non-work time is likely to 'saturate' rapidly and is likely to be difficult to 'bank' during a rest period. This is primarily because individuals find it difficult to extend sleep beyond 10-11 hrs irrespective of the amount of prior wakefulness (for review see Strogatz, 1986).

Thus, there is a significant experimental literature indicating that fatigue increases as a function of hours of prior wakefulness (Borbely, 1982; Daan *et. al.,* 1984). This increase is not, however, a simple monotonic function of hours of wakefulness. Rather, the function that links hours-of-wakefulness to fatigue levels shows a complex relationship in which there are significant linear and sinusoidal (circadian) components (Borbely, 1982; Folkard and Åkerstedt, 1991). In general, it is probably reasonable to argue that the fatigue value of a work period varies as a function of the duration and circadian timing of the work period. In simple terms, the longer the work period the more fatiguing it is likely to be. Furthermore, for any given duration of a work period the level of fatigue accumulated is likely to be greater when the work period occurs during the subjective night than during the subjective day.

Similarly, the recovery value of a non-work period is also likely to vary as a function of the duration and time at which it occurs. This is because the duration and quality of sleep (Czeisler *et al.,* 1980; for review see Strogatz, 1986) and, by inference, the 'recovery' value show a strong circadian component. Thus, the recovery value of a given rest period is likely to vary according to the time at which it occurred. For example, since the amount and quality of sleep varies as a function of the time of sleep onset (Czeisler *et al.*, 1980; Zulley *et al.*, 1981; Zulley and Wever, 1982), amount of sleep and, therefore, the recovery value of a 12h break during subjective night is likely to be greater than during subjective day.

Knowing the circadian timing and duration of work and non-work, the model will enable one to predict the amount of sleep an individual is able to obtain. This, in turn, provides the 'fatigue' and 'recovery' values for a specific work or non-work period. Given that the fatigue level of an individual can be viewed as an algebraic function of the 'fatigue' and 'recovery' functions, it is then possible to calculate the fatigue level for an individual on the basis of the shift schedule history of work and non-work periods. By recording an individual's hours-of-work, it should be possible to determine the work-related fatigue level at any particular point in time.

It is conceptually important to note that the relative contribution of each prior work and non-work period is unlikely to be equivalent. The relative contribution of a specific work or non-work period is likely to diminish as a function of the time distance from the shift in question. That is, the contribution of previous work and non-work periods to fatigue will tend to zero the further the shift is in the past. If we are interested in the fatigue level of an individual at a specific point in time, it is important to acknowledge that the timing of work and non-work periods months or years ago are unlikely to contribute any where near as strongly as a work periods in the last week. The nature of the function describing the relationship between prior work and non-work periods has, by definition, an empirically observable value. Unfortunately, the shape and dimensions of this 'window' are not experimentally known. For the purposes of this paper, we have arbitrarily defined values for this function. The function we have used is a linearly declining function which weights the current hour at a hundred percent and the same hour one week ago at zero percent. For more detailed information on the components of the model, see Fletcher and Dawson (1997).

If we assume that the level of work-related fatigue will reflect the timing and duration of the work and non-work periods and that these will sum as a function of the relative circadian time of the work and non-work periods it is possible to develop a simple but useful model. These general principles can be developed into a quantitative modelling tool by creating a 'token economy' and organisational 'budgets' in which employees acquire or lose fatigue tokens according to the timing and duration of work and non-work periods. In addition the 'value' of tokens will decline over time such that tokens recently acquired carry greater value than those from shifts further in the past. On this basis, the 'net worth' of the tokens that an individual holds at any point in time is a relative measure of the fatigue level for that individual at the time for which the value is calculated. Similarly, the fatigue levels of an organisation will reflect the aggregate value of fatigue tokens for all of the individual employees.

It is likely that the assumptions built into the model represent a large simplification of the actual processes which occur during work and non-work periods. However, it must be acknowledged that any model is just a simpler representation of a system used for predicting information. Its usefulness can only be determined by examining how closely the output of the model resembles the expected outputs.

To differentiate between schedules, the work-related fatigue model produces scores that are defined as being standard, moderate or high. Standard fatigue represents fatigue scores up to 40 points. That is, up to the level just above the maximum for the standard work week (Figure 1). Moderate fatigue scores represent fatigue scores up to 200% of the maximum fatigue scores produced for the standard work week, that is, up to 80 points. Most organisations in shiftworking industries use rosters which create high proportions of moderate fatigue. High fatigue scores are those over 200% of the maximum scores produced by the standard work week, that is, over 80 points.

The definition of high fatigue has not arbitrarily been set at 80 points. The level of fatigue at 80 points is the predicted level of fatigue achieved after 21-23 hours of continuous sleep deprivation following five days of work (0900 to 1700h) then two days off. This sleep deprivation protocol was carried out in a recent study observing performance loss due to sleep deprivation and performance loss due to alcohol intoxication (Dawson and Reid, 1997). Further analyses using the data from this study and the fatigue model suggests that the performance loss at 80 points represents performance loss at least as bad as an individual with a blood alcohol concentration of 0.05%. This level of performance loss due to alcohol would not be accepted in the workplace, therefore, whether this level of performance loss due to fatigue is appropriate is questionable.

The mathematical underpinning of the work-related fatigue model are presented in the Appendix.

RESULTS

The model attempts to illustrate work-related fatigue in a continuous manner. The output function of the model is derived from interactions between work and non-work input functions. In addition to this, the model utilises a transfer function which incorporates duration, circadian timing and recency of hours-of-work. Figures 1 to 8 illustrate the fatigue scores associated with each of the specific shift schedules outlined. Figures 1 to 4 show the fatigue scores associated with a range of standard work patterns. Figures 5 to 8 shows the scores for a range of rosters utilised in specific workplaces. Namely, rosters are shown for a police officer, a commercial pilot, a mining operator and a train driver

Fatigue scores are classified into three ranges: standard, moderate and high. These ranges represent relative fatigue scores for each hour of work. These three ranges reflect the fatigue score for each hour at relative scores of 0-100%, 100-200% or 200-300% of the peak fatigue score produced by a standard work week.

Overall, the fatigue scores for the rosters show considerable variation in the average and peak fatigue scores. Clearly, all 24 hour shift systems produce fatigue scores greater than the standard work week as do mixed start time schedules as seen from the train driver's scores. In addition, the aviation schedule produces peak fatigue scores greater than the standard work week despite the relatively small number of hours worked.

Standard work schedules

Figures 1 to 4 illustrate the fatigue scores associated with four standard work schedules. The first of these schedules is a standard 0900 to 1700h, five day per week roster (Figure 1). The second of these schedules is a two day on (0700 to 1900h), two night on (1900 to 0700h), four day off roster (Figure 2). Thirdly, Figure 3 illustrates a two day on (0700 to 1900h) two day off, two night on (1900 to 0700h), two day off schedule. The fourth of the standard schedules is five days on (0800 to 1600h), 2 off, five afternoons on (1600 to 0000h), two off, five nights on (0000 to 0800h), two off roster (Figure 4).

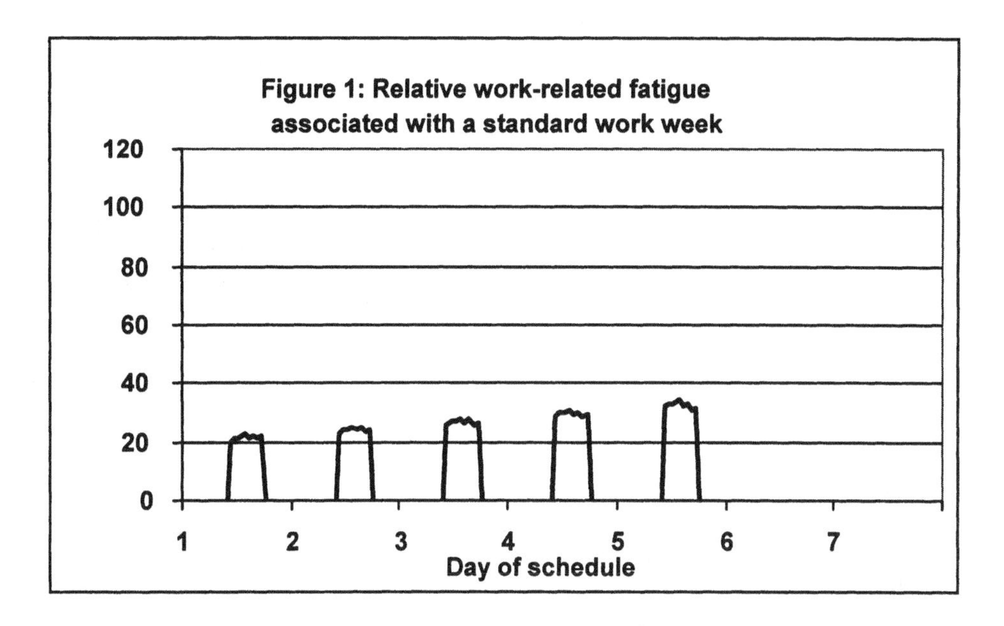

Figure 1: Relative work-related fatigue associated with a standard work week

Although the mean number of hours worked per week for the standard rosters in Figures 2 to 4 are similar to that for the standard work week (SWW), the range of values produced for the fatigue measures covers a much broader range. In general, any roster that covers hours outside 0900 to 1700h, five days per week, creates fatigue scores greater than the maximum for the SWW.

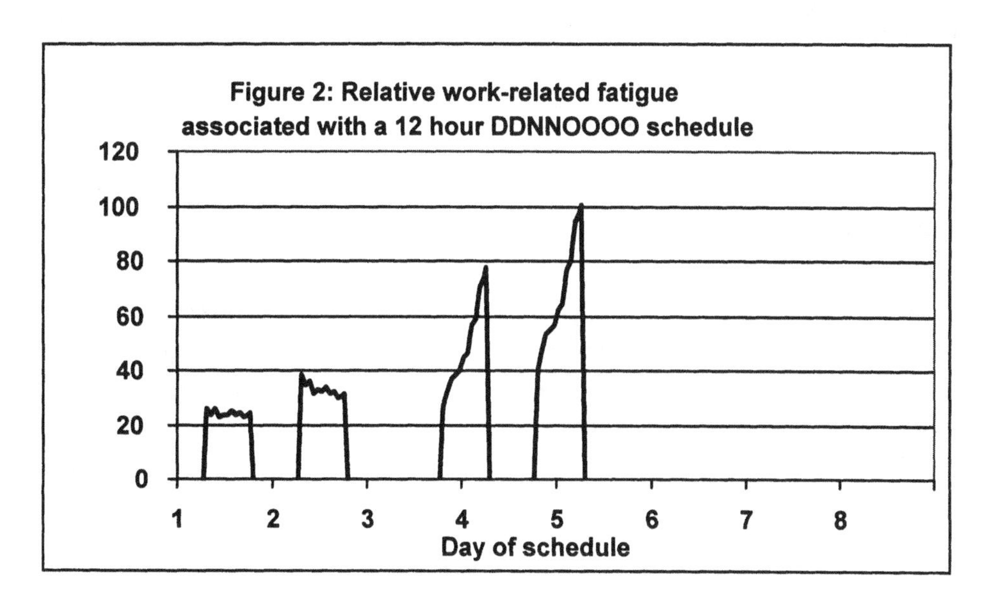

Figure 2: Relative work-related fatigue associated with a 12 hour DDNNOOOO schedule

For example, the roster illustrated in Figure 2 creates maximum fatigue scores greater than double those produced for the SWW. These greater than double scores are produced even thought the number of hours worked per week for the 12 hour system are only slightly higher than the SWW (42 vs 40 hours). Similarly, the 12 hour system illustrated in Figure 3 also produces maximum fatigue scores greater than double those produced for the SWW. However, the maximum scores for the roster in Figure 3 are approximately 20% lower than for the roster in Figure 2.

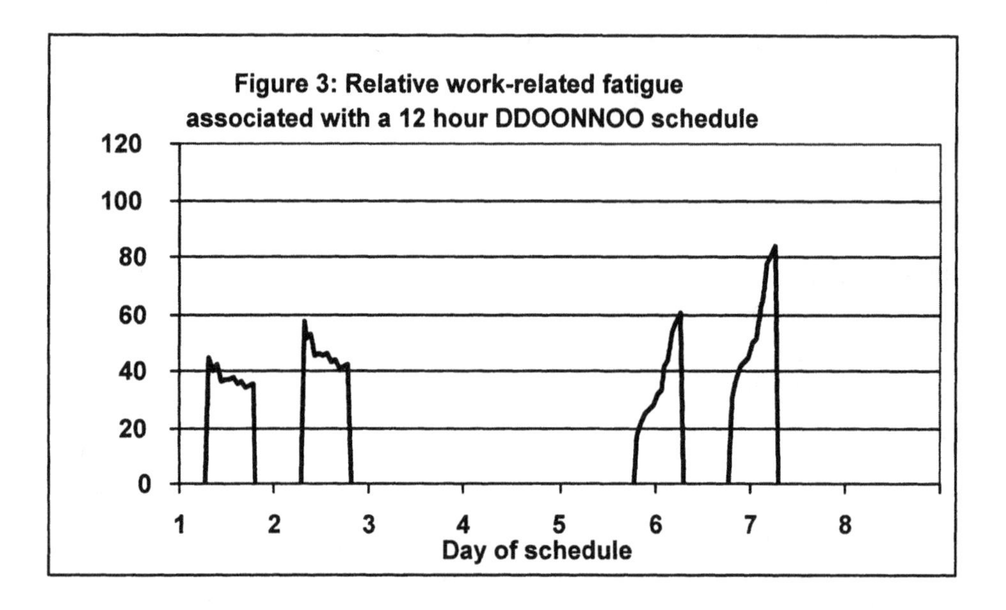

Figure 4 illustrates an 8 hour shift system with 24hr coverage. The number of hours worked per week is therefore identical to the SWW, however the timing of work periods rotate 8 hours on a weekly basis. Thus, a roster cycle is completed every three weeks. As with the night shift portions of the 12 hours rosters, the maximum fatigue scores produced are greater than double the maximum for the SWW. During the day and afternoon portions of the roster, however, the maximum scores are similar to the maximum obtained during a cycle of the SWW.

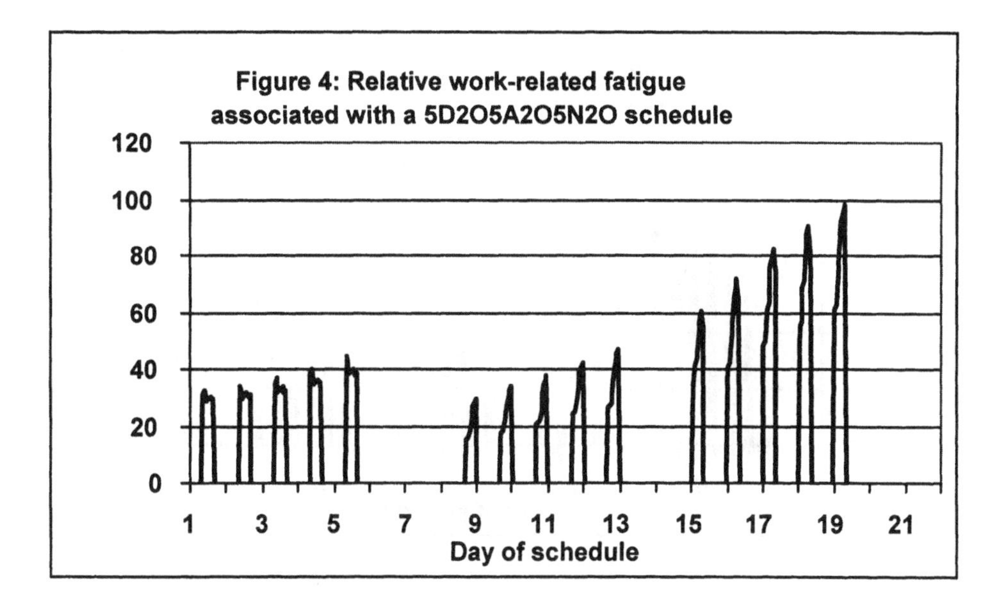

Figure 4: Relative work-related fatigue associated with a 5D2O5A2O5N2O schedule

Specific workplace rosters

Figures 5 to 8 illustrate the fatigue scores associated with rosters currently used in specific workplaces. The first of these schedules is a police officer's roster with varied start times and shift lengths (Figure 5). The second of these schedules is a commercial pilot's roster (Figure 6). Thirdly, Figure 7 illustrates a mining operator's roster. Finally, Figure 8 illustrates a train driver's work schedule.

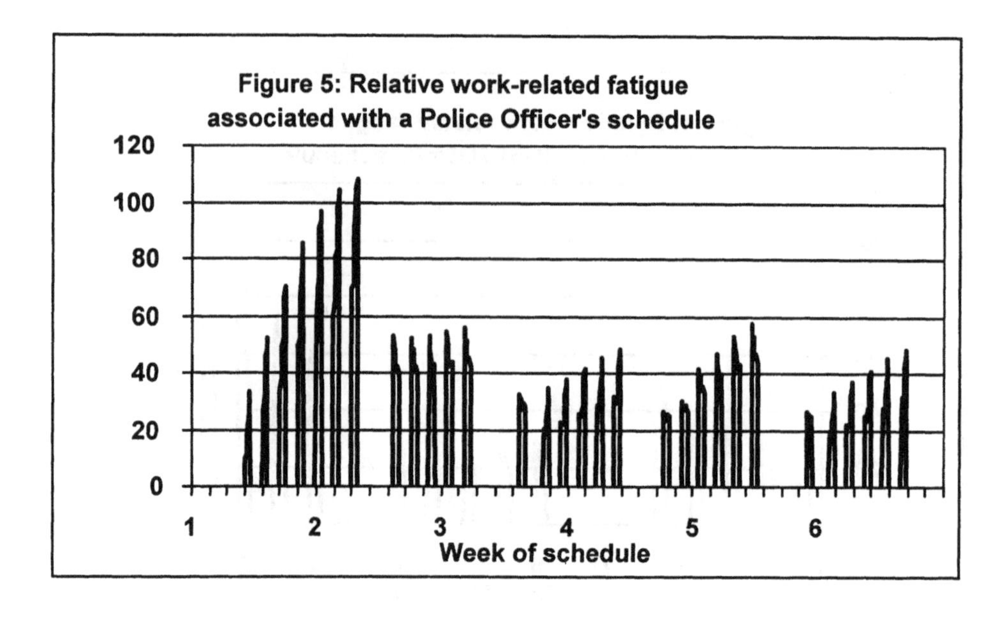

Figure 5: Relative work-related fatigue associated with a Police Officer's schedule

The schedule illustrated in Figure 5 is a schedule currently being used as a police officer's roster. The schedule uses four different shifts: day shift (0700 to 1500h), late morning shift (0830 to 1700h), afternoon shift (1500 to 2300h), and night shift (2300 to 0700h). As with the 8 hour roster in Figure 4, the portion of the police roster which creates high fatigue scores is the night shifts. Most of the night shifts are spent at fatigue scores greater than double the maximum produced for SWW. However, most of the non-night shifts are spent at scores at or below the maximum score for the SWW.

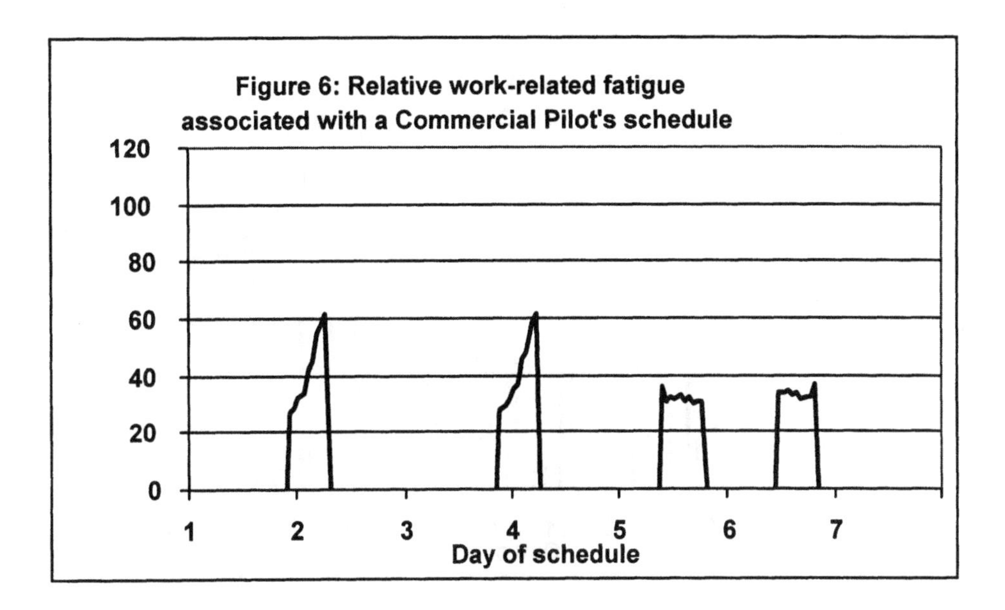

Figure 6 illustrates a commercial pilot's roster. This roster requires work to be carried out during four shifts per week. The first two of these are during night time hours (2200 to 0600h and 2100 to 0500h) and the second two during the day (1100 to 2100h and 1200 to 2200h). This roster requires pilots to work 34 hours per week, that is, 6 hours less than the SWW. However, most of these hours are spent at fatigue scores greater than the maximum produced during the SWW schedule. Once again, the maximum fatigue scores for this roster are produced during that work carried out during night time hours.

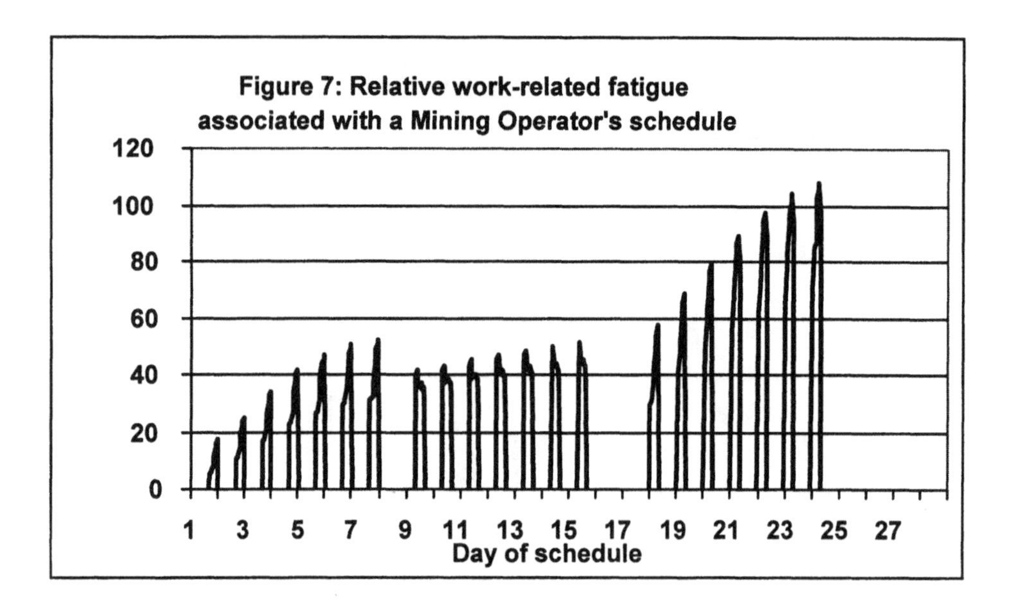

Similarly for the mining roster illustrated in Figure 7, the night hours on shift (0000 to 0800h) cause the greatest fatigue scores. The first night shift causes fatigue scores greater than any day (0800 to 1600h) or afternoon (1600 to 0000h) shift. The third to seventh night shifts all cause over 200% of the maximum SWW scores. As opposed to the mining roster, the train driver's roster depicted in Figure 8 does not have all of the high fatigue caused by night shifts. Some of the highest scores in the train driver's schedule are due to early morning shifts as seen on days 16 and 17. These two shifts started at 0200h and 0300h respectively and create scores as high as those produced during the night shifts on days 23 and 24. This schedule also shows that shifts worked after a break of two or more days is unlikely to produce scores greater than the maximum produced during the SWW.

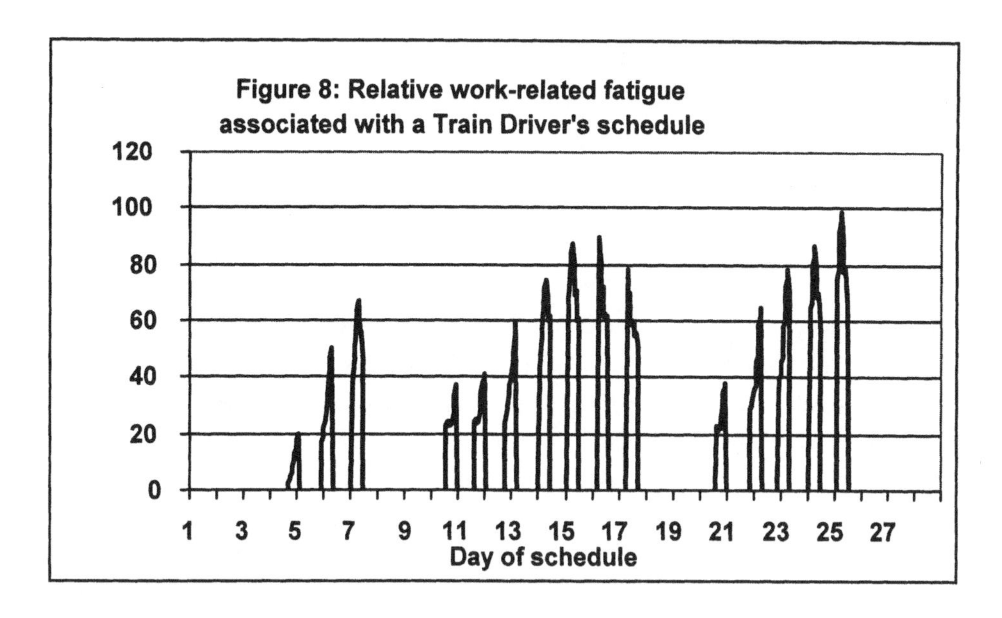

Figure 8: Relative work-related fatigue associated with a Train Driver's schedule

DISCUSSION

The current model constructs fatigue as a general input-output model modified by circadian, recovery and recency-of-work functions. Thus, we have been able to model work-related fatigue as a continuous output. Using only the timing and duration of work as an input, it is possible to calculate fatigue scores that allows us to discriminate between shift schedules. Moreover, by benchmarking schedules against a five day standard work week (SWW), it is possible to gain a perspective on the relative work-related fatigue scores associated with specific roster systems.

The fatigue scores calculated using the model are based only on the timing and duration of work and non-work periods in the roster schedules outlined above. This makes roster assessment simple from an organisational perspective since hours of work are usually already collected. Using such an approach, organisations can implement a sophisticated modelling approach without significantly altering workplace data collection. The current model is based on a theoretical distribution of the recovery function, however, commercial applications of this approach use recovery functions derived from specific individuals in specific workplaces under real operating conditions. Specific details of this methodology can be obtained from the authors.

The work-related fatigue model outlined in this paper is obviously a simplification of the actual processes that interact to modulate fatigue. Fatigue is a multi-dimensional outcome and it is obvious that no modelling methodology could ever incorporate all of the relevant factors. However, the information that this model provides allows simple comparisons of different rosters irrespective of amounts, duration or timing of hours-of-work. This model is useful in the sense that organisations can utilise information that is probably already captured, that is hours-of-work, and assess various outputs without undertaking a complex or expensive process.

The model does not make decisions on which roster schedules are most appropriate in specific workplaces, industries etc. What the model does, however, is provide information that can be useful when decisions about fatigue management need to be made. Therefore, it is capable of providing information that may allow the decision making process to be undertaken in a more informed manner. Ultimately, such a modelling approach may also be useful as a part of reporting processes for accidents, incidents etc. Tracking fatigue scores in relation to occupational health and safety data, absenteeism levels, employee illness days or other organisationally meaningful data would allow a clearer illustration of the relationship between hours-of-work and it's related costs. It could be foreseen that one part of a reporting process would require the capturing of a one week work history so that relative work-related fatigue levels prior to an incident can be assessed. To what extent work-related fatigue is a factor in incidents can also be critical in relation to legal issues where prior hours-of-work may influence outcomes and allocation of accountability.

If the results from the model calculations for each of the shifts are compared, a number of important differences can be observed. For example, comparisons of scores can be made between schedules with different numbers of work hours, different timing of work hours or a combination of the two. The major advantage of this approach is that it enables theoretical insights gained in laboratory studies to be applied to specific or potential roster systems without the need for extensive laboratory trials or actual implementations of a specific shift system. Furthermore, it provides the potential for organisations to evaluate the relationship between work-related fatigue and organisational performance indicators (For example, work performance, health and safety costs, absenteeism)

The distribution of work hours seems to have the largest impact on the fatigue scores produced by the model. By keeping the number of hours worked per week relatively stable, yet distributing hours in different ways, observations can be made regarding the importance of shift length and timing of hours of work. The best example of this is made by comparing the rosters in Figure 1 and 4. These rosters each cover 40 hours per week in the form of five consecutive eight hour shifts. However, the night shift component of the Figure 4 roster creates fatigue scores over 200% greater than the maximum scores produced in the roster in

Figure 1. Thus, compared to the time of day when hours are worked, the number of hours worked per week is not significant. It is unfortunate that most policies and awards only acknowledge number of hours per shift, number of hours per week and minimum break times. This is unfortunate as the scientific literature suggests that the time of day at which hours are worked may be the most important factor in fatigue and alertness management. The challenge is to distribute the hours of night work evenly through rosters to avoid large peaks in high fatigue.

Distribution of hours is even important in roster cycles with small hours of work per week. This is illustrated in the roster shown in Figure 6. This roster requires employees to work only 34 hours per week, however, the fatigue scores are all at or above the maximum for the SWW. This is the case even though the hours of work are evenly distributed evenly throughout the week. If hours are not distributed well, this can have a significant impact on maximal fatigue scores. This is shown best by comparing the rosters illustrated in Figures 2 and 3. These rosters have the same start times, the same shifts and the same number of hours per week. However, the distribution of hours is different. Inserting a break between the day and night shifts reduces the maximum fatigue by 20%.

Generally, the roster in Figure 2 would be preferred by employees even though the maximal fatigue is higher. The reason for this is that this roster gives a four day break every 8 days whereas the roster in Figure 3 gives a two day break every four days. Anecdotal evidence would suggest that most employees would prefer the four day break as it allows a larger block of time to pursue activities such as fishing, hunting, socialising, renovations or second jobs. From a fatigue perspective it is a shame that employees prefer longer breaks less often. This is unfortunate because the strategy that is best for recovery, that is shorter breaks more often, is not the preferred strategy for other activities. It is possible that education of employees to address strategies for recovery may lead to better decisions being made with regards to roster preferences.

In addition to the employees preferences, there may be many employers who choose to allow long blocks of night shifts and/or allow extra or extended shifts before, during or after a block of night shifts. As a strategy, this is also counter-intuitive from a fatigue perspective. Understandably, many employers just want the roster to be covered and do not give thought to whether the employee performing overtime or extra shifts is the least fatigued person for the job. To understand which employee/s will be least fatigued to work, we must understand the work history of all employees. It is in this respect that a modelling approach such as the one outlined in this chapter can help organisations be aware of the work-related fatigue of all employees.

In this capacity, such a modelling approach may provide organisational policy makers with a powerful tool. A multitude of organisational indicators could be linked to fatigue scores. For example, indicators which could be linked to fatigue scores include: absenteeism, employee performance, accidents and injuries, fuel efficiencies for tasks such as train or truck driving, sick leave or morale. An example in practice is that accident and injury reporting systems could require the work history for the week prior to any incident being recorded. The work history would then allow the work-related fatigue scores at the time of accident to be predicted. Over time, a relationship between high fatigue scores and accidents may be drawn out.

Similarly, fatigue scores could be linked to fuel usage of road or rail transport or mining operators. Organisations may find that fuel usage significantly increases at certain times of day or after certain levels of fatigue scores. Such information could clearly lead to considerable savings across organisations over time. In addition to the cost savings, it may eventuate that by managing high fatigue the alertness of employees is also improved. Such 'downstream' changes could potentially lead to improved morale, improved communication, further reductions in accidents and injuries and reduced family conflict. It is our belief that better management of fatigue and alertness may well hold benefits for both employers and employees. This understanding is essential for employers, unions and governments to appreciate the full impact of inappropriate hours-of-work and fatigue management policies. This is particularly true at a time in which longer and more flexible hours of work are seen as a guaranteed productivity gain.

REFERENCES

Åkerstedt, T., Arnetz, B. B. and Anderzén, (1990) I. Physicians during and following night call duty - 36 hour ambulatory recording of sleep. *Electroencephalogr. Clin. Neurophysiol.*, 76: 193-196.

Åkerstedt, T. and Folkard, S. A (1990) model of human sleepiness. In: J. A. Horne (Ed) *Sleep '90*. Pontenagel Press, Bochum, 310-313.

Åkerstedt, T., Kecklund, G. and Knutsson, (1991) A. Spectral analysis of sleep electroencephalography in rotating three-shift work. *Scand. J. Work Environ. Health*,17: 330-336.

Åkerstedt, T. (1994) Work injuries and time of day-national data. Work hours, sleepiness and accidents, Sept 8-10, 1994, *Stress Research Reports*, 248: 106.

Borbely, A. A. (1982) A two-process model of sleep regulation. *Hum. Neurobiol.*, 1: 195-204.

Czeisler, C. A., Weitzman, E. D., Moore-Ede, M. C., Zimmerman, J. C. and Knauer, R. S. (1980) Human sleep: its duration and organisation depend on its circadian phase. *Science*, 210: 1264-1267.

Czeisler, C. A., Dijk, D-J. and Duffy, J. F. (1994) Entrained phase of the circadian pacemaker serves to stabilize alertness and performance throughout the habitual waking day. In: R. D. Ogilvie and J. R. Harsh (Eds) *Sleep Onset: Normal and Abnormal Processes.* American Psychological Association, Washington D. C., 89-110.

Daan, S., Beersma, D. G. M. and Borbely, A. A. (1984) Timing of human sleep: recovery process gated by a circadian pacemaker. *Am. J. Physiol.*, 246: R161-R178.

Dawson, D. and Reid, K. (1997) Fatigue, alcohol and performance impairment. Nature, 388: 235.

Dijk, D-J., Duffy, J. F. and Czeisler, C. A. (1992) Circadian and sleep/wake dependent aspects of subjective alertness and cognitive performance. *J. Sleep Res.*, 1: 112-117.

Fletcher, A. and Dawson, D. (1997) A predictive model of work-related fatigue based on hours-of-work. *J. Occup. Health Safety*, 13(5): 471-485.

Folkard, S. and Monk, T. H. (1979) Shiftwork and performance. *Human Factors*, 21: 483-492.

Folkard, S. and Åkerstedt, T. (1991) A three process model of the regulation of alertness and sleepiness. In: R. Ogilvie and R. Broughton (Eds) *Sleep, Arousal and Performance: Problems and Promises*. Birkhaüser, Boston, 11-26.

Gander, P. H. and Graeber, R. C. (1987) Sleep in pilots flying short-haul commercial schedules. *Ergonomics*, 30: 1365-1377.

Gold, D. R., Rogacz, S., Bock, N. *et al.*, (1992) Rotating shift work, sleep and accidents related to sleepiness in hospital nurses. *Am. J. Public Health*, 82: 1011-1014.

Kecklund, G., Åkerstedt, T., Lowden, A. and von Heidenberg, C. (19940 Sleep and early morning work. *J. Sleep Res.*, 3, Suppl 1: 124.

Lauber, J. K. and Kayton, P. J. (1988) Sleepiness, circadian dysrhythmia, and fatigue in transportation system accidents. *Sleep*, 11: 503-512.

Laundry, B. R. and Lees, R. E. M. (1991) Industrial accident experience of one company on 8- and 12-hour shift systems. *J. Occup. Med.*, 33: 903-906.

Lauridsen, O. and Tonnesen, T. (1990) Injuries related to the aspects of shift working. A comparison of different offshore shift arrangements. *J. Occup. Accid.*, 12: 167-176.

Mitler, M. M., Dinges, D. F. and Dement, W. C. (1994) Sleep medicine, public policy, and public health. In: M. H. Kryger, T. Roth and W. C. Dement (Eds) *Principles and Practices of Sleep Medicine* (second edition). W. B. Saunders & Co., Philadelphia, 453-462.

Rosekind, M. R., Gander, P., Conwell. L. J. and Co, L. (in press) Crew factors in flight operations X: Alertness management in flight operations. *NASA Memorandum*.

Strogatz, S. H. (1986) *The Mathematical Structure of the Human Sleep-Wake Cycle.* Springer-Verlag, Berlin-Heidelberg.

Tepas, D. I., Walsh, J. D., Moss, P. D. and Armstrong, D. (1981) Polysomnographic correlates of shift worker performance in the laboratory. In: A. Reinberg, N. Vieux, P. Andlauer (Eds) *Night and Shift Work: Biological and Social Aspects.* Pergamon Press, Oxford, 179-186.

Tilley, A. J., Wilkinson, R. T. and Drud, M. (1981) Night and day shifts compared in terms of the quality and quantity of sleep recorded in the home and performance measured at work: a pilot study. In: A. Reinberg, N. Vieux, P. Andlauer (Eds) *Night and Shift Work. Biological and Social Aspects*. Pergamon Press, Oxford, 187-196.

Tilley, A. J., Wilkinson, R. T., Warren, P. S. G., Watson, W. B. and Drud, M. (1982) The sleep and performance of shift workers. *Human Factors*, 24: 624-641.

Torsvall, L., Åkerstedt, T. and Gilberg, M. (1981) Age, sleep and irregular work hours: a field study with EEG recording, catecholamine excretion, and self-ratings. *Scand. J. Work Environ. Health*, 7: 196-203.

Torsvall, L. and Åkerstedt, T. (1987) Sleepiness on the job: Continuously measured EEG changes in train drivers. *Electroencephalogr. Clin. Neurophysiol.*, 66: 502-511.

Torsvall, L., Åkerstedt, T., Gillander, K. and Knutsson, A. (1989) Sleep on the night shift: 24-hour EEG monitoring of spontaneous sleep/wake behaviour. *Psychophysiol.*, 26: 352-358.

Zulley, J., Wever, R. and Aschoff, J. (1981) The dependence of onset and duration of sleep on the circadian rhythm of rectal temperature. *Pflügers Arch.*, 391: 314-318.

Zulley, J. and Wever, R. (1982) Interaction between the sleep-wake cycle and the rhythm of rectal temperature. In: J. Aschoff, S. Daan and G. Groos (Eds) *Vertebrate Circadian Systems: Structure and Physiology*, Springer-Verlag, Berlin-Heidelberg, 253-261.

Appendix

Mathematical Description Of The Model

Intuitively fatigue and recovery functions would be:

$$\text{Fatigue}(x) = \text{Circadian}(x) \quad (1)$$

$$\text{Recovery}(x) = -\text{Scalar} \times \text{Circadian}(x) \quad (2)$$

where Circadian (x) is the score or measure on how hard an individual works during hour 'x'. The 'Scalar' determines the rate of fatigue discharge compared with the rate of fatigue accumulation. Note that 'Scalar' is a positive number.

Intermediate calculation

The workload model assumes that fatigue accumulates throughout the day and throughout the week. If their current fatigue level could be given a score based upon what has been worked previously, then the following recurrence relation can be derived:

$$\text{Score}(x) = \text{Score}(x-1) + \text{Fatigue}(x) \text{ (if hours '}x\text{' worked)}$$

$$\text{Score}(x) = \text{Score}(x-1) + \text{Recovery}(x) \text{ (otherwise)}$$

$$\text{with Score}(0) = 0 \quad (3)$$

This means that a person's working times can determine a score or fatigue profile. For a standard work week of 9:00am to 5:00pm Monday to Friday, with a normal Saturday and Sunday off:

$$\text{Score}(0) = \text{Score}(168) = 0 \qquad (4)$$

Equation 4 states that under normal conditions the person working the above shift will return the same fatigue level at the end of the week as they were at the beginning of the week.

A comparison of different shifts under the 'intermediate calculation' model will show the rate which the body's fatigue is accumulated and the differences in the final fatigue 'Score'.

Fatigue Calculation

The fatigue model does not accumulate in the same way as the 'intermediate calculation' model. Instead, a weighted moving average of previous fatigue and recovery scores is used. The calculation of the weights is independent for the fatigue and for the recovery. The weight for the fatigue is a function of the times and the hours worked in the previous week, and the weight for the recovery is a function of the times and hours not working in the previous week.

The functions (for fatigue and recovery) are taken as linearly weighted combinations of the worked or the non-worked hours.

Let : $W(x) = \text{Fatigue}(x)$ (if hour 'x' worked)

$W(x) = 0$ (otherwise)

$R(x) = -\text{Recovery}(x)$ (if hour 'x' not worked)

$R(x) = 0$ (otherwise)

Then : $\text{FWeight}(x)=(168.W(x)+167.W(x-1)+166.W(x-2)+....+2.W(x-166)+1.W(x-167))/168$

$\text{RWeight}(x)=(168.R(x)+167.R(x-1)+166.R(x-2)+....+2.R(x-166)+1.R(x-167))/168$

The score for the particular period is determined simply by multiplying the weight by the fatigue or recovery score for the period.

$\text{Score}(x) = \text{Fatigue}(x) \times \text{FWeight}(x)$ (if hours 'x' worked)

$\text{Score}(x) = \text{Recovery}(x) \times \text{RWeight}(x)$ (otherwise)

11

MANAGING FATIGUE BY DROWSINESS DETECTION: CAN TECHNOLOGICAL PROMISES BE REALIZED?

David F. Dinges, Ph.D. and Melissa M. Mallis
University of Pennsylvania School of Medicine, Philadelphia, Pennsylvania, USA

INTRODUCTION

This chapter draws on previous reports (Dinges, 1995b, 1997), as well as work in progress (Mallis et al., in preparation) to review key issues, and present illustrative data regarding the validation of technologies claiming to provide on-line monitoring of operator alertness and vigilance. Throughout the chapter the term "fatigue" is used to refer to the effects on performance capability of sleep loss, night work, prolonged work, or inadequate recovery, alone or in combination.

Although scientific and applied initiatives to develop on-line measures of alertness/drowsiness and hypovigilance have a long history (O'Hanlon & Kelley, 1974; Dinges & Graeber, 1989; Brookhuis, 1995), this area has undergone renewed interest and intensified activity especially in the USA (Rau, 1996; Knipling, 1996), and in Europe (Brown, 1995, 1997) in the past 3 years (see also Dinges, 1995a,b, 1996, 1997). Reasons for this "investment" in technology to manage fatigue-related performance impairment can be found in all modes of transportation, but especially in commerical motor vehicle operations. A number of arguments have been put forward to justify development of drowsiness detection technologies.

REASONS FOR FATIGUE-DETECTION TECHNOLOGIES

Below we summarize the main reasons that fatigue-detection technologies have become attractive as one way to prevent drowsy-driving crashes. These reasons are not necessarily the opinions of the authors, although we believe that drowsy-driving technology development has merit, as long as validity and reliability are demonstrated. We do not concern ourselves in this chapter with the often-voiced admonition that technology development is questionable because such drowsy-driving detection systems, even if valid and reliable, likely will encourage some drivers to continue driving in an impaired state due to the false sense of security the technology affords, or to coercion from employers, or to willful misuse of the technology (Evans, 1991; Brown, 1997). Misuse of a device or technology is a legitimate concern along with a range of other legal/ethical issues (see Dinges, 1997; Table 2 below), but a technology that can potentially enhance safety and save lives should not be prejudged based on speculations about users' ethics, whether realistic or exaggerated. (For example, most motor vehicles can be accelerated to dangerous speeds, but the fact that some drivers elect to speed is not used to justify placing acceleration governors on motor vehicles. Rather, education and laws exist to teach drivers how to properly use the acceleration potential of a motor vehicle.) Elsewhere, and below, we propose that education, legal, and policy standards for implementing fatigue-monitoring technologies be developed in parallel with the engineering and scientific development of the devices (Dinges, 1995b, 1997). This is essential, because the reasons for technology development in this area are compelling and indicate devices will be implemented whether or not their use is considered in public policy on road safety.

Fatigue-related crashes are common and serious. Since 1994, there has been growing evidence from industrialized countries that fatigue from varying combinations of sleep loss, night driving (i.e. circadian rhythms), and prolonged work time (i.e., wake time on task) contributes to substantial numbers of motor vehicle crashes. Although estimates and the methods they are based on vary widely, few dispute that the problem of drowsy driving is inadequately addressed. Fatigue has been estimated to be involved 2% to 23% of all crashes (cf., O'Hanlon, 1978; McDonald, 1984; Horne & Reyner, 1995; Knipling & Wang, 1995; Maycock, 1997); in 4% to 25% of single-vehicle crashes (cf., Wang & Knipling, 1994; Brown, 1995); in 10% to 40% of crashes on long motor ways (Shafer, 1993; Dinges 1995b); and in 15% of single-vehicle fatal truck crashes (Wang & Knipling, 1994). Fall asleep crashes are also very serious in terms of injury severity (Pack et al., 1995). In the USA, fatigue has been implicated as the most frequent contributor to crashes in which a truck driver was fatally injured (U.S. National Transportation Safety Board, 1990). Much of the focus on the putative role of sleepiness/drowsiness in traffic accidents has centered on single vehicle crashes, although there is no reason to believe that sleepiness is not also involved in multiple vehicle crashes. In the USA, single vehicle crashes in which no alcohol was involved account for more than a quarter of all fatal crashes, more than a quarter of all injury-only crashes, and more than a quarter of all property damage-only crashes (see Dinges, 1995b). While many factors can contribute to single vehicle non-alcohol-related crashes, the fact that they comprise 27% (i.e., 1.67 million crashes in 1993) of all motor vehicle crashes in the USA, suggests that fatigue may contribute to more of these crashes than current estimates allow.

Consequently, with so many persons driving fatigued, technologies that detect dangerous levels of sleepiness before a crash occurs are essential.

Subjective estimates of sleepiness are unreliable. Experiments have demonstrated that subjects cannot reliably predict when they are impaired to the point of having an uncontrolled sleep attack (i.e., microsleep) and/or a serious vigilance lapse (Dinges, 1989). Drivers know when they are experiencing sleepiness (Horne & Reyner, 1995), but they cannot necessarily translate those introspections into accurate predictions of how long their eyes are closed and whether they are missing signals, or when they will have an uncontrolled sleep onset while driving (Wylie et al., 1996; Brown, 1997). On the other hand, although self reports of sleepiness are highly influenced by contextual variables (Dinges, 1989, 1995b), drivers should know when they are experiencing heavy eyelids and head bobbing, which is likely past the point of impairment by drowsiness (Kribbs & Dinges, 1994). Hence, technology may offer the potential for an earlier and more reliable warning of performance-impairing sleepiness, before drowsiness leads to a catastrophic outcome.

Drowsiness-detection technology offers an alternative to proscriptive hours of service. Technology is viewed by some as a key component in a package of fatigue management options that can replace or at least put flexibility into federally-mandated proscriptive hours of service. For example, the current USA federal hours of service for commercial motor vehicle operators were written in 1939, and rely on work time as the primary determinant of fatigue (this is not unique to the trucking industry). It has been recognized for some time, however, that within limits, work duration accounts for only a modest proportion of accident risk (Hamelin, 1987). Thus, the current hours of service may not prevent many fatigue-related crashes, even when compliance is 100%. Fall-asleep crashes are more likely to occur during night driving and in sleep-deprived persons (e.g., Harris, 1977; Mitler et al., 1988; Pack et al., 1995). This is consistent with scientific studies in the past 30 years that have demonstrated that the level of waking alertness is regulated by two neurobiological forces that shape the time course of subjective fatigue and aspects of performance--the endogenous circadian rhythm and the need for sleep (Dinges, 1995b). When considered together and in combination with work hours, the product of these processes regulating fatigue and vigilance is nonlinear, temporally dynamic, and complex. This makes it complicated to derive regulatory schemes to prevent fatigue. Hence, technologies that monitor the driver's temporally dynamic state of alertness/drowsiness over time, are viewed as offering an advantage over proscriptive regulations. Such technologies may provide one of a number of ways to optimize safety through prevention of fatigue-related crashes, while permitting greater flexibility in work-rest scheduling to facilitate economic and related pragmatic goals, as well as drivers' personal choices.

Technological advances have made the goal feasible. Many technologies being developed for detecting drowsiness are miniaturized and unobtrusive (their durability and cost-effectiveness are less well established). Advances in electronics, optics, sensory arrays, data acquisition systems, algorithm development (e.g., neural nets), and other areas have made it far more likely that the goal of an affordable drowsiness detection system in a truck or automobile will be achieved and implemented in far less than the 10-20 years estimated by Brown (1995, 1997). In the USA, for example, there are currently many efforts underway at federal, industry, and entrepreneurial levels toward development of technologies for monitoring a driver's physiology or behavior in order to "manage" performance-impairment from fatigue in transportation. This marriage of technology and the human operator for drowsiness detection is part of a broader emphasis in the USA on development of "intelligent vehicle" and "driver condition warning" initiatives.

OPERATOR-CENTERED FATIGUE-MONITORING TECHNOLOGIES

Nearly all of the technologies currently being proposed to monitor on-line, the alertness or drowsiness or vigilance capability of a driver are in the prototypical development, validation testing, or early implementation stages. Their full effectiveness, practicality, and acceptance remain unproven scientifically and practically. Consequently, in this chapter we will not advocate for, endorse, or criticize any specific technology. However, we will demonstrate (below) the level of scientific standards that technologies must meet to warrant progression to the implementation stage. Reviewing the different categories of technologies being proposed for management of fatigue/sleepiness in transportation modes necessarily requires some categorizations. Technologies can be arbitrarily grouped by different criteria, but at least three broad categories of fatigue-related technologies (for detection and/or prevention) include: operator-centered technologies, system-centered technologies, and environmentally-oriented technologies (Dinges, 1997). Operator-centered technologies are the focus here.

Although many enthusiastic claims are made at meetings and in promotional materials regarding the merits of one type of operator-centered fatigue-monitoring technology or algorithm over another, such comparison data are only now coming to be generated (Mallis et al., in preparation). Moreover, the precise operational definition of what each technology is attempting to "detect" or the theoretical construct it purports to tap into, are often not well defined. In general, however, most technologies explicitly claim or imply detection of some aspect of either a heightened risk of operator error or outright impairment through one or more of the following hypothetical constructs: operator vigilance; operator attention/inattention; operator alertness/drowsiness; operator microsleeps; operator hypovigilancce; operator performance variability; or operator vulnerability to error. It remains a matter for scientific inquiry to what extent these constructs overlap empirically. However, no distinctions will be made among them in this chapter. Generic categories of operator-centered technologies currently being developed or marketed include, but are not limited to, the following four classes. This chapter is particularly concerned with the fourth category.

1. Readiness-to-perform and fitness-for-duty technologies

The final concept in the above list of fatigue constructs--vulnerability to error--bridges the distinction between technologies that afford on-line monitoring of the driver and those that involve probed evaluation or temporally discrete sampling of biobehavioral variables and that fall in the category of "fitness for duty." Fitness-for-duty or readiness-to-perform approaches, which are becoming popular replacements for urine screens for drugs and alcohol, can involve sampling aspects of performance capability or physiological responses. Because these tests are increasingly becoming briefer and more portable, the developers are seeking to extend their use beyond prediction of functional capability at the start of a given work cycle (i.e., prediction of relative risk over many hours), to prediction of capability in shorter time frames (e.g., whether someone is safe to extend work time at the end of a shift or duty period).

These technologies are intended to provide some behavioral or biological estimate of an operator's functional capability for work yet to be performed, relative to a standard, such as the operator's idiosyncratic function when unimpaired, or relative to a group norm (Gilliland & Schlegel, 1993). While some biologically-based (primarily ocular and pupilometric) technologies for fitness for duty are currently available ("EPS-100" by Eye Dynamics, Inc.; "FIT" by PMI, Inc.; "PUPILSCAN" by Fairville Medical Optics, Inc.), most of the technologies in this area are performance-based (Gilliland & Schlegel, 1993; Daecher, 1996). There are a large number of performance test batteries touted as candidates for readiness-to-perform and/or fitness-for-duty testing. Unfortunately, many of them are aptitude- and language-skill sensitive, and many have rather dramatic learning curves, making them less than ideal candidates for repeated usage in a diverse population. In addition, many have not been validated to be sensitive to fatigue, are not predicated on a model of human performance failure due to fatigue, and do not provide criteria by which to determine when someone is dysfunctional. On the other hand, there are a few simple performance tests that have been deployed (e.g., Rosekind et al., 1994; Dawson & Reid, 1997) that do not appear to have some of the aptitude- and language-sensitive problems of other tests, and that have been validated to be sensitive to sleep loss and circadian variation (Kribbs & Dinges, 1994; Dawson & Reid, 1997), suggesting that there is potential for this approach, both as a readiness-to-perform test and a way of probing functional capability while on the job.

2. Mathematical models of alertness dynamics joined with ambulatory technologies

This approach involves application of mathematical models that predict operator alertness/performance at different times based on interactions of sleep, circadian, and related temporal antecedents of fatigue (e.g., Akerstedt & Folkard, 1994; Belenky, 1997; Jewett, 1997). This is the subclass of operator-centered technologies that includes those devices that seek to monitor sources of fatigue, such as how much sleep an operator has obtained (via wrist actigraphy), and combine this information with a mathematical model that is designed to predict performance capability over a period of time and when future periods of increased fatigue/sleepiness will occur. However, like the other categories of technologies, precision and validation are critical criteria that must be met. A mathematical model that mis-estimates a cumulative performance decrement by only a few percentage points can lead to a gross miscalculation of alertness and performance capability over a work week. This is clearly a promising area, but much more work is needed.

3. Vehicle-based performance technologies

These technologies are directed at measuring the behavior of the transportation hardware systems under the control of the operator, such as truck lane deviation, or steering or speed variability, which are hypothesized to reflect identifiable alterations when a driver is fatigued (e.g., Wylie et al., 1996; Grace, 1996; King, 1996; Schwenk, 1996). The technologies are challenging to develop and implement owing to the complexity of driving behaviors under different conditions and the complexity of vehicle behavior relative to environmental conditions, but they offer the ultimate performance output--namely the behavior of the moving vehicle. They are less concerned with the condition of the operator than with the status of the vehicle. In addition to their face validity, they have many advantages (e.g., no wires, devices, or monitors on or aimed at an operator), but as with all technologies for preventing drowsy driving crashes, their scientific validity and cost-effectiveness remain to be demonstrated.

4. In-vehicle, on-line, operator status monitoring technologies

This category of fatigue-monitoring technologies includes a broad array of approaches, techniques, and algorithms. Technologies in this category seek to record some biobehavioral dimension(s) of an operator, such as a feature of the eyes, face, head, heart, brain electrical activity, etc., on-line (i.e., continuously, during driving). These are the most common and diverse of the fatigue-monitoring approaches. An ongoing review and categorization of technologies in this area reveals that there are currently more than 20 different initiatives in on-line biobehavioral monitoring in various stages of development (Mallis & Dinges, in preparation).

Table 1 presents examples of various classes and types of biobehavioral monitoring systems from which on-line driver recording technologies are being developed. Furthermore, within each class of measurement in Table 1 there are many possibilities, and even within a given technology/algorithm, there are often options. This diversity of measurement options is amply illustrated for by electroencephalogram (EEG) algorithms for detecting alertness. Since it has long been known that marked and characteristic changes occur in the EEG during transition from wake to sleep (Rechtschaffen & Kales, 1968), the EEG has been viewed as a standard for measuring alertness/drowsiness in the laboratory and in transportation operators (e.g., Brookhuis et al., 1986; Torsvall & Akerstedt, 1987; Brookhuis, 1995; Wylie et al., 1996). However, there are considerable disparities among current EEG fatigue-monitoring technologies. In addition to differing in the precise nature of their drowsiness/hypovigilance algorithm, EEG technologies can differ by the number and location of scalp electrodes from which they record; by the nature of the recording (e.g., monopolar vs. bipolar); by whether or not they also record and correct for eye movement (electrooculographic [EOG] activity); by many facets of signal processing characteristics; by whether their algorithms are standardized idiosyncratically for each subject; and by their recording hardware; to name but a few areas of difference. Consequently, when seeking to acquire a fatigue-monitoring technology based on EEG, it is best not to assume that different EEG drowsiness detectors are comparable. The better system is the one that has the best possible combination of scientific validation and practical utility. That remain to be determined even for EEG algorithms.

Table 1. Examples of biobehavioral measures used alone (or with other measures), as operator-centered technologies for on-line monitoring of alertness/vigilance.

Type of measurement	Examples of technologies
Video of the face (may include eyelid position, eye blinks, eye movements, pupillary activity, facial tone, direction of gaze, head movements)	• "PERCLOS" (Wierwille & Ellsworth, 1994; Wierwille, 1996) • Ford Motor Co. (UK) & HUSAT Res. Inst. (Richardson, 1995) • Nissan Research & Development, Inc. (Kaneda et al., 1994) • Toyota (Fukuda et al., 1995) • "Gaze Control System" (Stern, 1996)
Eye trackers	• "Eyegaze Systems" (LC Technologies; Lahoud, 1996) • "Eye Tracking System" (Applied Science Group, Inc.)
Wearable eyelid monitors	• "Alertness Monitor" (MTI Research, Inc.; MacLeod, 1996) • "Blinkometer" (IM Systems, Inc.) • "Nightcap" (Healthdyne Technologies; Stickgold et al., 1995) • "Eyelid activity measurement" (Leder et al., 1996) • "Stay-Awake Eye-Com Biosensor" (Torch, 1996)
Head movement detector	• "Proximity Array Sensing System" (Advanced Safety Concepts, Inc.; McIntosh, 1996)
EEG algorithms	• "Drowsiness detection" (Consolidated Research, Inc.) • "EEG algorithm adjusted by CTT" (Makeig & Jung, 1996) • "EEG spectral analysis" (Brookhuis, 1995) • "Quantitative EEG analysis" (Wylie et al., 1996)
ECG algorithms	• "MAP Process" (PALS Technology; Juszkiewicz, 1996)

STANDARDS FOR DROWSINESS DETECTION TECHNOLOGIES

If fatigue-monitoring technology development continues and is proposed as one piece of a programatic "fatigue management" alternative to proscriptive hours-of-service regulations, then technologies that allege to be effective must be shown to meet or exceed a range of criteria involving scientific, practical, and legal/ethical standards (Dinges, 1995b, 1996, 1997). A great deal of harm can be done if invalid and/or unreliable devices are quickly and uncritically implemented. In addition to the potential for increased crash risk, deployment of invalid and/or unreliable fatigue-detection technologies will result in wasted resources and provide only a false sense of security and fatigue management.

Table 2 summarizes the primary scientific and engineering criteria that technologies should meet to be maximally effective for monitoring operator vigilance (Dinges, 1997). The lack of answers to many of the scientific and engineering questions in Table 2 for most current technologies stems from the fact that many of them are being developed in a proprietary context, and their validation either is not attempted, or is not complete, or if complete is not available. There are many anecdotal claims and very few published validation studies. In addition to the lack of validation evidence, there is a dearth of information for many of the practical implementation questions in Table 2. Most technologies have not yet been systematically transitioned to operational test beds for evaluation. This is a critical second-stage hurdle to be overcome after validation is established. Practical questions of size, intrusiveness, cost, etc. can be addressed as the validity of a prototypical technology is tested.

The critical first step of establishing validity and reliability

The criteria outlined in Table 2 are achievable, but the progression should be from scientific / engineering validation, through the practical implementation phase, to the public-policy phase. Unfortunately, this has often not been the case. Excessive concern for selling technology to the user without first establishing the scientific validity and reliability of approaches is misguided and risky. For example, time, energy and resources spent on implementing a technology that is easily integrated into the work environment may be wasted if the device is later found not to detect vigilance errors, inattention, or drowsiness.

Once a technology has been developed to record a putative fatigue marker it must first be demonstrated to work. That is, it must actually record what it purports to record and do so consistently. Thus, an eyelid closure detection device must record eye lid closure and only eye lid closure and must not miss eyelid closures. After these engineering criteria are met, then scientific validity relative to fatigue/drowsiness/hypovigilance detection must be established. For example, in terms of devices that purport to detect a fatigued operator, basic data must be provided on whether the device detects what it alleges to detect (i.e., fatigue/drowsiness/hypovigilance--this is the validity standard), and whether it can detect it repeatedly (this is the reliability standard). Even if a device is valid and reliable, to be practically useful, it must meet additional standards of high sensitivity and high specificity. Thus, the device must detect all (or nearly all) fatigue events and fatigued operators (i.e., high sensitivity standard), without too many false alarms (i.e., high specificity standard). A device that has high sensitivity but low specificity will detect fatigue, but may give too many false alarms to be useful. In contrast, a device with low sensitivity but high specificity will give few false alarms, but it may miss too many fatigue events to be useful.

Given the growing federal support for technology development in fatigue management (the facilitators), the entrepreneurial zeal currently overtaking technology companies in this area (the vendors), and the escalating attractiveness of fatigue management technologies to transportation industries (the buyers), there is a risk of a rush toward widespread use of technologies that do not reliably detect fatigue. At this time, more proactive and coordinated efforts are urgently needed among relevant governmental agencies, transportation industries, and the scientific and engineering communities to ensure that promising technologies for fatigue management meet minimum standards for the criteria outlined in Table 2. There is, also a need to scientifically validate claims of the efficacy of on-board fatigue alarms and countermeasures delivered through contingency with fatigue-detection technologies (e.g., claims that certain odors will reverse fatigue for an unspecified period of time).

Table 2. Scientific, practical, and legal criteria and questions regarding the development and use of technologies for monitoring operator vigilance or impairment.

Scientific / Engineering criteria	
Validity	Does it measure what it purports to measure, both, operationally (e.g., eye blinks) and conceptually (e.g., hypovigilance)?
Reliability	Does it measure the same thing consistently?
Generalizability	Does it measure the same event (operationally and conceptually) in everyone?
Sensitivity	What proportion of the persons (or times within a given person) does it detect when reduced vigilance is actually present? (Does it miss some hypovigilance or some hypovigilant persons?)
Specificity	What proportion of the persons (or times within a given person) does it correctly identify safe vigilance when it is actually present? (How often does it false alarm?)
Practical / Implementation criteria	
Ease of use	Can nearly everyone use it correctly?
Acceptance	Will the target population use the technology?
Unobtrusiveness	Is the technology "transparent" or convenient for the user?
Robustness	Can the technology withstand heavy use and/or abuse?
Economical	Is the technology cost-effective?
Implementation	Operationally how is the technology to be used? (For example, does it only detect reduced vigilance conditions? Does it also alert the operator? If it alerts the operator, what is the nature of the alert? Does it trigger a broader countermeasure response?)
Legal / Policy criteria	
Purpose	What is the goal of implementing the technology? Is the use of the technology mandatory? If so, who mandates it and for what purpose?
Privacy	Who has access to any data acquired by the technology?
Enforcement	Is the technology to be used for enforcement, compliance, or advancement/demotion? If so, how is this to be accomplished?
Misuse potential	Can use of the technology lead to misuse (1) by the person being

	monitored (e.g., continuing to operate while impaired); (2) by the mandating entity (e.g., requiring an operation to continue when impairment is present).
Liability	Who is liable if the technology fails to detect impairment or if it is misused in association with an adverse event?

Experimental design for establishing scientific validity

It is not possible to establish scientific validity and reliability for a device that was developed to detect fatigue/drowsiness/hypovigilance without testing the technology in a controlled context in which the antecedents of fatigue (sleep loss and/or circadian time) are explicitly manipulated in an experiment, and the effects of this manipulation are explicitly measured. This requires an experiment with precise control of the independent variable (i.e., sleep loss and/or circadian time to induce a range of alertness/drowsiness) and precise measurement of dependent variables. The latter should include an outcome variable that is well established to co-vary with alertness level to serve as a validation criterion, as well as the variable measured by the device. A validation variable (preferably a vigilance performance variable) is critical in the experiment to confirm that what the device recorded was a meaningful covariate of the adverse effects of the induced fatigue. The implications of these experimental design criteria for validating fatigue-detection technologies are clear. Studies in which there is no controlled or reliable manipulation of the independent variable (i.e., fatigue) cannot provide evidence of fatigue-detection validity. Similarly, if the validation criterion variable is either inherently unreliable itself (e.g., self report of alertness) or a physiological variable of uncertain relationship to actual performance (e.g., Richardson, 1995), then it cannot be determined with certainty whether the device's output variable was changing in relationship to a reliable criertion of impairment from fatigue. To the best of our knowledge, only a very few studies of fatigue-monitoring technologies have actually used a performance criterion variable in conjunction with controlled sleep deprivation to validate their drowsiness detection algorithm (e.g., Wierwille & Ellsworth, 1994).

The arguments that laboratory validation studies of fatigue-detection technology should be avoided due to their necessarily artificial activities (relative to actual driving) or because their results have been ignored in the past (cf., Brown, 1995) misses the point of scientific validation and the circumstances under which it must be demonstrated. Scientific validity is the most basic standard upon which fatigue-detection is predicated. The controlled laboratory context is precisely where validity should first be established to ensure that the device measures a meaningful fatigue-induced change, especially since fatigue associated with sleep loss, circadian phase, and time-on-task is a brain-based phenomenon (Monk, 1991; Broughton & Ogilvie, 1992). Validation may be possible in some field-based controlled driving conditions, but only if error variance ("noise") from extraneous sources is controlled, and objective validation outcomes are recorded. In uncontrolled field driving trials, error variance can mask the validity of a technology, or create an apparent validity that is artificial and therefore not generalizable to other contexts. In contrast, if a device cannot be shown to be valid in well-designed laboratory experiments, where there is an optimal context for demonstrating that the independent variable (fatigue) will affect the device's dependent variable (alertness detection), there is little justification for transitioning the technology to field studies. If a device can be shown to be valid in well-designed laboratory experiments, then there is justification for transitioning it to field studies, where it will need to be evaluated to establish whether the added uncontrolled factors of a driving environment obscure its sensitivity or specificity. There are no shortcuts to this logical progression.

Experimental Validation Of Six Technologies To Detect Hypovigilance

In an effort to obtain measures of the scientific validity of a number of fatigue-detection technologies, we have recently completed a double-blind, controlled laboratory validation experiment on six of the technologies in Table 1. Psychomotor vigilance performance lapses were selected as the validation criterion variable for three reasons: (1) driving is fundamentally first and foremost a vigilance task requiring psychomotor reactions; (2) psychomotor vigilance has been validated to be very sensitive to fatigue from night work and sleep loss (Dinges et al., 1997); (3) hypovigilance while driving is the outcome most fatigue-detection technologies seek to identify. The technologies tested included a video-based scoring of eye closure by trained observers (PERCLOS; Wierwille & Ellsworth, 1994); two wearable eye-blink monitors (MTI Research, Inc.; IM Systems, Inc.); a head tracker device (Advanced Safety Concepts, Inc.); and two EEG algorithms (Consolidated Research, Inc.; Makeig & Jung, 1996). Fourteen healthy adult males remained awake in the lab for 42 hr, while working on a computerized test battery every 2 hr. Vigilance lapses each minute and every 20 minutes were used as the validation criteria. Each technology was time-locked to vigilance performance to test coherence between vigilance lapses and each technology's specific hypovigilance metric. Algorithms were applied to technology results by the respective vendors, who remained blind both to the lapse data and to time (i.e., the specific hour of continuous wakefulness at which each data file was acquired). These procedures were used to eliminate possible bias in drowsiness scores derived from the vendors.

The results of this trial are being published elsewhere, and therefore only a few key observations will be described here. While nearly all of the technologies were found to accurately predict lapses in at least one subject or a subset of subjects, only one technology consistently correlated at a high level with lapses within and between different subjects. Meeting the validation criterion both through high intra-subject and high inter-subject coherence is an important and highly promising outcome, since one of the more serious problems plaguing fatigue-detection and prevention is the large inter-subject differences in vulnerability to fatigue. For example, in the recent USA-Canada driver fatigue and alertness study, 14% of drivers accounted for 54% of all observed video-drowsiness episodes (Wylie et al., 1996). Some of the technologies proved to have inadequate predictive power or to be technically unreliable. While the technologies that performed poorly may not be able to predict performance outcomes in general, and fatigue-induced lapses in particular, there is also the possibility that their algorithms can be improved to enhance delectability. If such an retrospective enhancement were possible, a prospective re-validation trial would be necessary. More validation studies of this type are needed to sort out from the wide variety of biobehavioral fatigue monitors those that have the highest validity and reliability relative to actual performance capability.

THE NEED FOR LEGAL, ETHICAL, AND PUBLIC POLICY STANDARDS

Finally, the legal and policy questions in Table 2 are largely not considered by the technology developers, beyond what is necessary to meet a given standard of safety in product development. However, these issues must be confronted if fatigue-detection technologies are to be located in the workplace. One of these issues concerns who has control over the detection technology. Another concerns the privacy rights of the individual being monitored, and the confidentiality issues of the information acquired. Related to the confidentiality question, are issues of enforcement and use of punitive contingencies when fatigue is detected. For example, from an enforcement perspective, should fatigue be viewed in the same way impairment from alcohol and drugs are viewed (Dawson & Reid, 1997)? What role should fatigue monitoring/detection have in assessing regulatory compliance versus providing feedback and/or education to an operator? What are the practices and policies for repeated detection of fatigue in an operator? If fatigue detection is involuntary and subject to enforcement contingencies, will operators accept it or seek to disable it? The answers to these and related questions are not obvious. Public policy discussion of these issues should begin now, while different technologies for fatigue detection in all modes of transportation are being developed, to allow legal policy to guide integration of the use of these technologies.

THE ULTIMATE CHALLENGE IS HUMAN

It appears that within the not-too-distant future some technologies will be deployed to prevent or limit certain catastrophic outcomes due to fatigued performance while driving. However, as reviewed above, there is much validation work yet to do to achieve this goal, and to determine how to use the most promising technologies in the face of large inter-individual differences in response to sleep loss and night work. While there is clearly potential for improving the safety margin with technologies, such a development should not be a substitute for setting standards for the functional capability of a transportation operator. Technologies may eventually prevent or limit certain catastrophic outcomes due to fatigued performance, but technologies are not substitutes for setting societal standards for the functional capability of an operator. On the other hand, technologies can help establish and maintain adherence to that standard if they are developed and used in a valid and responsible manner.

ACKNOWLEDGMENT

The substantive evaluation upon which this article was based was supported by contract DTNH22-93-D-07007 from the U.S. Department of Transportation, and in part by grant F49620-95-1-0388 from the U.S. Air Force Office of Scientific Research, by cooperative agreement NCC-2-599 from U.S. National Aeronautics and Space Administration, by grant NR04281 from the National Institutes of Health, U.S. Public Health Service, and by the Institute for Experimental Psychiatry.

REFERENCES

Akerstedt, T. and Folkard, S. (1994) Prediction of intentional and unintentional sleep onset in Ogilvie, R.D. and Harsh, J.R. (eds) Sleep Onset: Normal and Abnormal, American Psychological Association; Washington, D.C.

Belenky, G. (1997) Sustaining performance during continuous operation: The U.S. Army's sleep management system. International Conference Proceedings on Managing Fatigue in Transporation, Tampa, FL

Brookhuis, K.A. et al. (1986) EEG energy-density spectra and driving performance under the influence of some antidepressant drugs in O'Hanlon, J.F. and de Gier, J.J. (eds) Drugs and Driving, Taylor & Francis; London

Brookhuis, K. (1995) Driver impairment monitoring by physiological measures in Hartley, L. (ed) Fatigue & Driving: Driver Impairment, Driver Fatigue and Driving Simulation. Taylor & Francis; London

Broughton, R.J. and Ogilvie, R. (eds) (1992) Sleep, Arousal and Performance. Birkhauser-Boston, Inc.: Cambridge

Brown, I.D. (1995) Methodological issues in driver fatigue research in Hartley, L. (ed) Fatigue & Driving: Driver Impairment, Driver Fatigue and Driving Simulation. Taylor & Francis; London

Brown, I.D. (1997) Prospects for technological countermeasures against driver fatigue, Accident Analysis and Prevention, 29, 525-531

Daecher, C. (1996) Fitness for duty testing in the trucking environment in Proceedings of the Technological Conference on Enhancing Commercial Motor Vehicle Driver Vigilance, American Trucking Associations Foundation, FHWA, NHTSA, McLean, VA, 31-34

Dawson D. and Reid K. (1997) Fatigue, alcohol and performance impairment. Nature, 388, 235

Dinges, D.F. and Graeber, R.C. (1989) Crew fatigue monitoring. Flight Safety Digest, 8, 65-75

Dinges, D.F. (1989) The nature of sleepiness: Causes, contexts and consequences in Stunkard, A., Baum, A. (Eds) Perspectives in Behavioral Medicine: Eating, Sleeping and Sex. Lawrence Erlbaum; Hillsdale

Dinges, D.F. (1995a) Technology / Scheduling Approaches in Managing Fatigue in Transportation: Promoting Safety and Productivity. Proceedings from the multi-modal symposium co-sponsored by the National Transportation Safety Board and NASA Ames Research Center, 53-58

Dinges, D.F. (1995b) An overview of sleepiness and accidents. Journal of Sleep Research, 4, Suppl. 2, 4-14

Dinges, D.F. (1996) Validation of psychophysiological monitors. Proceedings of the Technological Conference on Enhancing Commercial Motor Vehicle Driver Vigilance, American Trucking Associations Foundation, FHWA, NHTSA, McLean, VA, 35-41

Dinges, D.F. (1997) The promise and challenges of technologies for monitoring operator viiglance. Proceedings of the International Conference on Managing Fatigue in Transportation, American Trucking Associations Foundation, Tampa, FL, 77-86

Dinges, D.F. et al. (1997) Cumulative sleepiness, mood disturbance, and psychomotor vigilance performance decrements during a week of sleep restricted to 4-5 hours per night. Sleep, 20, 267-277

Evans, L. (1991) Traffic Safety and the Driver. Van Nostrand Reinhold: Ncw York

Fukuda, J. et al. (1995) Development of driver's drowsiness detection technology. Toyota Technical Review, 45, 34-40

Gilliland K. and Schlegel, R.E. (1993) Readiness to perform: A critical analysis of the concept and current practices. Office of Aviation Medicine, Federal Aviation Administration, NTIS No. AD-A269 379

Grace, R. (1996) Field testing a prototype drowsy driver detection and warning system in Proceedings of the Technological Conference on Enhancing Commercial Motor Vehicle Driver Vigilance, American Trucking Associations Foundation, FHWA, NHTSA, McLean, VA, 27-30

Hamelin, P. (1987) Lorry drivers' time habits in work and their involvement in traffic accidents. Ergonomics, 30, 1323-33

Harris, W. (1977) Fatigue, circadian rhythm and truck accidents in Mackie, R.R. (ed) Vigilance: Theory, Operational Performance and Physiological Correlates. Plenum Press; New York

Horne, J.A. and Reyner, L.A. (1995) Sleep related vehicle accidents. British Medical Journal, 310, 565-567

Jewett, M.E. (1997) Models of circadian and homeostatic regulation of human performance and alertness. Doctoral Thesis, Harvard University, Cambridge, MA.

Juszkiewicz, K. (1996) Correlation of ultra-slow electrophysiological signals with performance. Proceedings of the Technological Conference on Enhancing Commercial Motor Vehicle Driver Vigilance, American Trucking Associations Foundation, FHWA, NHTSA, McLean, VA, 79

Kaneda, M. et al. (1994) Development of a Drowsiness Warning System. Proceedings of the 14th International Technical Conference on Enhanced Safety of Vehicles, Munich, Germany

King, D. (1996) Correlation of steering behavior with heavy truck driver fatigue. Proceedings of the Technological Conference on Enhancing Commercial Motor Vehicle Driver Vigilance, American Trucking Associations Foundation, FHWA, NHTSA, McLean, VA, 69

Knipling, R.R and Wang, J.S. (1995) Revised estimates of the US drowsy driver crash problem size based on general estimates system case reviews. 39th Annual Proceedings, Association for the Advancement of Automotive Medicine, Chicago

Knipling, R. (1996) The promise of technology for fatigue management: The Federal Highway Administration perspective. Proceedings of the Technological Conference on Enhancing Commercial Motor Vehicle Driver Vigilance, American Trucking Associations Foundation, FHWA, NHTSA, McLean, VA, 8-13

Kribbs, N.B. and Dinges, D.F. (1994) Vigilance decrement and sleepiness in Ogilvie, R. and Harsh, J. (eds) Sleep Onset: Normal and Abnormal Processes. American Psychological Association; Washington, D.C.

Lahoud, J.A. (1996) Technologies that have potential for driver vigilance studies. Proceedings of the Technological Conference on Enhancing Commercial Motor Vehicle Driver Vigilance, American Trucking Associations Foundation, FHWA, NHTSA, McLean, VA, 67

Leder, R.S. et al. (1996) Eyelid activity measurements: A new retroreflective sensor. Sleep Research, 25, 509

MacLeod, E. (1996) Personal Alertness Monitor. Proceedings of the Technological Conference on Enhancing Commercial Motor Vehicle Driver Vigilance, American Trucking Associations Foundation, FHWA, NHTSA, McLean, VA, 47-52

Makeig, S. and Jung, T-P. (1995) Changes in alertness are a principal component of variance in the EEG spectrum. NeuroReport, 7, 213-216

Maycock, G. (1997) Sleepiness and driving: The experience of U.K. drivers, Accident Analysis and Prevention, 29, 453-462

McDonald, N. (1984) Fatigue, Safety and the Truck Driver. Taylor & Francis: London

McIntosh, B. (1996) Capacitive Proximity Sensor. Proceedings of the Technological Conference on Enhancing Commercial Motor Vehicle Driver Vigilance, American Trucking Associations Foundation, FHWA, NHTSA, McLean, VA, 75

Mitler, M.A. et al. (1988) Catastrophes, sleep and public policy: Consensus report of a committee for the Association of Professional Sleep Societies. Sleep, 11, 100-109

Monk, T.H. (ed) (1991) Sleep, Sleepiness and Performance. John Wiley and Sons, Ltd.: Chichester

O'Hanlon, J.F. and Kelley, G.R. (1974) A psychophysiological evaluation of devices for preventing lane drift and run-off-road accidents, Federal Highway Administration, Report No. 1736-F

O'Hanlon, J.F. (1978) What is the extent of the driving fatigue problem? in Driving Fatigue in Road Traffic Accidents, Brussels: Commission of the European Communities Report No. EUR6065EN, 19-25

Pack, A.I. et al. (1995) Characteristics of crashes attributed to the driver having fallen asleep. Accident Analysis and Prevention, 27, 769-775

Rau, P. (1996) The National Highway Traffic Safety Administration's Drowsy Driver Technology Program. Proceedings of the Technological Conference on Enhancing Commercial Motor Vehicle Driver Vigilance, American Trucking Associations Foundation, FHWA, NHTSA, McLean, VA, 14-16

Rechtschaffen, A. and Kales, A. (eds) (1968) A Manual of Standardized Terminology, Techniques and Scoring System for Sleep Stages of Human Subjects. U.S. Department of Health, Education and Welfare, Public Health Service, Bethesda, MD

Richardson, J.H. (1995) The development of a driver alertness monitoring system in Hartley, L. (ed) Fatigue & Driving: Driver Impairment, Driver Fatigue and Driving Simulation. Taylor & Francis; London

Rosekind, M.R. et al. (1994) Crew factors in flight operations IX: Effects of planned cockpit rest on crew performance and alertness in long-haul operations, NASA Technical Memorandum 108839

Schwenk, S. (1996) TRW Driver Alertness Warning System (DAWS). Proceedings of the Technological Conference on Enhancing Commercial Motor Vehicle Driver Vigilance, American Trucking Associations Foundation, FHWA, NHTSA, McLean, VA, 76-77

Shafer, J.H. (1993) The decline of fatigue related accidents on the NYS thruway. Proceedings of the Highway Safety Forum on Fatigue, Sleep Disorders and Traffic Safety, Albany, NY

Stickgold, R. et al. (1995) Nightcap detection of decreased vigilance. Sleep Research, 24, 500

Stern, J.A. (1996) Alertness and states of attention. Proceedings of the Technological Conference on Enhancing Commercial Motor Vehicle Driver Vigilance, American Trucking Associations Foundation, FHWA, NHTSA, McLean, VA, 72-74

Torsvall, L. and Akerstedt, T. (1987) Sleepiness on the job: Continuously measured EEG in train drivers. Electroencephalography and Clinical Neurophysiology, 66, 502-511

Torch, W.C. (1996) The Stay-Awake Eye-Com Biosensor & Communicator in Proceedings of the Technological Conference on Enhancing Commercial Motor Vehicle Driver Vigilance, American Trucking Associations Foundation, FHWA, NHTSA, McLean, VA, 68

U.S. National Transportation Safety Board (1990) Safety Study: Fatigue, Alcohol, Other Drugs, and Medical Factors in Fatal-to-the-Driver Heavy Truck Crashes, Vol. 1, Washington, D.C., NTSB/SS-90-01

Wang, J.S. and Knipling, R.R. (1994) Single-Vehicle roadway departure crashes: Problem size assessment and statistical description. US Department of Transportation, National Highway Traffic Safety Administration

Wierwille, W.W. and Ellsworth, L.A. (1994) Evaluation of driver drowsiness by trained raters. Accident Analysis and Prevention, 26, 571-581

Wierwille, W.W. (1996) Driver status in Proceedings of the Technological Conference on Enhancing Commercial Motor Vehicle Driver Vigilance, American Trucking Associations Foundation, FHWA, NHTSA, McLean, VA, 17-26

Wylie, C.D. et al. (1996) Commercial Motor Vehicle Driver Fatigue and Alertness Study: Project Report, U.S. Department of Transportation Report No. FHWA-MC-97-002

12

QUANTITATIVE SIMILARITY BETWEEN THE COGNITIVE PSYCHOMOTOR PERFORMANCE DECREMENT ASSOCIATED WITH SUSTAINED WAKEFULNESS AND ALCOHOL INTOXICATION.

Drew Dawson, Nicole Lamond, Katharine Donkin and Kathryn Reid.The Centre for Sleep Research, The Queen Elizabeth Hospital, Woodville Rd Woodville SA 5011, Australia.

INTRODUCTION

Since the industrial revolution shiftwork has become an increasingly common work practice. It has been estimated that 15-20% of the working population in industrialised countries are currently employed on some form of non-standard work schedule (Knauth, 1993; Baker, 1980). While the economic benefits of shiftwork are self evident (Harrington, 1978), the benefits are accompanied by significant health and social costs (Mitler at al, 1988; Moore-Ede et al, 1985; Spelton et al, 1993). Research studies over the last 20 years have clearly identified shiftwork as an occupational health and safety risk factor (Akerstedt, 1995a).

Reduced opportunity for sleep and reduced sleep quality are generally considered to be the major risk factors associated with shiftwork related accidents (Mitler et al, 1988; Leger, 1994; Akerstedt et al, 1994). Not surprisingly, the combination of these factors leads to increased fatigue, lowered levels of alertness and impaired performance on a variety of cognitive psychomotor performance tasks (Harrington, 1978).

Experimental studies have shown that sustained wakefulness (SW) impairs several components of performance including hand-eye co-ordination, decision-making, memory, cognition, visual search performance and speed and accuracy of responding (Linde et al, 1992; Fiorica et al, 1968; Babkoff et al, 1988). In addition to cognitive factors, affective components of behaviour such as motivation, and mood are altered as the duration of SW increases (Babkoff et al, 1988; Bohl, 1993).

From the studies cited above it is clear that there is a general consensus that cognitive psychomotor performance is impaired by the sleep disruption and extended wakefulness associated with shiftwork (Akerstedt et al, 1994). Moreover, this performance impairment is associated with an increased risk of accident (Dinges, 1995).

Surprisingly, however, policy makers in western industrialised countries have generally not legislated to manage and control fatigue in a manner commensurate with the statistical risks associated with it. This attitude is in stark contrast to the response to alcohol-related performance impairment. Policy makers and the community have frequently proscribed work and/or the operation of dangerous equipment under the influence of alcohol. Given that the effects of SW are qualitatively similar to the effects of even moderate alcohol intoxication (Klein et al, 1970), it is paradoxical that fatigue-related performance impairment has not been subject to similar levels of regulatory intervention. This failure to address the occupational, health and safety impact of fatigue may, in part, reflect a failure to provide policy makers with a readily understood index of the relative risk associated with sleep loss and fatigue.

The current studies sought to express the impairment associated with fatigue equivalent to those currently accepted by policy makers and the community. That is, by expressing the performance impairment as its equivalent level of alcohol intoxication. By expressing the performance impairment associated with fatigue in terms of its equivalent BAC it is hoped to provide an easily-grasped index of comparative impairment.

METHODS

Study One

Subjects

Forty subjects (27 male; 13 female) gave informed consent to participate in the study. Subject ranged from 18 years to 32 years of age (mean 21.1 (± 3.7) years). The subjects selected were recruited using advertisements placed around the University of Adelaide. Volunteers were required to complete a general health questionnaire prior to the study. Subjects who had a current health problem, a history of psychiatric or sleep disorders were excluded. Subjects who smoked cigarettes or who were taking medication known to interact with alcohol or affect sleep patterns were also excluded. Subjects who did not drink alcohol, or who habitually consumed more than 6 standard drinks per day were excluded.

Procedure

All investigations were conducted at the Centre for Sleep Research at the Queen Elizabeth Hospital. Subjects participated in a randomised cross-over design involving two experimental conditions,

1. A sustained wakefulness condition (SW).

2. An alcohol condition (A).

The two conditions were administered at least one week apart to allow subjects time to recover.

See Figure 1 for a schematic representation of the experimental protocol.

A previous pilot study for this protocol (Dawson et al, 1995) indicated that there was no performance decrement associated with the placebo condition, and that all subjects could correctly identify whether they were intoxicated or not. Since all subjects were regular social drinkers (4-8 drinks/week), and therefore experienced in the effects of alcohol a placebo condition was not included in this protocol.

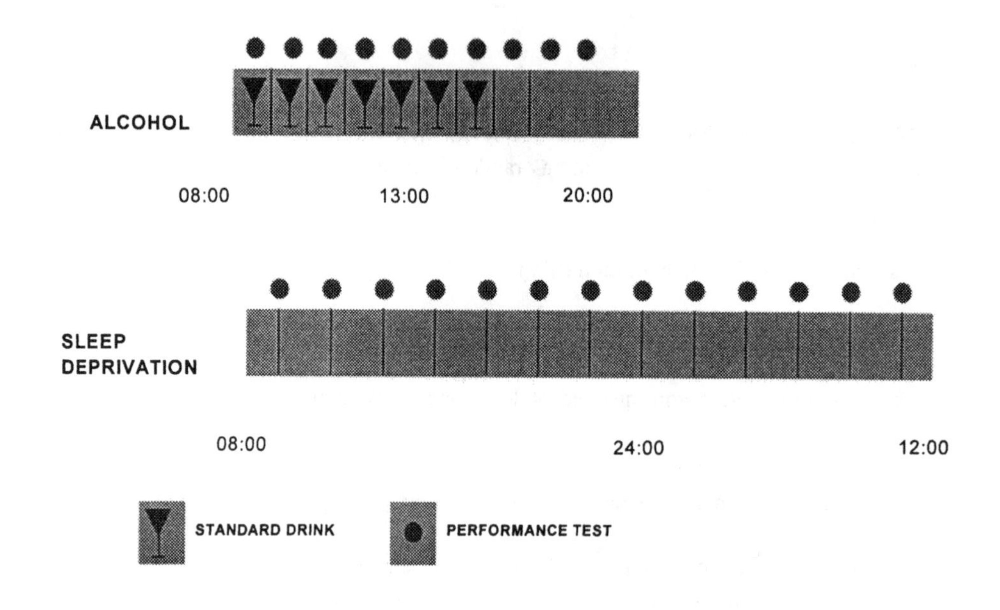

Figure 1. Schematic representation of the protocol for the sustained wakefulness (SW) and alcohol (A) experimental conditions. The alcohol condition commenced at 0800 hours. Subjects consumed 10mg of ethanol in orange juice every half hour until 1600h or until they reached a BAC of 0.10%. Every 30 minutes, subjects were breathalysed, completed three performance tests and then, if necessary, consumed another alcoholic beverage. The sleep deprivation condition commenced at 0800. Subjects completed three performance tests every 30 minutes until 1200h the following day.

Training Session

Subjects arrived at the sleep laboratory at 8:30 p.m. on the night prior to the commencement of each study period. They were required to complete 40 OSPAT tests to familiarise themselves with the assessment procedure and to minimise improvement in performance resulting from learning.

Sustained Wakefulness Condition

Subjects arrived at the sleep laboratory at 8:30 p.m. on the night prior to the commencement of the study period and completed a training session before going to sleep at approximately 11:00 p.m. Subjects were woken at 7 a.m. the following morning, after breakfast at approximately 7:45 a.m., 9 practice OSPAT tests were completed. Subjects then completed three performance tests at half hourly intervals from 8:00 a.m. until 12:00 p.m. the following day. In between tests, subjects were allowed to read, watch television and play games. Careful monitoring by research staff ensured wakefulness over the entire 28 hour period.

Alcohol Condition

Subjects arrived at the sleep laboratory at 8:30 p.m. on the night prior to the commencement of the study period and were required to complete a training session before going to sleep at approximately 11:00 p.m. Subjects were woken at 7 a.m. the following morning, after breakfast at approximately 7:45 a.m, 9 performance tests were completed. From 8:00 a.m. subjects underwent a breath test, completed three OSPAT tests and consumed an alcoholic drink at half hourly intervals. If a BAC of 0.1% was reached no further alcohol was given. Subjects were not informed of the BAC at anytime during the test period. All drink consumption and performance testing ceased at 4:00 p.m., but subjects were required to stay in the sleep laboratory under supervision until their BAC returned to 0%.

Subjects ate standard hospital meals during the study, although food and drinks containing caffeine were prohibited. Subjects were required to sit quietly and watch television or play boardgames during their time in the laboratory. Subjects were not permitted to exercise, shower or bath.

Equipment

Cognitive psychomotor performance

Cognitive psychomotor performance was measured using the Occupational Safety Performance Assessment Test (OSPAT). OSPAT is a unpredictable tracking task that subjects perform on a computer workstation. In simple terms, the task required subjects to keep a randomly moving cursor in the centre of three concentric circles, using a standard trackball. After the cursor is 'centred' the cursor moves to a random position away from the centre and the subject is required to 're-centre' the cursor. Subjects were seated in front of the workstation in an isolated room, free of distraction and were instructed to manipulate the track-ball using their dominant hand. Subjects completed three one-minute tests in each testing session and received no feedback between tests in order to avoid the knowledge of results affecting performance levels.

A global performance measure for each test is determined by summing the 'error' distance between the cursor and target and the rate at which the subject adapted to the random changes. This measure indicates how "well" the subject performed the task.

Blood alcohol

During the alcohol condition subjects were given alcohol loaded drinks consisting of 95% ethanol and orange juice at a rate designed to increase their BAC to 0.10% over a 4-6h period. Prior to all breath tests subjects were required to rinse their mouths with water. A standard calibrated breathalyser was used to estimate blood alcohol concentration (BAC) (Lion Alcolmeter S-D2, Wales). The breathalyser was accurate to 0.005% BAC.

Study Two

Subjects

Eight subjects (5 male; 3 female) gave informed consent to participate in the study. Subjects ranged from 19 years to 25 years of age (mean 20.25 (± 1.28) years). Method of recruitment and exclusion criteria were the same as those employed in study one.

Procedure

Subjects participated in a two-condition protocol similar to that in study one. In addition to the two experimental conditions, subjects attended a separate training day.

Training Session

During the week prior to commencement of the experimental conditions, subjects were required to attend the lab for a training session to familiarise themselves with the tasks used to assess performance and to minimise improvement in performance resulting from learning. They were required to complete each test until their performance reached a plateau.

Sustained Wakefulness Condition

Subjects arrived at the sleep laboratory at 8:00 p.m. on the night prior to commencement of the study. Prior to retiring at approximately 11:00 p.m., subjects completed a re-training session to reacquaint themselves with the performance tasks. During this session, they completed practice tests for each of the performance tasks in the study. Subjects were woken at 7:00 a.m. the following morning and a baseline testing session was completed at 8:00 a.m. Subjects then completed a testing session every hour, until 11:00 p.m. the following day.

Alcohol Condition

Subjects arrived at the sleep laboratory at 8:00 p.m. on the night prior to commencement of the experimental period. Before retiring for the night, at approximately 11:00 p.m., subjects completed a practice session as outlined above. Subjects were woken at 7:00 a.m. the following morning and completed a baseline testing session at 8:00 a.m. From 9:00 a.m. subjects completed a testing session every hour. As in study one, subjects underwent a breath test and consumed an alcoholic beverage at half hourly intervals.

Equipment

Cognitive psychomotor performance

Cognitive psychomotor performance was measured using a standardised computer based test battery (Worksafe Integrated Test Battery). The apparatus for the battery consists of an IBM compatible computer, microprocessor unit, response boxes and computer monitor. The test battery is based on a standard information processing model (Wickens, 1984). According to this model there are seven key information processing functions (see Figure 2).

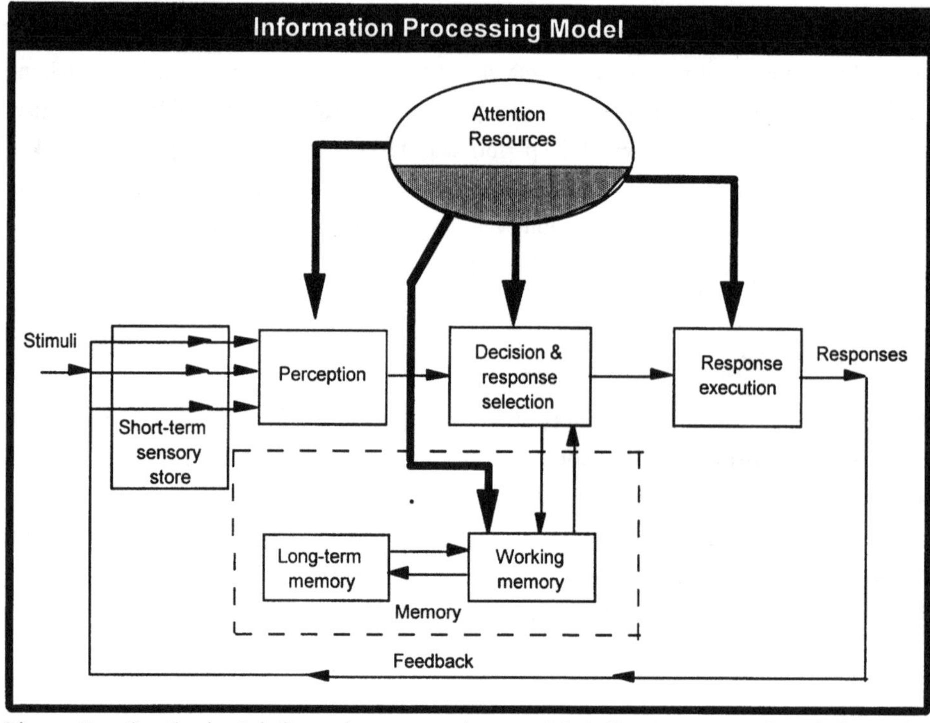

Figure 2. Synthesised information processing model indicating seven key information processing functions. Adapted from Wickens (1984).

The test battery software permits any combination of 12 visual and/or auditory tasks. The four tasks used in this study were (in order of complexity),

1. Simple Reaction Time (SRT)
2. Predictable Tracking (PT)
3. Vigilance (VIG)
4. Grammatical Reasoning (GR)

Each of these tasks was used to measure specific components of cognitive psychomotor performance according to the combinations outlined in Table 1.

	Simple Reaction Time	Predictable Tracking	Vigilance	Grammatical Reasoning
Short-term sensory store				
Perception	P	P	P	
Decision and Response selection			P	P
Working memory				P
Long-term memory				P
Attention resources		P	P	P
Response execution	P	P	P	

Table 1. Components of cognitive functioning required for performance of each test.
P: Primary information processing component skill(s) assessed by the task.

Performance Tasks

Simple Reaction Time

Simple reaction time is an unpredictable task that measures both reaction time and response time. Subjects were instructed to depress the home button on a tri-button unit. Then, using the same finger, they were required to depress the left hand response button when a stimulus was observed, returning to, and depressing, the home button afterwards.

Predictable Tracking

In the predictable tracking task , subjects were required, with the use of a joystick, to keep a cursor on, or as close as possible to, a target box. In this task, the whole track the target box is going to make was revealed to the subject in the set-up lap.

Vigilance

Vigilance is an unpredictable task that measures both accuracy and response time. To begin this task, subjects were instructed to have their hand hovering over the display area, ready to press any of the six black buttons or the single red button. Subjects were instructed to press the black button corresponding to the illuminated light, if only one light was illuminated, and to press the red button if two lights were illuminated simultaneously.

Grammatical Reasoning

Grammatical reasoning, the most complex of the tasks, measures accuracy, response time and reaction time. Using the same tri-box as SRT, the task began after the subject depressed the home button. Subjects were instructed to keep their finger on the home button, until a decision as to the truth/falsity of a specific statement (displayed on the monitor) had been made, and then to press the left (true) or right (false) button accordingly, using the same finger. After responding, subjects were required to return to and depress the home button, to initiate the next statement. Subjects were instructed to concentrate on accuracy rather than speed.

ANALYSIS OF RESULTS

Cognitive psychomotor performance data was analysed using relative performance. That is, each individual's performance was expressed relative to their personal baseline. In study one, the baseline measure was calculated by averaging the scores of each individuals nine practice trials carried prior to the first 8:00 a.m. performance test on the first of the two counterbalanced experimental conditions. In the second study the scores of each individuals 8:00 a.m. test session were used as the baseline measure.

Figure 3 indicates that subjects in the first study rapidly mastered the performance test during the practice session. There was little variation in mean relative performance after completing

5 tests and by 25 tests subjects had reached a clear performance plateau. Similar findings were observed in the second study.

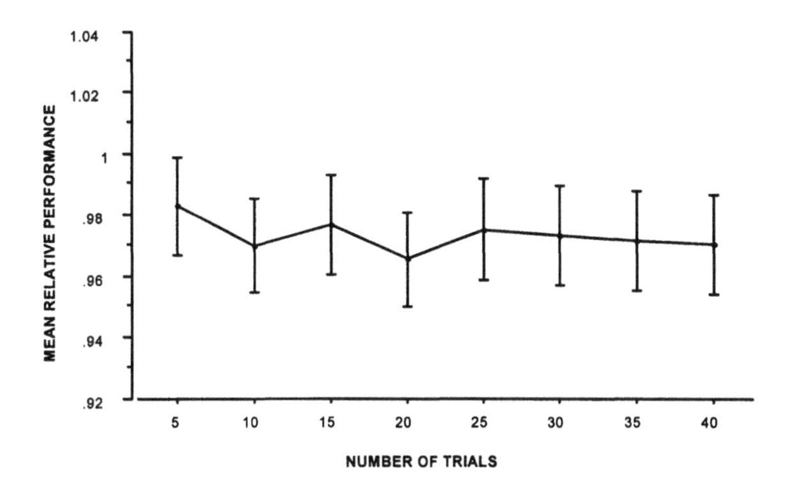

Figure 3. Mean relative performance for training session trials in the night prior to the first experimental condition (Study one). Error bars indicate ± s.e.m.

Alcohol Intoxication

To determine the relative effect of alcohol on performance, mean relative performance scores for all subjects were collapsed into 0.005% BAC intervals to determine the average performance decrement per unit increase of BAC. The linear relationship between increasing BAC and performance impairment was analysed by regressing mean relative performance against BAC for each 0.005% interval. Figure 4, shows the regression line between estimated BAC and mean relative performance in the alcohol condition.

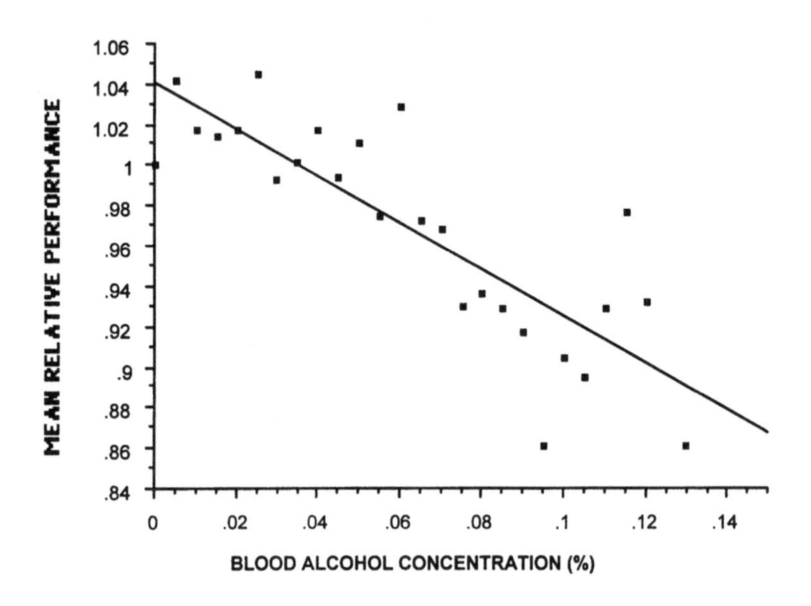

Figure 4. Scatter plot and linear regression of mean relative performance levels against blood alcohol concentrations between 0.00-0.13%.

The regression analysis indicated a significant linear correlation between subject's mean BAC and mean relative performance (Table 2). It was found that for each 0.01% increase in BAC, performance decreased by 1.16%. Thus, at a mean BAC of 0.10% mean performance decreased, on average, by 11.6%.

PERFORMANCE TEST	DF	F	P	R^2
Study One				
Ospat	1,24	54.4	< 0.05	0.69
Study Two				
Simple Reaction Time	1,7	13.41	<0.05	0.66
Predictable Tracking	1,7	9.61	<0.05	0.58
Vigilance				
Response Time	1,8	5.14	N.S.	0.39
Response Variability	1,6	1.63	N.S.	0.17
Grammatical Reasoning				
Response Time	1,6	7.33	<0.05	0.45
Accuracy	1,6	7,27	<0.05	0.39

Table 2. Regression analyses between mean relative performance and mean BAC.

Sustained Wakefulness

Performance in the SW condition was analysed by averaging performance data into two-hourly bins across the 28 hours of the study. Since there is a strong non-linear (circadian) component to the performance data and shiftworkers do not typically spend less than 10 or more than 26 hours awake (Australian Bureau of Statistics, 1993), the linear performance decrement per hour of wakefulness, was calculated using a linear regression between the tenth and twenty-sixth hour of wakefulness. This was methodologically appropriate since analysis of the performance data across this period shows a significant linear component ($p<.05$) and a non-significant non-linear component. Figure 5. illustrates this relationship plotting mean relative performance (from study one) against hours of wakefulness between the tenth and twenty-sixth hours.

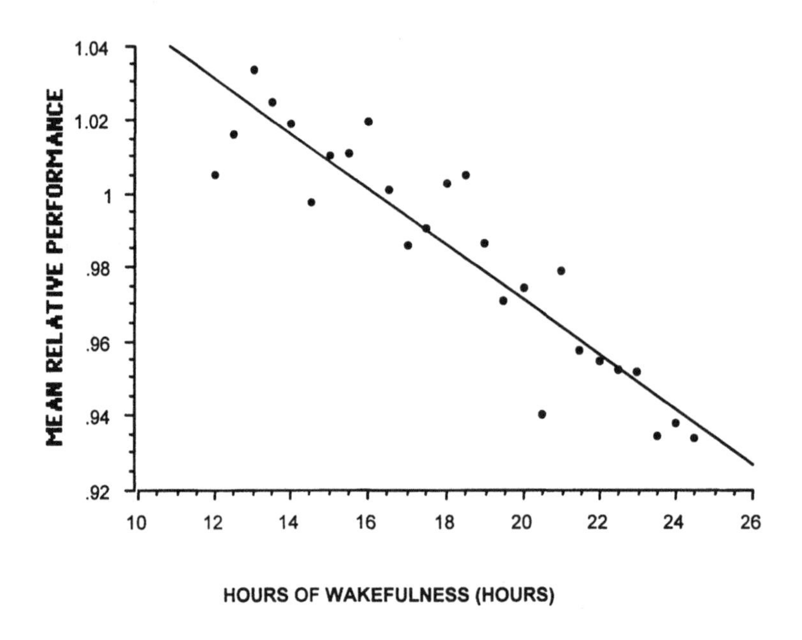

Figure 5. Scatterplot and linear regression of mean relative performance levels against prior wakefulness between the 10th and 26th hour of sustained wakefulness.

Regression analysis revealed a significant linear correlation between mean relative performance and hours of wakefulness (Table 3). Between the tenth and twenty-sixth hours of wakefulness, performance relative to baseline decreased by 0.74%/h.

PERFORMANCE TEST	DF	F	P	R^2
Study One				
Ospat	1,24	132.9	< 0.05	0.92
Study Two				
Simple Reaction Time	1,6	14.1	<0.05	0.70
Predictable Tracking	1,6	6.0	<0.05	0.5
Vigilance				
Response Time	1,6	22.53	<0.05	0.79
Response Variability	1,6	16.91	<0.05	0.74
Grammatical Reasoning				
Response Time	1,6	44.24	<0.05	0.88
Accuracy	1,6	14.42	<0.05	0.71

Table 3. Regression analyses between mean relative performance and hours of wakefulness.

The results discussed above illustrate the effects of SW and alcohol intoxication on cognitive psychomotor performance. However, the aim of the present study was to express the effects of fatigue as a blood alcohol equivalent. Figures 6 to 9 illustrate the comparative effects of SW and alcohol consumption on performance by plotting mean relative performance and BAC against hours of wakefulness.

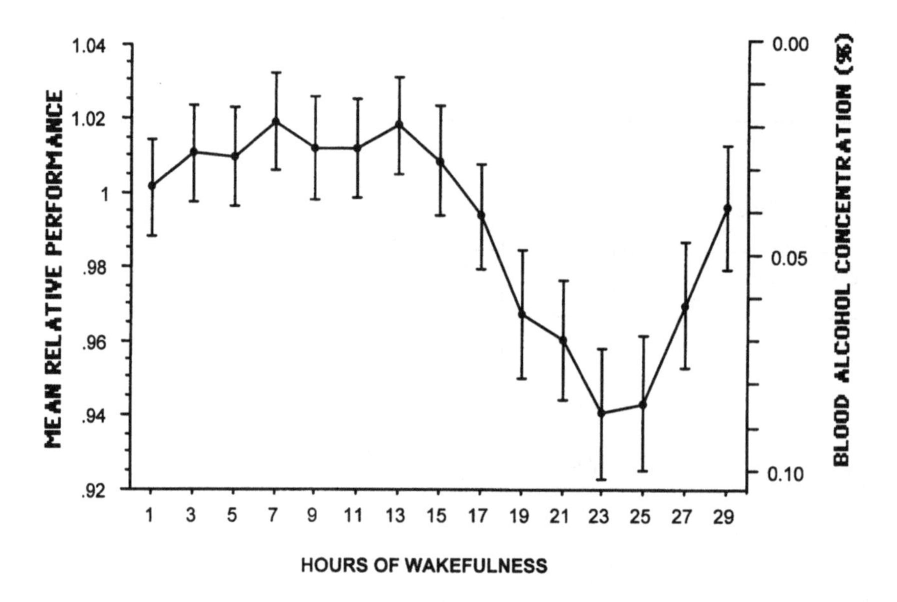

Figure 6. Performance (OSPAT) in the SW condition expressed as mean relative performance on the left hand axis and the %BAC equivalent on the right hand axis. Error bars indicate ± one s.e.m.

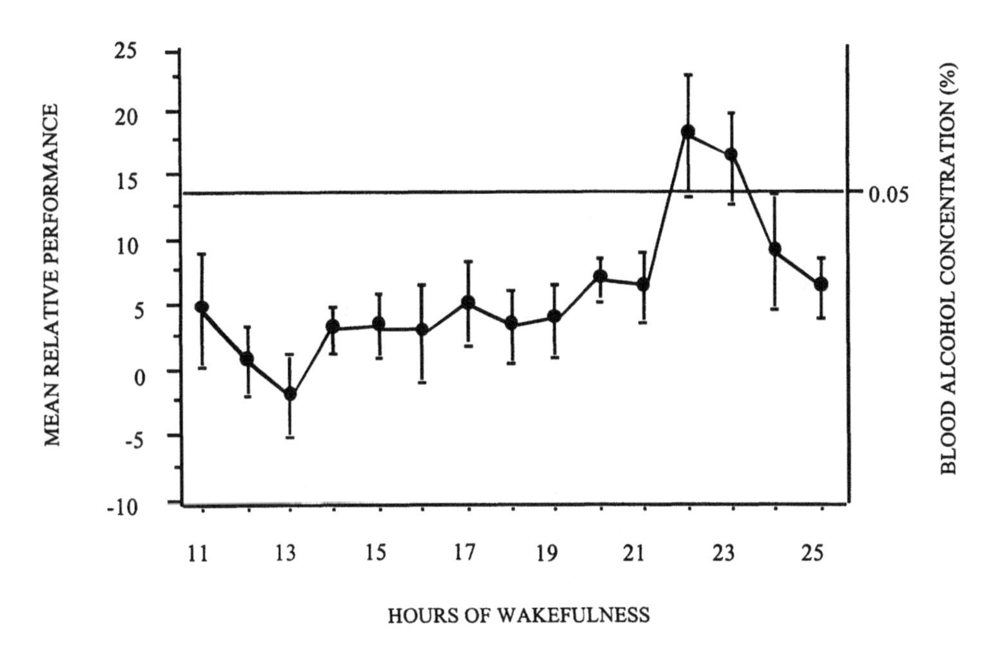

Figure 7. Performance (SRT) in the SW condition expressed as mean relative performance on the left hand axis and the %BAC equivalent on the right hand axis. Error bars indicate ± one s.e.m.

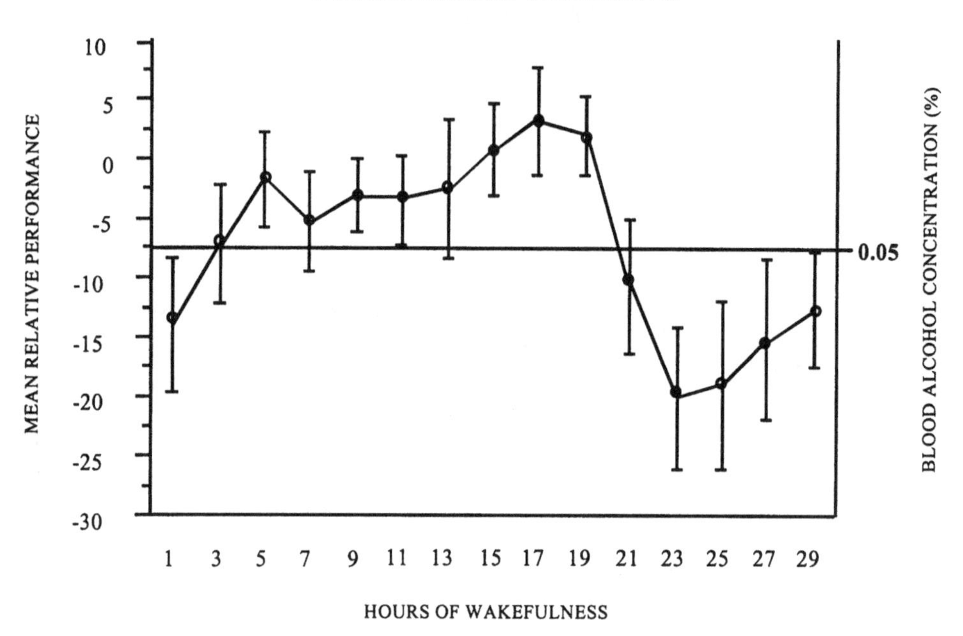

Figure 8. Performance (PT) in the SW condition expressed as mean relative performance on the left hand axis and the %BAC equivalent on the right hand axis. Error bars indicate ± one s.e.m.

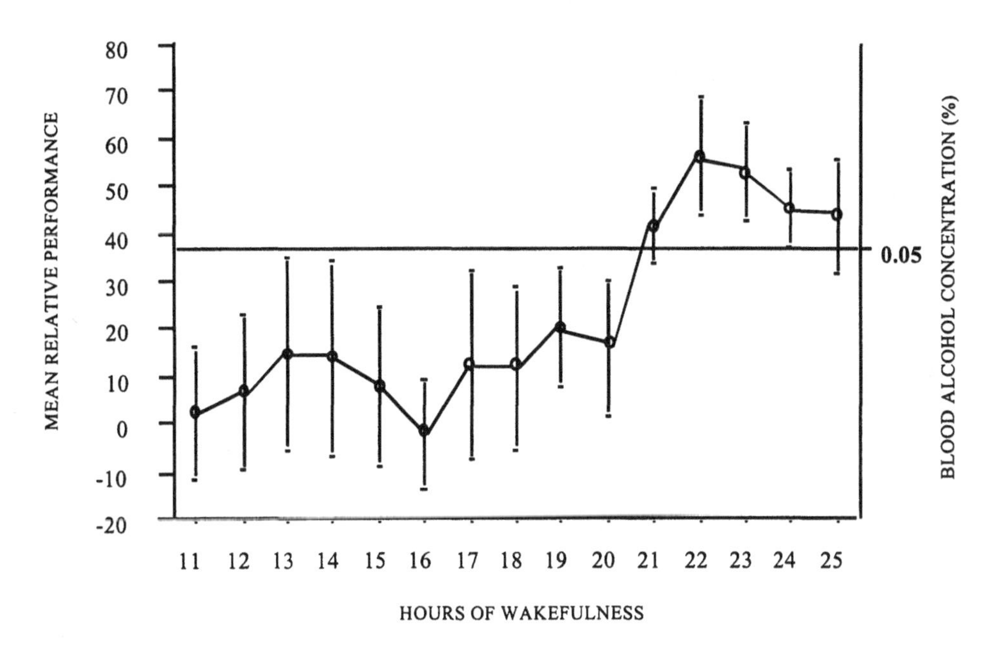

Figure 9. Performance (GR accuracy) in the SW condition expressed as mean relative performance on the left hand axis and the %BAC equivalent on the right hand axis. Error bars indicate ± one s.e.m.

By equating the two rates at which performance declines, (i.e. % decline / hour of wakefulness and % decline / BAC) it was calculated that the performance decrement for each hour of wakefulness was equivalent to the performance decrement observed with a 0.04% rise in BAC (study one data). Therefore, after 24 hours of SW cognitive psychomotor performance in study one decreased to a level equivalent to the performance observed at a BAC of 0.096%. While in the second study, after 24, 21, 21 and 13 hours of SW, performance on tasks of simple reaction time, predictable tracking and grammatical reasoning (accuracy and response time), respectively, decreased to a level equivalent to the performance observed at a BAC of 0.05%.

DISCUSSION

Cognitive psychomotor performance levels for all tests except for vigilance decreased significantly in the alcohol condition. Similarly, cognitive psychomotor performance levels decreased significantly for all performance tests in the SW conditions. Comparison of the two effects indicated that moderate levels of sustained wakefulness produce performance decrements comparable to those observed at moderate levels of alcohol intoxication in social drinkers.

In the alcohol condition increasing blood alcohol concentrations were associated with a significant linear decline in cognitive psychomotor performance. For example, in study one mean relative performance in the alcohol condition was impaired by approximately 5.8% at a BAC of 0.05% and by 11.6% at a BAC of 0.10. Overall, mean relative performance declined by approximately 1.16% per 0.01% BAC. These results are consistent with previous findings that suggest that cognitive psychomotor performance declines linearly with increasing intoxication between 0.0-0.075% BAC (Billings et al, 1991).

It is important to note that there was no decrease in mean relative performance up until a BAC of 0.03%. This is similar to the findings of Wilkinson and Colquhoun (1968) who also reported an increase in performance on a choice serial reaction test up until a BAC of 0.032%. This result is thought to reflect the fact that alcohol acts as a stimulant at low blood alcohol concentrations.

In contrast, cognitive psychomotor performance in the SW condition showed a more complex relationship. Mean relative performance showed three distinct phases. In the first phase (0-10 hours) performance remained relatively stable at a plateau. In the second phase (10-26 hours) performance declined linearly. During the third interval (26-28 hours) mean relative performance increased again presumably reflecting the well reported circadian variation in cognitive psychomotor performance (Folkard et al, 1993).

Since few shiftworkers remain awake for less than 10 or more than 26 hours between shifts (Australian Bureau of Statistics, 1993), the comparative analysis focussed on the second phase. Between the 10th and 26th hours mean relative performance, showed a strong linear decline of approximately 0.74 % per hour. The performance decline observed between hours 10 and 26 is consistent with previous studies, documenting cognitive psychomotor performance decreases for periods of sustained wakefulness between 12 and 86 hours (Linde et al, 1992; Storer et al, 1989; Fiorica et al, 1968).

While the results in each of the individual experimental conditions have, in and of themselves been previously established (Linde et al, 1992; Storer et al, 1989; Wang et al, 1992; Gustafon, 1986; Roache et al, 1992) equating the effects is relatively novel.

The results of this comparison indicate that the effects of 10-26 hours of SW from 1800-1000 hours, and moderate alcohol consumption have quantitatively similar effects on cognitive psychomotor performance. Although there are previous anecdotal reports indicating qualitative similarities between fatigue and alcohol intoxication (Klein et al, 1970; Kleitman, 1939), these studies establish the quantitative similarities of the two forms of impairment. In study one, equating the performance impairment between the 10th and 26th hour indicated a mean BAC equivalent of approximately 0.05% after 18 hours and 0.096% after 24 hours. If the results of this study were generalised to an applied setting they suggest that between 0300h and 0800h on the first night shift a shiftworker would show a cognitive psychomotor performance decrement similar to or greater than the legally proscribed BAC for many industrialised countries.

The second study further expanded on these findings. The results of the comparisons indicate that sleep deprivation effects specific components of performance differently, dependent on their relative degree of complexity. That is to say, sustained wakefulness effects more complex cognitive psychomotor abilities before simpler abilities. In accordance with the Information Processing Model earlier referred to, the simplest measure of performance incorporated in this study, simple reaction time, required only perception and response excecution functions. It was found that 24 hours of sleep deprivation were necessary to produce a performance decrement comparable to that associated with a BAC of 0.05%. Whereas for performance on the predictable tracking task, a slightly more complex task that also requires attention resource functions, a decrement equivalent to that of BAC of 0.05% was observed after 21 hours of sustained wakefulness.

Similarly, 21 hours of sleep deprivation produced a decrement in performance on the grammatical reasoning task equivalent to that associated with a BAC of 0.05%. It is interesting to note, however, that a decrement in the speed component of grammatical reasoning, equivalent to that associated with a BAC of 0.05%, was observed after only 13 hours of sustained wakefulness (graph not shown). While this may at first contradict the suggestion that more complex abilities are affected sooner by sustained deprivation, it must be remembered that subjects were told to concentrate on accuracy in this task, rather than speed. Indeed, the apparent speed-accuracy trade-off observed in the grammatical reasoning task is similar to that found in previous studies (Craig and Condon, 1985).

The data from both studies supports the idea that sustained wakefulness may carry a risk comparable with moderate alcohol intoxication since approximately 50% of shiftworkers on 8 hour shift patterns typically spend at least 24 hours awake on the first night shift in a roster (Knauth et al, 1981). Furthermore, the highest level of impairment observed in this study (~0.096% BAC) would occur at the end of a typical night shift (i.e. 0600-0900h) and would frequently coincide with the trip home for many shiftworkers.

While the results of this study clearly illustrate the comparative risks associated with sustained wakefulness for the first night shift, these results may underestimate the effect of night work in many real world settings. Previous research suggests that the performance impairment associated with shiftwork may be even greater on subsequent night shifts because of the reduced recuperative value of poor daytime sleep (Akerstedt, 1995b). Several studies have reported that the performance decrements, reduced alertness and fatigue reported by night shift workers is greater on the second and third night shift (Tilley et al, 1981). If this is the case, then it may be reasonable to assume that the alcohol impairment equivalent on these nights may be even greater than reported here for the first night.

However, it is not a simple process of calculating the performance decrement for hours of wakefulness, since shiftworkers may be sleeping at different times of the day and night. In addition they may have accumulated sleep loss from night one to three of their work schedule. Therefore, it may be useful to use longer experimental protocols to model actual shift schedules and establish the BAC equivalence for the performance decrement associated with the fatigue that can accumulate over a sequence of night shifts.

Taken together, the results from this study support the idea that the performance impairment and, by inference, the risk associated with sustained wakefulness across the night are not insubstantial and are quantitatively similar to those observed for moderate alcohol intoxication in social drinkers.

References

Akerstedt, T. (1995a) Work hours, sleepiness and accidents. *Journal of Sleep Research*, 4, (Suppl 2): 1-3.

Akerstedt, T. (1995b) Work hours, sleepiness and the underlying mechanisms. *Journal of Sleep Research*, , 4 (Suppl 2): 15-22.

Akerstedt, T., Czeisler, C., Dinges, D. F. and Horne, J. A. (1994) Accidents and sleepiness: a consensus statement from the International Conference on work hours, sleepiness and accidents, Stockholm. *Journal of Sleep Research*, 3: 195.

Angus, R. and Heslegrave, R. (1985) Effects of sleep loss on sustained cognitive performance during a command and control simulation. *Behaviour Research Methods, Instruments and Computers*, 17: 604-613.

Babkoff, H., Mikulincer, M., Caspy, T. and Kempinski, D. (1988) The topology of performance curves during 72 hours of sleep loss: A memory and search task. *The Quarterly Journal of Experimental Psychology*, 324: 737-756.

Baker, D. The use and health consequences of shiftwork. *International Journal of Health Services*, 1980, 10 (3): 405-419.

Billings, C., Demosthenes, T., White, T., O'Hara, D. (1991) Effects of alcohol on pilot performance in simulated flight. *Aviat. Space Environ. Med.*, 6 (3): 233-5.

Bohl, P. (1993) Predicting mood change on night shift. *Ergonomics*, 36 (1-3): 125-133.

Dawson, D., Wagner, R. (1995) Comparing the effects of sleep deprivation and alcohol consumption on cognitive psychomotor performance. *Sleep Research*, 24A: 425.

Dinges, D. (1995) An overview of sleepiness and accidents. *Journal of Sleep Research*, 4, (Suppl 2): 4-14.

Fiorica, V., Higgins, E., Iampietro, P., Lategola, M. and Davis, W. (1968) Physiological responses of men during sleep deprivation. *Journal of Applied Physiology*, 24(2): 167-176.

Folkard, S. and Tottersdell, P. (1993) Dissecting circadian performance rhythms: Implications for shift work. *Ergonomics*, 36(1-3): 283-288.

Gustafson, R. (1986) Alcohol and vigilance performance: Effect of small doses of alcohol on simple visual reaction time. *Perceptual and Motor Skills*, 62: 951-955.

Harrington, J. (1978) Shiftwork and Health: A Critical Review of the Literature. *Report to the Medical Advisory Service, UK Health and Safety Executive*, Her Majesty's Stationery Office, London

Klein, K., Bruner, H., and Holtman, J. (1970) Circadian rhythm of pilots efficiency and effects of multiple time zone travel. *Aerospace Medicine*, 41: 125-132.

Kleitman, N. (1939) Deprivation of sleep. In Kleitman, N. *Sleep and Wakelfulness.* Chicago: The University of Chicago Press pp 223.

Knauth, P. (1993) The design of shift systems. *Ergonomics*, 36(1-3): 283-288.

Knauth, P., Rutenfranz, J. (1980) Duration of sleep related to the type of shiftwork. In Reinberg, A., Vieux, N., Andlauer, P. (Eds.) *Advances in the Biosciences vol 30. Night and Shiftwork Biological and Social Aspects*. Pergamon Press Oxford, New York .pp 161-168.

Leger, D. (1994) The cost of sleep related accidents: A report for the national commission on sleep disorders research. *Sleep*, 17(1): 84-93.

Linde, L. and Bergstrom, M. (1992) The effect of one night without sleep on problem solving and recall. *Psychological Research*, 54: 127-136.

Mitler, M., Carskadon, M., Czeisler, C., Dement, W., Dinges, D., Graeber, R. (1988) Catastrophes, sleep, and public policy: Consensus report. *Sleep*, 11(1): 100-109.

Moore-Ede, G., Richardson, G. (1985) Medical implications of shift work. *Annual Review of Medicine*, 36: 607-617.

Roache, J., Gingrich, D., Landis, R., Severs, W., Pogash, D., Londardi, L. Kantner, A. (1992) Effects of alcohol intoxication on risk strategy, and error rate in visuomotor performance. *Journal of Applied Psychology*, 72(4): 515-524.

Spelton, E., Barton, J. and Folkard, S. (1993) Have we underestimated shift worker problems? Evidence from a reminiscence study. *Ergonomics*, 36(1-3): 307-312.

Storer, J., Floyd, H., Gill, W., Giusti, C. and Ginsberg, H. (1989) Effects of sleep deprivation on cognitive ability and skills of paediatric residents. *Academic Medicine*, 64: 29-32.

Tilley, A., Wilkinson, R., Drud, M. (1980) Night and day shifts compared in terms of the quality and quantity of sleep recorded in the home and performance measured at work: a pilot study. In Reinberg, A., Vieux, N., Andlauer, P. (Eds.) *Advances in the Biosciences vol 30. Night and Shiftwork Biological and Social Aspects*. Pergamon Press Oxford, New York, pp 187-196.

Wang, M., Fitzhugh, E., Christina, R. (1992) Psychomotor and visual performance under time course effect of alcohol. *Perceptual and Motor Skills*, 75: 1095-1106.

Wickens, C. (1984) Engineering psychology and human performance. Columbus: C.E. Merill.

Wilkinson, R., Colquhoun, W. (1968) Interaction of alcohol with incentive and with sleep deprivation. *Journal of Experimental Psychology*, 76(4): 623-629.

Working Arrangements Australia. (1993) Australia Bureau of Statistics..Cat # 6342.0.

13

Does Regulating Driving Hours Improve Safety?

Narelle Haworth, Monash University Accident Research Centre, Melbourne

Introduction

Most developed countries have regulations which limit driving hours for drivers of heavy vehicles. The regulations commonly include a limitation on the maximum number of hours that can be driven per day or per week and specifications relating to the length and timing of rest periods. Some jurisdictions limit driving hours (that is, hours behind the wheel) whereas others limit working hours (which may include loading, paperwork, waiting etc.).

Much of the argument for regulations governing driving hours is on the basis of ensuring safety standards. The underlying assumption is that limiting the hours of driving per day and per session results in drivers who are more alert and are, therefore, involved in fewer crashes. "Prescriptive regulation of driving hours is based on the premise that the regulation is effective in limiting driving or working periods and that these limitations are effective in reducing fatigue and hence fatigue related crashes" (Moore & Moore, 1996, p. 547).

Worsford states that "in Britain and Europe it is generally accepted that there must be legal limits on driving hours that the public authorities can enforce. There are a number of good reasons for this:

- there are high density populated areas;
- there is a high road vehicle population and traffic flows;
- the road freight industry is highly competitive and there must be an effective mechanism to ensure operating standards are not allowed to slip below a level that results in safety risk;
- for social reasons it is accepted that drivers should not work excessive hours behind the wheel." (NRTC, 1996, p.24)

That the majority of road users killed and injured in truck crashes are not truck occupants probably provides public (and thus political) support for restrictions on truck drivers including driving hours regulations. "Whilst it is a misguided perception this type of regulation gives comfort to the community that fatigue is controlled" (Mahon, 1996, p.520).

Recent Developments

In the last decade, the appropriateness of the current driving hours regulations has been debated in both the United States and Australia. Three types of developments have ensued: moves to change the provisions of the regulations, development of alternatives to driving hours regulations (driver alertness monitoring or fatigue management programs) and calls for the abolition osf the regulations.

In the US, an Advance Notice of Proposed Rulemaking on the Commercial Vehicle Hours of Service (HOS) rules was issued in November 1996 which may result in the first major changes to these rules in nearly 60 years. "The rulemaking addresses the potential for both conventional HOS rules and performance-based alternatives, with the latter including both fleet management and individual monitoring approaches to performance-based regulation" (US DOT, 1997, p.5).

Efforts to modify and standardise driving hours regulations in Australia have been underway since the mid-1980s (Moore & Moore, 1996). The proposed changes have related to changes to daily maximum hours of driving and possible differentials between remote and non-remote areas. Lack of agreement among jurisdictions has hindered these proposals.

Many observers see changes to hours of service regulations as an interim measure which will be replaced by future technological approaches to reduction of driver fatigue. The Office of Motor Carriers of the US Department of Transportation has stated that "although OMC has set its long-term sights on performance-based rules, the current focus is on prescriptive HOS rules" (US DOT, 1997, p.6).

Background To These Developments

There have been a number of important issues in the debate about the appropriateness of current driving hours regulations. Some of these are related to safety:

- whether the provisions of the driving hours legislation are such as to ensure adequate control of fatigue

- the future availability of driver alertness monitoring as an alternative
- fatigue management (formal or informal) as an alternative with safety and productivity aims

Safety is not the only issue, however, in the debate on driving hours regulations. In their Technical Working Paper No. 5, the National Road Transport Commission acknowledged that "the issue of driver working hours is contentious. The question raises trade-offs between productivity, road safety and occupational health objectives." (1992, p.1). Moore and Moore (1996) claim that "the imposition of regulations which, because they cannot be effectively enforced, adversely effect (sic) the competitive position of law-abiding operators is undesirable" (p.552). They also note that the overall cost in reduced productivity of complying with driving hours regulations has never been measured.

Adequacy of the provisions for controlling fatigue

The current driving hours regulations were framed at a time when little was known about the causes of fatigue and so they do not take into consideration many of the factors that are now known to be important. Feyer and Williamson (1996) note that current driving hours regulations have three critical shortcomings: they place limits on consecutive hours of work and rest irrespective of the time of day, they are not derived from empirical research basis and do not take into account inter- and intra-driver variability. "The omission of important factors affecting alertness levels, such as time of day, activity during rest breaks and prior activity, means that these regulations are incapable of being completely effective, even if problems related to enforcement were solved" (Haworth, 1995, p.46).

The interpretation and application of the regulations by a highly competitive industry has led to problems. The regulations are framed in a proscriptive manner. For example, driving more than five hours without a break of at least 30 minutes is proscribed in many states of Australia. Yet the interpretation, and even the general term used by many, is that of **prescriptive** driving hours. The interpretation is that drivers **should** drive for five hours, then take a break of 30 minutes and then drive for another five hours, etc. This does not take into consideration that some drivers at some times may not be sufficiently alert to drive safely during a five-hour period. Mahon notes that "the driving hours regime by its nature trained operators to not only think that any driver should be capable of 12 hours a day, 6 days a week but further if they only extracted this amount of hours out of each driver per week then they were meeting their fatigue responsibilities irrespective of the state of health of that driver" (Mahon, 1996, p.520).

Fatigue management as an alternative

Formal or informal fatigue management has been strongly proposed, and is being trialed, as an alternative to driving hours regulations. The Queensland Fatigue Management legislation has as its objective to ensure "that the driver of any heavy vehicle was always in a fit state of health and well being to safely drive that vehicle" (Mahon, 1996, p.520).

Moore and Moore (1996) argue that "in an unregulated situation, drivers and operators/employers have strong private incentives to manage fatigue. Poor management of fatigue by any of the parties in the transport chain can lead to inefficient practices and unsafe outcomes" (p.545). However, these incentives may not be sufficiently strong to ensure safe practices are adopted. Drivers and operators/employers may have even stronger incentives to undertake behaviours which are not compatible with managing fatigue. These incentives relate to the length of trips and times that cargo is available to be loaded and unloaded, etc.

Difficulties In Answering The Question

Moore and Moore (1996) concluded "we wonder whether the introduction of prescriptive hours could be justified if the burden of proof lay with those seeking the new regulation" (p.550). This section of the paper seeks to describe the difficulties in answering the question posed by Moore and Moore: Does regulating driving hours improve safety?

What is "safer"?

The specification of what is considered "safer" needs to be made. While some past comparisons have used the number (or rate) of fatal crashes involving trucks as a safety measure, this ignores the larger number of crashes that involve injury. If "safer" is defined as resulting in a lower total road trauma cost, then this may not necessarily be reflected in fewer fatigue-related crashes of heavy vehicles. It may be that fatigue-related crashes of heavy vehicles in rural areas (largely night-time and single-vehicle) result in less road trauma cost than other truck crashes in urban areas (largely multi-vehicle and frequently involving unprotected road users). Thus the net effects on road trauma costs of reducing fatiguing night-time rural travel by heavy vehicles, at the expense of reducing injury crashes in urban areas, would have to be carefully considered.

What should be compared?

In attempting to answer the question of whether regulating driving hours improves safety, one of the major decisions is the choice of what situations should be compared. One could compare

1. hypothetical complete compliance with current regulations with hypothetical behaviour if there were no regulations, or

2. current level of compliance with current regulations with hypothetical behaviour if there were no regulations.

The first comparison answers the question: "Could driving hours regulations improve safety?", the second answers the question, "Do current driving hours regulations improve safety?" The latter is probably of more interest.

An alternative would be to estimate in monetary and safety terms what it would cost to change the currently unregulated areas to (a) hypothetical complete compliance with current regulations or (b) a level of compliance with current regulations similar to that which occurs in other states.

The real answer would need to consider what all sectors of industry would do if there were no driving hours regulations. Perhaps constraints such as geography and delivery times would become the major determinants of driving hours. Would unions seek to bring in their own replacement set of regulations?

Possible comparisons and their limitations

Both of the comparisons above involved measurement of "hypothetical behaviour if there are no regulations". Clearly, measurement of hypothetical behaviour would be difficult (more so than measurement of actual behaviour). So, in reality, what could be compared? The following comparisons are outlined and then their limitations discussed:

- regulated driving hours area versus unregulated area
- regulated driving hours versus fatigue management in same area
- current regulated driving hours versus improved regulations

Comparing regulated driving hours area versus unregulated area. Hartley, Penna, Corry and Feyer (undated) state that "Australia is the ideal place to make comparisons on the effectiveness of HSR [hours of service regulations], because as noted earlier, some states implement HSR and some states do not. The rationale for comparing states is simple, if HSR regulations have been effective then one would expect to see lower fatigue related accidents and incidents in those states that control hours of service as compared to those that do not." (p.34)

Hartley, Arnold, Penna, Hochstadt, Corry and Feyer (1996) examined the prevalence of fatigue in the Western Australian transport industry (a non-regulated state) and compared the findings to similar studies carried out on Eastern states drivers who operate under a regulated hours regime. Hartley et al concluded that if Western Australia is compared with states that regulate driving and working hours, there appears to be little difference in the proportion of crashes that are fatigue related.

Yet this comparison is flawed "as the results are aggregates (covering urban, rural and remote areas and all types of operations), it is not possible to draw conclusions about the effectiveness of prescriptive driving hours" (Moore & Moore, p.555). In addition, fatigue-related incidents are more likely to have occurred without a resulting crash in the more remote areas of Western Australia and the Northern Territory. To quote Haworth (1995)

> Crashes are often the combination of characteristics of the driver and characteristics of the driving environment. The driving conditions result in a level of impairment of performance of the driver. Depending on the driving environment, a crash may or may not occur........Because of the longer trips and because of factors such as the smaller number of other vehicles, driving in a remote area is more monotonous and so is likely to lead to higher levels of driver fatigue. The probability of dozing off, of leaving the roadway or drifting across the road is higher in these areas. The probability of having a crash would be much higher if it was not for the lack of other vehicles, poles, etc. Thus in remote areas, we have the strange phenomenon that the very characteristics which increase fatigue, reduce the risk of impaired driving resulting in a crash. (p.44-45)

The same level of fatigue in a populated area would be more likely to have resulted in a crash.

In the same study, Hartley et al (1996) reported that 30% of the Western Australian drivers surveyed exceeded 72 hours per week compared with 25 to 46% of drivers in regulated states

(Williamson et al, 1992). This finding is more compelling evidence that regulating driving hours may not improve safety than the crash comparisons.

Regulated driving hours versus fatigue management in the same area. One possibility would be to take fatigue management practices as an example of what would happen if driving hours were not regulated. However, the pilot nature of the program in Queensland and the self-selecting companies involved and the desire to show that it works would likely result in a bias in any comparisons.

Mahon states that "we cannot be conclusive that the Fatigue Management Program will reduce the total of fatigue related accidents nevertheless we know:

- what the inputs are
- the stresses, hours and other matters that make a person fatigued

The output is that a certain amount of fatigue is generated in the system, from which the outcome is fatigue related accidents. If a process is put in place to manage and countermand those influences and reduce the amount of inputs in the system, it is reasonable to conclude that on a balance of probabilities the amount of fatigue in the system will be reduced and thereby reduce the number of fatigue related accidents." (Mahon, 1996, p.525)

Current regulated driving hours versus improved regulations. As mentioned earlier, there are current US (and Australian) moves to improve driving hours regulations, at least as an interim move until performance monitoring is fully developed. It is possible that improved driving hours regulations could improve safety, if the hours were reset in such a way that time of day and other effects on driver performance were acknowledged. There is the potential to include some flexibility in the hours (for example, the French regulations allow the trip to continue if within a certain time from the destination) and technology could be incorporated to improve the effectiveness of enforcement.

What experimental studies tell us

Experimental studies have provided considerable information on the factors influencing the development of driver fatigue but have shed little light on whether regulating driving hours improves safety. Those studies which examined driving hours were conducted under a range of constraints (legal, industrial, ethical) which restricted their usefulness in answering this question. Williamson et al (1994) compared driver performance on trips driven according to driving hours regulations or not. They found no differences in the amount of fatigue experienced on regulated and unregulated trips. Unfortunately, only one unregulated trip was driven per driver and it is possible that their stopping behaviour on that trip was determined to a large extent by habit. In addition, the trip was only about 12 hours long which limited the flexibility possible.

The Driver Fatigue and Alertness Study (Wylie, Shultz, Miller, Mitler and Mackie, 1996) compared driver performance on 10 and 13 hour workdays. It confirmed earlier findings that time of day was the strongest factor affecting driver alertness and fatigue. The number of hours or days of driving were not strong or consistent predictors of driver fatigue.

Research Strategies

There are two types of approaches to research in this area: research strategies that can give relatively precise answers about restricted situations from which you may not be able to generalise or research strategies which will give imprecise, generalisable results.

An example of the precise study, with lack of generality, is the California agricultural example reported in Heavy Duty Trucking (Ryder and Whistler, 1995). In order to retain its federal highway funding, California enacted legislation in 1992 to limit intrastate hours of service to 80 hours in eight days. However 112 hours in eight days was allowed for drivers hauling agricultural commodities. The legislation led to an overall 22 percent decrease in productivity in the trucking industry and a 25 percent decrease in driver income. There were no fatigue-related crashes in the short (four month) period of the 112 hour limit for the agricultural hauliers.

Comparing the fatigue management practices and crash rates of similar operations in regulated and unregulated areas would also offer precise results with limited generality.

An example of the imprecise, generalisable study would be surveying a representative sample of operators in regulated areas of Australia about what they would do if regulation were removed. Productivity changes could be calculated but some general rules would have to be applied in order to calculate effects on road trauma. (e.g. doubling risk after 8 hours and so on).

Three possible methods of answering the question of whether driving hours improve safety - which vary along the precision-generality dimension - are presented in the remainder of this section.

Development of a model to estimate safety effects of differing (or no) driving hours regulations

A simulation model could be developed to estimate the safety (and cost) implications of having no driving hours regulations or different types of regulations. The model would require as input certain parameters about where trips are and origin-destination distances (that should not change), when freight is required at destination (time of day and time of week), and certain assumptions (based on earlier research) about the crash risks associated with driving after a certain time on task and at a certain time of day. The model could output scenarios and related cost estimates (in both cost of transport and cost of crashes terms).

The input parameters could be devised from currently available truck flow information and by surveying trucking companies about what they would do if there were no regulations or by examining what the companies in the Fatigue Management Programs are advocating (the latter might give a very positive bias to the comparisons, however, as noted earlier).

Proposed study of relationship between violation of driving hours and crash history

This proposed study builds on the methodology used in earlier US studies by Hertz (1991) and Beilock (1995) who measured the time taken for particular heavy vehicles to travel between locations and calculated driving times from speed estimates. It is proposed to record the registration numbers and times of arrival of heavy vehicles at particular points and to measure travel times and derive driving hours from these times. In NSW this could be very efficiently done by the use of SAFE-T-CAM records if access was provided. Careful choice of the survey points would be needed to minimise the likelihood that changes of drivers occurred between the points.

The registration numbers (without driving hours) would then be forwarded to the relevant road authority who would be requested to provide details of crashes involving that vehicle within a given period (perhaps the last five years).

This study would show whether vehicles driven by drivers who were judged to have violated driving hours regulations were involved in more crashes than those driven by drivers who were judged to have complied with the regulations. As such it compares the safety of complying versus not complying with driving hours regulations within a regulated area.

It would be possible to extend the study to unregulated areas and compare the driving times with those specified in the regulated areas. Extension of the study to international collaboration would also be of interest.

Clearly, the feasibility of this study would depend on the extent of cooperation provided by the relevant road authorities.

Proposed study of risk of crashing associated with violation of driving hours

This proposal is for a case-control study of truck crashes, similar to that reported by Jones and Stein (1987). Researchers would interview truck drivers involved in crashes and then interview truck drivers passing the site at the same time of day and week one week later. This would control for road environment and traffic factors in the crashes. Comparisons of the hours driven and other risk factors for the two groups would be made. Police assistance in stopping trucks (but not interviewing) would be most helpful, otherwise there could be difficulties in stopping the comparison trucks near the crash site and other methods of recruiting a control sample might have to be developed.

This study could be conducted in both regulated and unregulated areas. It could also be extended to international collaboration which would allow the required number of crashes to be collected more quickly and would allow sharing of the costs of the study.

CONCLUSIONS

While improving safety by reducing fatigue-related crashes has been a major aim of the regulation of driving hours, the ability of these regulations to achieve this aim has been questioned in the last decade. In addition, the development of alternatives to driving hours regulations - driver alertness monitoring or fatigue management programs - have led to an increasing debate about the usefulness in both safety and productivity terms of driving hours regulations.

There are many difficulties in answering the question of whether driving hours regulations improve safety. The findings of many of the "natural experiments" - such as the comparison of crashes in regulated and unregulated areas - are difficult to interpret and experimental studies have not yet addressed the issues. Three possible research strategies are proposed: development of a model, examination of crash history and a case-control study of truck crashes.

ACKNOWLEDGMENT

The preparation of this paper was supported by the National Road Transport Commission. The views expressed in the paper are those of the author and do not necessarily reflect those of the Commission.

REFERENCES

Beilock, R. (1995). Schedule-induced hours of service and speed limit violations among tractor-trailer drivers. Accident Analysis and Prevention, 27, 33-42.

Feyer, A.-M. and Williamson, A. M. (1995). Managing driver fatigue in the long-distance road transport industry: interim report of a national research programme. In L. Hartley (Ed.), Fatigue and driving. Driver impairment, driver fatigue and driving simulation. London: Taylor and Francis. (pp.25-32)

Hartley, L.R. and Arnold, P.K. (1995). Trends and themes in driver behaviour research: the Fremantle Conference on 'Acquisition of skill and impairment in drivers'. In L. Hartley (Ed.), Fatigue and driving. Driver impairment, driver fatigue and driving simulation. London: Taylor and Francis. (pp.3-13)

Hartley, L.R., Arnold, P.K., Penna, F., Hochstadt, Corry, A. and Feyer, A.M. (undated). Fatigue in the Western Australian transport industry. Part One. The principle and comparative findings (Report No.117). Institute for Research in Safety & Transport, Murdoch University, Western Australia.

Hartley, L.R., Penna, F., Corry, A. and Feyer, A.M. (undated). Comprehensive review of fatigue research.. (Report No.116). Institute for Research in Safety & Transport, Murdoch University, Western Australia.

Haworth, N. (1995). The role of fatigue research in setting driving hours regulations. In L. Hartley (Ed.), Fatigue and driving. Driver impairment, driver fatigue and driving simulation. London: Taylor and Francis. (pp.41-47)

Hertz, R.P. (1991). Hours of service violations among tractor-trailer drivers. Accident Analysis and Prevention, 23, 29-36.

Jones, I.S. and Stein, H.S. (1987). Effect of driver hours of service on tractor-trailer crash involvement. Washington, D.C.: Insurance Institute for Highway Safety.

Mahon, G.L. (1996). Fatigue & transportation: Engineering, enforcement and education solutions. In L. Hartley (Ed.), Proceedings of The Second International Conference on Fatigue and Transportation: engineering, enforcement and education solutions. Fremantle February 11-16 1996. Canning Bridge: Promaco Conventions. (pp.517-543).

Moore, B. and Moore, J. (1996). Prescriptive driving hours: The credibility gap. In L. Hartley (Ed.), Proceedings of The Second International Conference on Fatigue and Transportation: engineering, enforcement and education solutions. Fremantle February 11-16 1996. Canning Bridge: Promaco Conventions. (pp.545-560).

National Road Transport Commission. (1992) Driver working hours (Technical Working Paper No. 5). Melbourne: National Road Transport Commission

National Road Transport Commission. (1996). Use of tachographs in the enforcement of truck driving hours: The British and European experience. (Technical Working Paper No.23). Prepared by Frank Worsford. Melbourne: National Road Transport Commission.

Ryder, A. and Whistler, D. (1995). Sleepy drivers: How big the threat? Heavy Duty Trucking, March 1995, pp.69-74.

United States Department of Transportation. Federal Highway Administration. Office of Motor Carriers. (1997). Summary of driver fatigue programs. December 1997. Washington, D.C.: US Department of Transportation.

Williamson, A.M., Feyer, A.M., Coumarelos, C. and Jenkins, T.. (1992). Strategies to combat fatigue in the long distance road transport industry - The industry perspectives (Report No.CR108). Canberra: Federal Office of Road Safety.

Williamson, A.M., Feyer, A.M., Friswell, R. and Leslie, D. (1994). Strategies to combat fatigue in the long distance road transport industry - Stage II: Evaluation of alternative work practices (Report No.CR144). Canberra: Federal Office of Road Safety.

Wylie, C.D., Shultz, T., Miller, J.C., Mitler, M.M. and Mackie, R.R. (1996). Commercial Motor Vehicle Driver Fatigue and Alertness Study: Project Report (FHWA-MC-97-002). Washington, D.C.: Federal Highway Administration.

14

PRESCRIPTIVE DRIVING HOURS: THE NEXT STEP

Barry Moore, National Road Transport Commission, PO Box 13105, Law Courts, Melbourne, Victoria, 8010

The views expressed in this paper are those of the author and not necessarily of the Australian National Road Transport Commission.

INTRODUCTION

The regulatory approach to the control of fatigue in drivers of heavy vehicles in eastern Australia is based on the application of prescriptive limits to hours of driving. This same approach is used in Europe, the United States, Canada and New Zealand. In developing uniform regulation of hours of driving and work for the regulated areas of Australia[1], some improvements to this prescriptive regime have been made and there has been debate over the place of formal fatigue management programs in a fatigue control strategy.[2]

This paper addresses the issue of whether the provision of additional flexibility within a prescriptive regime will make it a more effective approach to the control of fatigue in heavy vehicle drivers.

1 Restrictions on driving hours have not been applied in Western Australia, Northern Territory or Australian Capital Territory and have not been enforced in Tasmania. Under the national Regulations, Western Australia and Northern Territory will form a non-regulated zone. The Regulations will apply in the New South Wales, Victoria, Queensland, South Australia, Tasmania and the Australian Capital Territory. Tasmania has stated that it will seek approval from Ministerial Council for Road Transport not to implement requirements relating to logbooks.

2 For discussion of the national Road Transport Reform (Truck Driving Hours) Regulations see NRTC (1998)

BACKGROUND

Many concerns have been expressed about the effectiveness of prescriptive hours in managing driver fatigue. However no comprehensive alternative has yet been demonstrated. Technological approaches to the measurement of fatigue have not yet been widely implemented in operational situations and the regulatory use of fatigue management programs is still at an early stage of development.

In Australia, there has been a strong focus on the development of fatigue management as an alternative to prescriptive hours of driving. The purpose of this paper is to stimulate discussion of whether the prescriptive approach can be modified to more effectively achieve intended road safety outcomes.

In the populous areas of Australia there is a perception of strong public support for prescriptive restrictions on hours of driving for drivers of heavy vehicles. It is difficult for a regulatory authority to explain to the average motorist that truck drivers are permitted to drive beyond a daily limit which appears to those outside the industry to be high, whether it be 10, 12 or 14 hours. Politicians are understandably averse to the possible consequences of easing controls which have been in place for decades, whether or not they can be demonstrated to have had beneficial effects.

An indication of the public concerns over relaxing the regulation of driving hours for heavy vehicles is provided by a news release issued by the NRMA, a New South Wales based motoring association, during the development of the national policy (NRMA 1996):

> *"The telephone survey, which involved 800 people from Sydney, Melbourne and Brisbane, indicated that the community would not accept the NRTC proposal and indeed believe that the maximum driving time should be much closer to nine hours. ...*
>
> *"Eighty three per cent of those thought the time a truck driver can spend behind the wheel should be regulated by law. Seventy-six per cent disagreed with the first phase of the NRTC proposal, which would allow a truck driver to work an 18 hour day, including 12 hours behind the wheel with a minimum of two half hour breaks. Seventy-five per cent disagreed with the first phase of the proposal, which would increase the permissible driving hours to 14, as long as drivers had regular health checks and are better educated about driver fatigue.*

"The survey results suggest that industry and NRTC perceptions of acceptable truck industry work practices are very different from the community at large. People do not accept that the trucking industry should be allowed to extend working hours in return for the adoption of good working practices, such as ensuring the health of drivers and providing fatigue education. The industry has an obligation to employ safe work practices, including acceptable working hours."

It appears that a similar situation applies in other developed countries:

- *"... in Britain and Europe it is generally accepted that there must be legal limits on driving hours that the public authorities can enforce."* (NRTC, 1996, p24, quoted in Haworth 1998);
- *"The Office of Motor Carriers of the US Department of Transportation has stated that 'although OMC has set its long-term sights on performance-based rules, the current focus is on prescriptive HOS* [hours of service] *rules'"* (US DoT, 1997, p6, quoted in Haworth 1998)

The prescriptive approach is based on the proposition that longer hours of driving will lead to increased driver fatigue and to worse safety outcomes.

Research results suggest that many other factors must be taken into account.[3] Recent studies highlight time-of-day effects and the importance of the quantity and quality of rest. Currently regulated prescriptive driving hours in Australia and North America provide for rest periods but take no account of quality of rest or time-of-day effects.

Two pieces of research which have been undertaken in North America in recent years are of particular relevance.

The **Driver Fatigue and Alertness Study** was recently conducted by the US Department of Transportation, Transport Canada and the Trucking Research Institute (Department of Transportation (US) 1996). The study involved 80 drivers divided into four contrasting schedule types: 10 hour daytime, 10 hour rotating start, 13 hour night-time start and 13 hour daytime start. The study included 360 trips and about 4,000 hours driving. The 10 hour schedules were driven in the US and the 13 hours schedules in Canada. Major findings of the study were:

---

3 A brief review of research is included in Moore and Moore (1996). More comprehensive reviews include Hartley et al (1997).

- *The strongest and most consistent factor influencing driver fatigue and alertness in this study was time of day. Drowsiness ... was markedly greater during night driving than during daytime driving. Peak drowsiness occurred during the 8 hours from late evening until dawn.* (p7)
- *Hours of driving (time-on-task) was not a strong or consistent predictor of observed fatigue. Most notably, there was no difference in the amount (prevalence) of drowsiness observed on video records of comparable daytime segments of the 10-hour and the 13-hour trips.* (p8)
- *There was some evidence of cumulative fatigue across days of driving ...However, cumulative number of trips was neither a strong nor consistent of fatigue...* (p8)

An earlier study conducted by the United States National Transportation Safety Board concluded that:

> *The most critical factors in predicting fatigue-related accidents ... are the duration of the most recent sleep period, the amount of sleep in the past 24 hours, and split sleep patterns.* (National Transportation Safety Board 1995, p51)

The Board recommended that driver working hour regulations be amended to enable drivers to obtain additional hours of sleep.

Research evidence and developments in fatigue management suggest that it would be desirable to modify prescriptive approaches to take into account or allow for:

- trip preparation
- quality and quantity of rest
- cumulative effects
- circadian rhythms
- individual differences.

Due to the fundamental limitations of the prescriptive approach, it is unlikely that the addition of more flexibility would allow all of these factors to be addressed. The issue to be investigated is whether the added flexibility would allow worthwhile improvements in some areas.

EFFECTS OF PRESCRIPTIVE HOURS

A major problem in the prescriptive approach is that it implies a world which is black or white: fatigued or alert, safe or unsafe, drive or not drive; whereas the real world includes every situation between these extremes. The implication with prescriptive hours is that one particular combination of hours of driving over a cycle represents the optimal bounds in all circumstances. Up to these bounds the risks to society are acceptable - beyond these bounds the risks are unacceptable.

Of course, we all know that the world rarely works like that. Degrees of risk increase (perhaps at an increasing rate) as hours of work increase. Boundaries are rarely hard and fast, or independent of circumstances. We also know that there is considerable variation between individuals.[4]

Among the more obvious difficulties caused by the "hard edges" of prescriptive hours are:

- a need to drive for an additional period in order to finish a trip

    There may be many reasons to exceed current prescriptive periods of driving in this way, including:
    - a driver getting home to enjoy better quality rest
    - perishable commodities
    - animal welfare (providing livestock water, food and rest)
    - the delivery of inputs crucial to just-in-time operations
    - vehicle breakdown
    - floods, bushfires or other natural or man made disasters.

- longer term operations which would benefit from flexibility

    One example would be an operation based on two drivers per truck, which may require longer hours per day, but may be unaffected by a total over a cycle. Do we have research evidence to say that 15 hours of work three days a week (or even three days in succession) should be prevented by law when 14 hours for six or 10 days in succession is permitted? A more extreme example is of work for one hour per day for more than

---

4 One very experienced and well respected Australian long distance truck driver has been heard to complain bitterly about the requirement for a minimum daily period of continuous rest of six hours. This driver argues that he can effectively manage his fatigue using shorter but more frequent periods of sleep.

12 days in succession - not permitted under current or proposed prescriptive schemes in Australia.

Anomalies resulting from the rigid application of prescriptive hours were discussed in Moore and Moore (1996)

"Any prescriptive regulation of human behaviour creates anomalies. Examples of the types of anomalies created by prescriptive regulation of driving hours are:

- ***type of load***
  some loads are subject to more urgent deadlines, while delays to others have direct welfare effects (eg livestock)

- ***circumstances***
  there is an argument for relaxation of requirements in the event of unusual circumstances (eg bushfire or flood)

- ***proximity to destination***
  a driver obeying the rules may be required to pull over and rest in an unfavourable location when close to a destination which provides a more suitable environment for rest (eg home).

"Some of these anomalies can be addressed in exemptions, but excessive reliance on exemptions can introduce an undesirable degree of arbitrariness into decision-making.

"The practical reality of exemptions is that they tend to break down the whole fabric of control. If exemptions are given for livestock why not for other perishable goods? If for bushfire and flood why not heatwave? If for proximity of 20 km from the end of the journey why not for 25 km or for similar proximity to an intermediate destination? The notion that it is 'safe' to exceed the prescribed limit in certain circumstances begs the question why it is not just as safe under other or all circumstances. This argument can be followed by the assertion that exemption should be given on a regional basis or on the basis of traffic density or almost any pretext. Regulatory control should be seen to be equitable and exemptions are of their nature inequitable.

"It is very difficult to use a regulation to force a driver to rest or sleep. The best that can be done is to provide time away from driving, when sleep may be taken. If the time or place is inappropriate, the rest may be of poor quality and the ability of the driver to obtain quality rest in a preferable time or place may be reduced. The net result could be negative."

In an environment where enforcement is incomplete, one effect of prescriptive regulation is to reward operators who breach the rules in relation to those who comply. Allowing choice of options which reduce the worst anomalies of prescriptive schemes, without reducing the effectiveness of the approach, would, at least in part, remove this anomaly and reduce the competitive disadvantage faced by more "honest" operators and drivers.

THE OPTIONS

There are many problems in the use of prescriptive hours as an approach to the control of driver fatigue. Some of these problems may be quite fundamental and their solution would require an entirely different approach, ie full fatigue management.

Over the longer term, it may be that approaches to the regulation of driver alertness will be dominated by technical measurement of fatigue and attempts to control fatigue precursors, through fatigue management schemes. In the shorter term, however, it may be useful to explore options for greater flexibility within prescriptive approaches.

Fatigue detection

Research into the physical measurement of fatigue is developing rapidly and will be considered by others at this conference. The premise of this paper is that the widespread application of these technologies will not occur for some years yet.

Fatigue management

The fatigue management approach attempts to control a comprehensive set of fatigue precursors.

A fatigue management scheme is under development in Australia by Queensland Transport in conjunction with the Road Transport Forum. Pilot schemes are also under consideration for North America. Widespread application of this approach is still some time away. The Australian scheme has been under development for five years. The first operator was accredited in 1995 and only two further operators have now been accredited.

In Australia, an effective scheme must allow travel through all regulated States. One of the reasons for the slow expansion of the Australian pilot scheme is the concern of one jurisdiction to allow little extension of maximum daily hours of driving and work beyond what is currently available through the prescriptive regime.

Even if this opposition were removed, expansion of the scheme will be gradual due to resource constraints, the need to carefully develop internal guidelines, the need to educate additional operators, the need for thorough evaluation at every stage and the need to win public acceptance of the approach.

Rigid prescription

Control of driver fatigue based on the rigid prescription of hours of driving and/or work has the limitations discussed above. In fact, there is an argument that prescriptive regulation could be counter-productive if it creates an impression that compliance with the requirements equates to adequate management of fatigue.

However, there is little doubt that the prescriptive approach will not be removed in populous areas of Australia, at least in the short to medium term. The situation probably applies in other developed economies, including Canada and the United States.

It must also be accepted that, for many road transport activities, the nature of the task means that driver fatigue is not an issue and current prescriptive arrangements do not impose operational constraints.

Flexible prescription

It may be possible to remove some of the anomalies of prescriptive driving hours regimes by allowing limited and controlled flexibility. This may allow the safety objectives of this form of regulation to be more effectively achieved whilst reducing or removing some undesirable side-effects. This option is the subject of this paper.

FLEXIBLE PRESCRIPTION

The major flexibility which can be injected into a prescriptive scheme involves changing the "value" of driving time. This would involve allowing additional hours of driving or work at "high risk" times, in exchange for greater reductions in hours of driving or at "low risk" times. In effect, this would allow smaller amounts of higher-risk time to be traded for larger amounts of lower-risk time and would increase the potential for rest over cycle. The "exchange rate" or relative risk premium would be determined on the basis of industry and regulatory experience and research evidence. This principle could be applied to either the duration or the time-of-day of the driving period.

Increased flexibility based on time-shifting may be politically attractive in that increases in some driving periods can be more than compensated by increased availability of rest, through reductions in driving hours across a cycle. The increases in some periods of driving could be traded against longer absolute and proportional increases in the time available for rest, and potential improvements in the quality of rest.

Duration of driving periods

Examples of extensions in the duration of driving periods include:

- Allowing increased hours on specific days in return for reduced total hours, ie increased potential for effective rest.
    - This could allow for longer hours on the last day of a trip, so long as the total weekly driving hours were reduced. For example, on one day in each cycle, the number of hours could exceed the usual maximum (subject to the retention of an upper bound for the day) in return for a reduction of a multiple of this amount in the weekly total. A high multiplier may be required to allow for adequate rest.
    - An alternative would be to allow the additional period of driving, so long as the hours in the preceding and subsequent periods had been sufficiently low. For example, a driver could be allowed an additional hour on a specific day, so long as he/she had made provision for additional rest on the day or days before and/or after.

- Allowing the deferral of short rest breaks. In the regulated States of Australia, rest breaks totalling 30 minutes are required for each five hours of driving. On some occasions, a suitable location for a rest break may lie outside the permitted driving range. It would be possible to devise a system that would permit a longer period between short rest breaks, so long as the breaks were longer. For example, rest of 40 minutes in a six hour period could be required as an alternative to rest of 30 minutes in a five hour period. It would also be possible to limit the number of times this option could be chosen over a driving cycle.

Time-of-day effects

Time-of-day effects could also be taken into account by changing the "value" of periods of driving at different times of the day, ie placing a higher risk premium on these periods. Under prescriptive schemes, there is currently no distinction drawn on the basis of the time of day at which driving takes place. This is despite evidence of strong time-of-day effects of fatigue related crashes. Under a more flexible approach, the risk weighting placed on active periods could vary depending on the time of day at which they occurred. Driving during this period could be linked to a requirement for increased provision for rest.

Introduction of a weighting for circadian effects would present particular difficulties. Drivers currently have unrestricted access to these hours. The development of flexibility in this direction may require the offer of additional driving hours elsewhere to "compensate" for the loss of access to the high-risk periods. Operators could object strongly to any reductions in hours of driving and work currently available under prescriptive regimes.

Other

Once the notion of imposing differential risk premia in hours driven at different times of day or different times in a shift had been established, the range of options available would be very wide.

Examples include:

- imposing a premium on hours later in a shift; and
- imposing the risk premium after repeated occurrence
  - For example the first two night driving shifts in a cycle could be premium free and a premium imposed on subsequent night driving shifts. This possibility would have to be assessed against research evidence on the effects of occasional night-shifts against regular night-shifts.
  - one extended shift in a cycle could be premium-free, but subsequent extended shifts could attract a high premium.

Use of geographical positioning devices (perhaps implanted in the skull of the driver) could even allow the quality of rest to be differentiated on the basis of its location, though it may be a little early to pursue that avenue just now. The world may not yet be ready for the spectacle of groups of experts on the quality of rest assigning weightings to rest undertaken in different circumstances. Would rest at home with one's spouse attract a higher or lower rating than rest alone in a sleeper cab?

Implementation

Initial steps towards differential valuation of driving and working time would be small. An initial scheme should be kept simple in order to maximise the chance of acceptance and simplify enforcement. A likely first injection of flexibility would be to allow a single extended shift in a driving cycle to allow for an emergency, or a return home at the end of a trip to benefit from improved quality of rest.

An attraction for policy makers in time-shifting is that the increased flexibility may be politically acceptable. A policy based on some increase in specific periods of driving or work in exchange for higher absolute and proportional increases in potential rest could be "marketable" to the public.

This form of exchange could be attractive to a growing number of operators. Some operators are moving towards increasing the ratio of drivers per truck and reducing the hours of driving and work for each driver. Reducing the total number of hours, in return for increased length of some shifts, may be an attractive option for these operators.

Record Keeping

In Australia the compliance methods currently applied also work against the introduction of flexibility into the prescriptive approach. Recording of driving hours for long-distance operations in Australia currently relies almost entirely on log books and is given effect through roadside enforcement. For trips close to base, management records are permitted and in one State a limited exemption permits the use of management records by one long-distance operator [6].

The recently agreed national Regulations provide for electronic driver-specific monitoring devices as a voluntary alternative to logbooks. A performance standard for these devices is currently under development. Use of these devices will result in the availability of more comprehensive records at the premises of operators and will facilitate a shift in the emphasis of enforcement towards audit of these records.

In a recording system restricted to logbooks, only limited flexibility is feasible. Examples may include the inclusion of a longer shift in return for shorter shifts on the days before and after and/or lower aggregate hours over a cycle.

In the absence of some form of electronic recording, more sophisticated examples of flexibility would present difficulties for drivers, operators and enforcement officers. Options could be added to driver specific monitoring devices, electronic tachographs or vehicle management systems to allow the recording of hours and the presentation of options to drivers and operators. It should be possible to program these devices to warn drivers and/or operators when the driver is approaching certain limits. These limits could be based on either the standard hours or the flexibility option chosen by the driver/operator. Drivers and operators should also be able to query the devices on the effect of a range of driving options.

In any form of time-shifting or differential risk premia, external validation of time-of-day would be required. This could be incorporated into electronic recording devices, either through the signal which is incorporated into global positioning systems or by some other means.

The adoption of an electronic tachograph in Europe may provide opportunities in this area.

LINKS TO OTHER APPROACHES

Access to increased flexibility could be linked to the implementation of other fatigue reduction strategies. Australia's Transitional Fatigue Management Scheme (TFMS) is an example of this. Under the TFMS, drivers and operators are given the flexibility to run the same number of hours of driving as in the base prescriptive scheme but over a longer cycle (two weeks as compared to one week in the standard scheme) in return for the adoption by drivers and operators of limited aspects of fatigue management, including training in fatigue management for drivers and operational staff, driver health checks and the implementation of management procedures to monitor and manage driving hours.

Other elements of a more comprehensive fatigue management strategy could be included if beneficial results could be expected.

An another example of this type of approach is the recently announced exemption from some aspects of prescriptive hours for livestock carriers in Queensland. Under the proposed Livestock Welfare Exemption Scheme, the 12 hour daily limit could be extended to 16 hours, so long as the extension does not occur on consecutive days. The exemption does not change total weekly hours of driving and contains conditions including driver health checks and the availability of approved sleeper berths.

There appears to be no reason why this approach could not be made more widely available, perhaps in combination with a reduction in total driving hours over the driving cycle. This would then be an example of a trade-off of increased hours at specific times for increased potential for rest over the driving cycle.

SOME ARGUMENTS AGAINST INCREASED FLEXIBILITY

Fatigue Management Programs

Some will argue that more flexibility should not be allowed in current prescriptive schemes because of a possible lessening of the attraction of full fatigue management as an alternative.

Full fatigue management, at least in its initial stages, imposes a significant burden on operators in the sense that it involves considerable cultural and operational change and may require significant establishment costs. Whilst there is a strong case that the approach results in longer term benefits to both the operator and society, these effects have not yet been demonstrated for large numbers of operators.

We cannot yet be confident on when, or whether, the approach will be widely available. The slow expansion of the pilot demonstrates the difficulty of achieving such a fundamental change in the regulatory approach to driver fatigue. Over the medium term, and perhaps even in the longer term, it is likely that fatigue management schemes will be a voluntary alternative to more prescriptive regulation.

A choice could be offered from a menu of rigid prescription, flexible prescription and full fatigue management. The recently agreed national policy in Australia offers these options: a Prescriptive Hours Scheme, the Transitional Fatigue Management Scheme (a form of flexible prescription) and a Fatigue Management Scheme (included as an exemption to the prescriptive scheme pending further development of the pilot).

If the prescriptive approach is to be maintained beyond the short term, there may be significant safety and productivity benefits in allowing increased flexibility.

Risk-weighting

It may be argued that any exchange between high-risk and low-risk periods of activity is arbitrary. The response is that to take no account of risk differentials is to choose an exchange rate of 1:1. This is equally arbitrary and in conflict with research results.

Consideration should be given to the process used to determine risk premia for different periods of time - either time-of-day or duration of a shift. The process chosen should be based on the best available research and result in an outcome which has reasonable credibility, both with the road transport industry and other road users.

Exemptions to Rigid Prescription

It could be argued that many of the cases presented are extreme and could be addressed by exemptions. However, if the examples cited represent acceptable patterns of work and rest, it may be preferable to allow them through a flexible scheme. The advantage of a more structured approach is that all eligible applicants would be allowed the same outcome, rather than relying on individual decisions which may be inconsistent between locations and over time.

A further disadvantage of exemptions is that the process of application and assessment is intensive in the time of drivers/operators and regulatory authorities.

Another consideration in a federal system is that it may be easier to achieve mutual recognition between jurisdictions within an accepted framework of flexible prescription, than through exemptions which allow for more subjectivity in decision making. (Although the recent experience in attempting to achieve uniformity in a prescriptive scheme in Australia may suggest otherwise.)

Conclusion

The rigidities imposed by current prescriptive hours schemes for the control of fatigue in drivers of heavy vehicles impose operational inefficiencies and work against the objectives of the regulation. However, this form of regulation is deeply embedded in the regulatory culture of populous areas in many developed countries and is likely to be an important element of regulatory attempts to control fatigue for at least the medium term. For this reason, increases in the flexibility of prescriptive schemes to improve their effectiveness and reduce their operational impacts is a worthwhile goal.

Steps which can be taken in this direction include allowing longer periods of driving on limited occasions in return for increased provision for rest over a driving cycle. This may allow productivity benefits as well as resulting in the improved road safety outcomes which are the objective of prescriptive regulation.

Flexibility in prescriptive hours could also be used to impose a risk premium on the hours driven at times in the day which have been identified by research as high risk. A particular difficulty in the introduction of this form of flexibility is that it would require a reduction in existing entitlements to drive in high risk periods. The provision of higher aggregate hours to compensate for this loss may be considered unacceptable on safety grounds.

The next stage in the evolution of regulatory approaches to the control of fatigue in drivers of heavy vehicles may be to allow drivers a choice of options: a base case of rigid prescription against options of flexible prescription and full fatigue management.

The development of the option of flexible prescription may require:

- the availability of more sophisticated forms of record-keeping;
- the establishment of credible processes to determine the appropriate risk premia for different periods of time; and
- public acceptance of the validity of the approach.

The addition of flexibility to current schemes for the prescriptive regulation of driving hours could lead to increased productivity in road transport and improved safety outcomes. Methods of modifying existing prescriptive schemes should be explored.

REFERENCES

Department of Transportation (US), Transport Canada, American Trucking Associations Foundation (1996) Commercial Motor Vehicle Driver Fatigue and Alertness Study, Executive Summary (November)

Hartley, L.R, Arnold, P.K., Penna, F., Corry, A. And Feyer, A.M. (1997). Comprehensive review of fatigue research. Institute for Research in Safety and Transport, Murdoch University, Western Australia

Haworth, N. (1998). Does regulating driving hours improve safety? In L.R. Hartley (Ed), Fatigue and Transportation, London: Elsevier.

Moore, B. and Moore, J. (1996) Prescriptive Driving Hours: The Credibility Gap. In L. Hartley (Ed.), Proceedings of the Second International Conference on Fatigue and Transportation: engineering, enforcement and education solutions. Fremantle February 11-16 1996. Canning Bridge: Promaco Conventions. (pp.545-560)

National Road Transport Commission. (1996). Use of tachographs in the enforcement of truck driving hours: The British and European experience. (Technical Working Paper No 23). Prepared by Frank Worsfold. Melbourne: National Road Transport Commission.

National Road Transport Commission. (1998, forthcoming). Regulatory Impact Statement: Truck Driving Hours

National Transportation Safety Board (1995) Factors that Affect Fatigue in Heavy Truck Accidents. Safety Study NTSB/SS-95/01, Washington D.C. (January)

NRMA (1996) News Release: Community against increase in truck hours, National Roads and Motorists Association (18 March)

United States Department of Transportation. Federal Highway Administration. Office of Motor Carriers. (1997). Summary of driver fatigue programs. December 1997. Washington, D.C.: US Department of Transportation.

15

RISK-TAKING AND FATIGUE IN TAXI DRIVERS

James R. Dalziel & R. F. Soames Job, Department of Psychology, University of Sydney, Australia

INTRODUCTION

There has been considerable interest in the problems of driver fatigue in recent years, and many road safety researchers currently consider fatigue to be a problem of approximately equal importance to drink-driving and speeding. While driver fatigue may have devastating consequences, particularly in the case of single vehicle "run-off-road" accidents resulting from "driver asleep" (Fell & Black, 1996), it is difficult to identify. Fatigue leaves no obvious biological traces that can be used to externally identify fatigued drivers (such as blood alcohol levels can be used to identify drink-drivers), nor does driving while fatigued always show obvious driving behaviour aberrations (such as with speeding) until the moments immediately prior to a fatigue-related incident.

As Brown (1994) has argued, fatigue is primarily a subjectively experienced phenomenon, and its hazards include not just physiological impairment (such as slower reaction time) but also psychological impairment in the form of worsened mood and impaired judgement. In addition, fatigue impairment, like alcohol impairment, may have the insidious effect of reducing meta-cognitive abilities to evaluate one's impairment, hence increasing risk due to the inability of drivers to realistically assess their own driving performance. Fatigue can result from lack of sleep, circadian rhythm disruption (NASA, 1996) and from prolonged performance of a task (Crawford, 1961) - all of which are potentially relevant to taxi drivers due to the nature of their work (Dalziel & Job, 1997a).

Prior to our first study of taxi drivers in 1993 (Dalziel & Job, 1994, 1997a), no study of the fatigue issues associated with taxi driving had been presented in the road safety literature. In the 1997 Dalziel & Job paper, we indicated several issues of interest to fatigue research based on the study of taxi drivers: many taxi drivers work long hours per week (average hours of driving as a taxi driver was 50 hours, and total average work per week was 59 hours); only a relatively small percentage report having ever fallen asleep at the wheel (5-12%, depending on definition); a significant negative correlation was observed between accident rate and total average break time per shift (although this relationship may be complicated by employment and personality variables); and that optimism bias regarding the ability to "drive safely while very tired" was significantly less than other "skill" abilities such as the ability to "swerve around a sudden road hazard". Since the time of the original study, Corfitsen (1993) has examined fatigue and reaction time among night shift taxi drivers, and these results may be compared with those of young male drivers (Corfitsen, 1994).

The present study, as part of a more general study of taxi driver road safety, sought to extend the fatigue findings of the 1993 study. In the first of these developments, a question regarding accidents while driving home after completing a taxi shift was included, as was a question on the existence of sleeping problems (in addition to the "asleep at the wheel" question used previously). Second, a question regarding driving while very tired was included in both optimism bias and risk-taking scales, and hence the relationship between these particular questions and scales can be compared with other variables, including accident involvement. Third, it is possible to examine the fatigue-related factors that predicted accidents, going beyond the negative correlation observed between total breaks length and accident rate observed in the 1993 study. By attempting to build a more comprehensive predictive model of accident involvement, it was possible to examine the inter-relationships between breaks and other fatigue-related factors (such as average hours per shift), as well as between other potentially relevant variables, such as optimism bias, risk-taking, aggression and sensation seeking. This proposal provides a basis for examining the speculation that fatigue issues may be related to aspects of personality or motivation (Dalziel & Job, 1997a).

METHOD

Surveys were distributed to Sydney taxi drivers at taxi ranks throughout the metropolitan region, and some surveys were distributed during each possible shift period. Surveys were returned using an attached prepaid envelope, and drivers were offered $10 compensation for time lost from their shift while completing the survey. Further details of the methods of this study are described in Dalziel & Job (1997b). The working conditions survey questions relevant to fatigue issues were:

Q7. How many breaks do you normally take during a shift (for gas, a meal, etc.), and how long are each of these?
Q14. Apart from when you are stationary at a rank, have you ever fallen asleep at the wheel (even just for a few seconds) while driving a cab?
(1) Yes(2) No
If yes, how many times would this have happened during 1995 & 1996?
Q15. Have you ever had an accident while driving home after a shift that was at least partly the result of tiredness?
(1) Yes(2) No (3) Don't drive home after shift
Q16. Do you have Sleep Apnea, chronic snoring, or any other major sleeping difficulties?
(1) Yes(2) No

The reason for the qualifier regarding falling asleep on ranks is that feedback on the 1993 survey indicated that some drivers "nod off" for brief periods during quiet times of their shift when the taxi is stationary (and often with the engine turned off) while waiting on a taxi rank.

The question included in the optimism scale was the same as reported in Dalziel and Job (1997a), that is, "How able would you be to do the following actions compared to an average taxi driver (same age and sex as yourself): To drive safety when very tired?" The question included in the risk-taking scale was "How often do you: Keep driving even when very tired?" Further details concerning the construction and psychometric properties of these scales is contained in Dalziel & Job (1997b).

RESULTS

Of 165 respondents, 163 were male, and 2 were female. Driver ages ranged from 21 to 68 years, with a mean of 43. Time holding a driving license ranged from 2 to 44 years, with a mean of 22, and time holding a taxi license ranged from 1 month to 35 years, with a mean of 10 years. Drivers with less than 50 shifts of experience (14 drivers, including both female drivers) were excluded from analyses that required some degree of experience as a taxi driver in order to be meaningful (such as frequency of falling asleep at the wheel and accident involvement).

The average time spent driving a taxi per week was 51 hours, with a range from 9 to 101 hours. Of this, 64% of drivers were on the road for at least 50 hours per week. When other (non-taxi driving) work is included, the average taxi driver works 58 hours per week, with a range from 16 to 112 hours. Of this, 9% of drivers worked less than 40 hours per week, and 38% of drivers worked more than 60 hours per week. Further working pattern details may be found in Dalziel & Job (1997b).

Fatigue problems

Of the 151 drivers with sufficient exposure and experience who returned surveys, 35 of them (23%) indicated that they had fallen asleep at the wheel at some stage during their taxi driving career. Of these 35 drivers, 27 had fallen asleep at the wheel more than once, and 14 indicated that they suffered from some form of sleeping disorder. Drivers answering yes to the "asleep at the wheel" question were asked to indicate the total number of times they had ever fallen asleep at the wheel in the past two years: responses were positively skewed, and ranged from 1 to 30 with a mean of 5.

For the question regarding sleeping disorders, 27 drivers (18%) indicated they experienced "sleep apnea, chronic snoring, or other major sleeping difficulties". Thirteen (50%) of these 27 drivers with sleeping disorders reported falling asleep at the wheel (one driver did not answer the second question), whereas only 22 (18%) of the remaining 121 drivers who reported no sleeping disorders had fallen asleep at the wheel, indicating that drivers with sleeping disorders appear significantly more prone to falling asleep at the wheel than those without sleeping disorders (chi-square = 11.9, $df = 1$, $p<.001$).

For the question regarding whether or not drivers had ever had an accident while driving home after work that was at least partially attributable to tiredness, five drivers (3%) indicated that this had occurred, whereas 80% had not experienced this event, and a further 17% indicated that they did not drive home after work. Of these five drivers, four of them had also reported falling asleep at the wheel while driving a taxi, and two of them indicated that they also suffered from sleeping disorders. When the answers to all three of these questions are combined (falling asleep at the wheel, sleeping disorders and tiredness-related accidents when driving home after a taxi shift), 33% of drivers surveyed acknowledged some form of fatigue-related difficulty.

Breaks

The average number of breaks taken per shift was 2, with a range from 1 to 5 (33% take one break, 34% take two breaks, 24% take three breaks, 5% take four breaks, and 4% take five or more breaks). The average total break time is 41 minutes per shift, with a range from 1 to 160 minutes. The average length of the first break is 22 minutes, followed by 19 minutes for the second, 16 minutes for the third, and 11 minutes each for the fourth and fifth breaks (although only 14 drivers reported taking more than 3 breaks per shift). Where drivers have only one break, this break is significantly longer than the first break of drivers who have three or more breaks (one break mean length = 25 minutes, three or more breaks, 1st break mean length = 16 minutes, $t = 2.9$, $p<.01$). If drivers take several breaks, the second and subsequent breaks are normally of equivalent length to, or shorter than, the first break. This indicates that, in general, drivers tend to take one long break or several shorted breaks during their shifts.

Fatigue, optimism bias and risk-taking

The average score on the optimism bias question concerning the ability to drive safely when very tired was not significantly different from the designated average value of four (mean = 4.24, $t = 1.58$, $p = .06$, 1 tailed test) indicating that taxi drivers as a group do not exhibit the typically strong optimistic bias concerning the issue of driving safely when very tired, compared to other questions. There was a significant difference between scores on this question and the other optimism bias questions (mean difference = .7, $t = 6.14$, $p<.001$). However, this difference was not due to the fatigue question being a separate construct to the others, as the correlation between scores on this question and total optimism bias scale scores was .75, and all questions clearly loaded on a single factor (see Dalziel & Job, 1997b for further details). None of the other variables examined in this study were significant predictors of scores on this question.

The mean score on the fatigue risk-taking question was 1, with a range of 0 to 5 (higher scores indicate greater frequency of risk-taking, 1 indicates "hardly ever"). As with the optimism bias questions, the fatigue question was not a separate construct to other risk-taking questions, but rather was correlated with overall scale scores ($r = .59$), and all risk-taking questions clearly loaded on a single factor (it should be noted that optimism bias and risk-taking factors were separate and unrelated, and the resultant scales were not correlated: $r = -.03$). There was no significant correlation between the individual fatigue and optimism questions ($r = .08$), nor was their a significant correlation between the fatigue optimism question and the overall risk-taking scale ($r = -.01$), or between the fatigue risk-taking question and the optimism scale ($r = -.09$). Overall risk-taking was significantly correlated with a number of other variables, including age, time holding a car and taxi license, driving style, aggression, and sensation seeking. The pattern of correlations between the individual fatigue question and these other variables was the same as for the whole scale.

There was no significant difference between drivers who reported falling asleep at the wheel ("asleep") and drives who reported not falling asleep at the wheel ("non-asleep") on the fatigue optimism bias question (mean "non-asleep" = 4.0, mean "asleep" = 4.3; $F(1, 143) = .94$, $p>.05$). However, there was a significant difference between "asleep" and "non-asleep" drivers for the risk-taking fatigue item (mean "non-asleep" = .8, mean "asleep" = 1.5; $F(1, 147) = 13.58$, $p<.001$). This difference in risk-taking between "non-asleep" and "asleep" drivers was significant for the overall risk-taking scale score (mean risk-taking scale score for "non-asleep" = 9.6, "asleep" = 13.1; $F(1,147) = 8.1$, $p<.01$).

To eliminate the possibility that this effect was due to a confounding of general risk-taking with the fatigue risk-taking question (ie, the overall scale score differences were due only to differences on the fatigue question), the risk-taking scale was re-analysed with the fatigue question removed. A significant difference was still observed based on the remaining nine items ("non-asleep" mean = 8.8, "asleep" risk-taking mean = 11.5; $F(1,147) = 6.48$, $p<.05$). There were no significant differences between the "asleep" and "non-asleep" groups on optimism bias, anger, sensation seeking, average number of breaks or average total break time. Finally, there was no significant different in risk-taking for the nine item scale between drivers with sleeping disorder and those without (mean sleeping disorders risk-taking = 11.0, mean non sleeping disorders risk-taking = 9.1, $F(1,147) = 2.5$, $p=.11$).

Fatigue and accidents

A total of 121 accidents were reported for the 2 year period studied, with an average of .8 per driver, ranging from 0 to 4. Seventy four driver (49%) had been involved in at least one accident. None of the accidents reported appeared to be due to "driver asleep", and only two of the accident descriptions provided by drivers mentioned tiredness.

There was no significant correlation between average total break time and accident rate ($r = -.08$, $p>.05$), despite our earlier finding of a correlation between this variable and accident rate (Dalziel & Job, 1997a), and it was not a significant predictor of accidents in the overall model (see below). There was also no significant difference in average break time by employment type.

Using regression analysis, four factors were determined to be predictors of accident involvement: anger, risk-taking, vehicle type and average number of hours per shift, and together these accounted for 21% of the variance. Of these, average number of hours per shift was the only variable directly related to fatigue issues. Average hours per shift was not significantly correlated with any of the variables examined, such as risk-taking, anger, age, time holding a car or taxi license, sensation seeking.

A significant difference in accident rate between "asleep at the wheel" ("asleep") and non-asleep at the wheel ("non-asleep") drivers was observed (mean accident rate for "asleep" = 1.2, "non-asleep" = .7; $F(1,147) = 5.41$, $p<.05$) but there were also significant differences for these two groups on both risk-taking (see above) and average hours per shift (mean average shift length for "asleep" = 11.5, "non-asleep" = 10.7; $F(1,144) = 4.47$, $p<.05$). When entered into the model to predict accident rate, asleep at the wheel did not make a significant contribution once risk-taking and average shift length had been taken into account, and therefore was discarded from the model.

DISCUSSION

The findings presented here provide further information about the role of fatigue-related variables in the experiences of taxi drivers, and present some new findings concerning the connections among these variables. Fatigue is clearly an issue of relevance to taxi drivers, as the findings here concerning sleeping problems, accidents and shift length suggest. However, the results do not simply indicate that fatigue is an independent factor in road safety, but rather that it is part of a complex web of relationships related, in particular, to risk-taking.

The relationship between taxi driver fatigue and accident rate appears complex. While drivers who have fallen asleep at the wheel at least once during their careers have higher accident rates than those who have not, fatigue alone may not be the major determinant of this finding, due to the relationship between this variable and both risk-taking and shift length. Increased shift length, in addition to being an exposure variable, may be considered a fatigue-related variable, for as shift length increases beyond 11-12 hours, even by only small amounts, the risk of involvement in an accident increase considerably (Folkard, 1996), probably due to the effects of exhaustion. This argument is supported by other evidence presented in Dalziel & Job (1997b) regarding increased accident rates towards the end of weekend night shifts - traditionally the longest shifts in the taxi driving week (based on New South Wales Roads and Traffic Authority data of all Sydney metropolitan taxi accidents during 1993-1995). These findings suggest two important questions: why drivers would put themselves at risk of the effects of fatigue by continuing to drive beyond average shift lengths, and why would drivers with experience of falling asleep at the wheel have higher, rather than lower, levels of risk-taking on both the fatigue question, and the overall risk-taking scale?

The answer may be that a relationship exists between fatigue and risk-taking that deserves further study. As noted in the introduction, fatigue may lead to greater accident involvement due to the inability of the fatigued individual to realistically assess their performance. However, it is also conceivable that fatigue may be the result of a more general disposition to take more risks when driving. The significantly higher risk-taking levels of drivers who have fallen asleep at the wheel may actually be the underlying <u>cause</u> of continued driving while fatigued, rather than being unrelated to fatigue incidents. That is, risk-taking while driving may not be related only to specific risky driving manoeuvres such as running a red light (Morgan & Job, 1995), but also to more general behaviours such as continuing to drive in spite of tiredness. If this were the case, drivers who exhibit low levels of risk-taking while driving would be likely to stop driving when they first begin to feel the effects of tiredness, while drivers who exhibit high levels of risk-taking while driving may continue to drive despite tiredness. This would explain the finding that overall risk-taking is higher in drivers who have experienced falling asleep at the wheel.

It should be noted, however, that this increased risk-taking is not the product of optimism bias concerning an individual's driving abilities. In the present study, optimism bias about general driving abilities appears unrelated to actual risk-taking on the road, which argues against the theory that increased optimism bias causes risk-taking, and hence more accidents (at least for experienced drivers). This finding holds for the specific issue of fatigue as well: drivers who are optimistic about their ability to drive safely while very tired are no more likely to continue driving when very tired, or to fall asleep at the wheel than drivers who are not optimistic about this ability. This finding should also be viewed in the light of the overall lack of optimism among taxi drivers concerning the ability to drive safety when fatigued, compared to other skills-based abilities.

Driver countermeasures concerning fatigue

Drivers who experience some form of sleeping disorder (such as sleep apnea, chronic snoring or other major sleeping disturbances) are significantly more likely to have fallen asleep at the wheel than those who do not. Approximately 50% of drivers with sleeping disorders have fallen asleep at the wheel, whereas only 14% of drivers without sleeping disorders have fallen asleep at the wheel. While this 50% is a relatively small number of total drivers (14 of 151), presumably due to the low prevalence of sleeping disorders overall, it indicates that road safety countermeasures regarding fatigue could appropriately be targeted at this specific group (taxi drivers with sleeping disorders), due to the fact that half of them have fallen asleep at the wheel at some stage, and that many fall asleep at the wheel a number of times. These countermeasures could include further driver education, enforcement of regulated shift times and/or break periods, penalties for excessive time periods on the road, and technological fatigue countermeasures (as these devices become sufficiently advanced).

The findings concerning the relationship between fatigue incidents and risk-taking is also an important findings for countermeasures concerning fatigue. In the case of taxi drivers, it appears that most drivers have a realistic (rather than optimistic) assessment of their ability to drive safely when very tired. Even where individual drivers do have an optimistic view of their abilities, this does not appear to be the predictor of fatigue incidents. Rather, drivers who generally take more risks while driving are more likely to experience the effects of fatigue. Thus, in countermeasures directed at taxi drivers concerning fatigue, it is not necessary to try to convince drivers that fatigue is a genuine problem that may affect them - taxi drivers seem to already acknowledge this issue (as indicated by the difference in optimism scores between fatigue and other skill-based abilities). However, to attempt to reduce the effects of fatigue on drivers, a more general approach to reducing risk-taking while driving is needed. This needs to address two potential causes of risk-taking: personality (anger and sensation seeking) and motivational factors (such as the need for earnings). Each of these appears to contribute to risk-taking among taxi drivers, and hence if the effects of these factors can be reduced, then overall risk-taking may be lowered, which would lead to reduced problems with fatigue.

Finally, as one of the causes of driving behaviours which lead to fatigue (such as continuing to drive while tired) appears to be a general predisposition to risk-taking while driving, this finding may be of value to all road safety research on fatigue, not just research in relation to taxi drivers.

REFERENCES

Brown, I. D. (1994). Driver fatigue. Human Factors, 36, 298-314.

Corfitsen, M. T. (1993). Tiredness and visual reaction time among nighttime cab drivers: A roadside survey. Accident Analysis and Prevention, 25, 667-673.

Corfitsen, M. T. (1994). Tiredness and visual reaction time among young male nighttime drivers: A roadside survey. Accident Analysis and Prevention, 26, 617-624.

Crawford, A. (1961). Fatigue and driving. Ergonomics, 4, 143-154.

Dalziel, J. R. & Job, R. F. S. (1994). A reconsideration of the relationship between driving experience, accident record and optimism bias. Presented at the 23rd International Congress of Applied Psychology, Madrid, Spain.

Dalziel, J. R., & Job, R. F. S. (1997a). Motor vehicle accidents, fatigue and optimism bias in taxi drivers, Accident Analysis and Prevention, 29, 489-494.

Dalziel, J. R. & Job, R. F. S. (1997b). Taxi Drivers and Road Safety. Report to the Federal Office of Road Safety, Australia.

Fell, D., & Black, B. (1996). Driver fatigue in the city. Proceedings of the Second International Conference on Fatigue and Transportation: Engineering, enforcement and education solutions. Canning Bridge: Promaco.

Folkard, S. (1996). Black Times: Temporal Determinants of Transport Safety. Presentation at the 2nd International Fatigue and Transport Conference, Fremantle, Australia.

Morgan, G.A. & Job, R.F.S. (1995). Red light cameras: drivers knowledge, attitudes and behaviours. In: D. Kenny & R.F.S. Job (Eds.). Australia's Adolescents: A Health Psychology Perspective. (pp 144-150) Armidale, NSW: New England University Press.

National Aeronautics and Space Administration (1995). Fatigue Resource Directory. Proceedings of the Second International Conference on Fatigue and Transportation: Engineering, enforcement and education solutions. Canning Bridge: Promaco.

16

UPDATE ON THE U.S. FHWA COMMERCIAL DRIVER FATIGUE RESEARCH & TECHNOLOGY, RULEMAKING, EDUCATION/OUTREACH, AND ENFORCEMENT PROGRAM

Ronald R. Knipling, Ph.D., Chief, Research Division, Office of Motor Carrier Research and Standards, Federal Highway Administration, U.S. Department of Transportation, Washington, DC

INTRODUCTION

The FHWA of the U.S. Department of Transportation regulates and supports the Nation's interstate commercial motor carrier industry. The mission of the FHWA's OMC is to help move people, goods, and commercial vehicles on our Nation's highways in the most efficient, economical, and crash-free manner possible. OMC is a leader in identifying and promoting new technologies which enhance motor carrier performance and safety. In order to accomplish our charge, OMC has placed a priority on human factors research, with a special attention given to driver fatigue and hours-of-service.

Driver fatigue is a safety issue of special concern to commercial motor vehicle transportation. CMV drivers may drive up to 10 hours continuously before taking a break, often drive at night, and sometimes have irregular and unpredictable work schedules. Much of their mileage is compiled during long trips on Interstate and other four-lane roadways. Because of their far greater mileage exposure and other factors, commercial drivers' risk of being involved in a fatigue-related crash is far greater than that of non-commercial drivers -- even though CMV drivers represent only about 4 percent of the drivers involved in known fatigue-related crashes and their rate of involvement per mile traveled is no greater than that of non-commercial drivers. In addition, other "competing" crash factors such as alcohol, speeding, and other unsafe driving acts are less common among commercial drivers and thus less important *relative* to fatigue.

The 1995 FHWA-sponsored Truck and Bus Safety Summit and other industry conferences have identified driver fatigue as the top-priority commercial motor vehicle safety issue. The OMC supports this designation and has allocated its resources accordingly. Driver fatigue is a primary focus in four major OMC program areas: research and technology, rulemaking, education/outreach, and enforcement/consultation.

Research And Technology (R&T)

Driver drowsiness/fatigue is the dominant human factors research issue in the OMC R&T program. Altogether, OMC has more than 20 recently-completed, ongoing, or planned R&T projects relating to driver drowsiness/fatigue and HOS. These are described below.

Completed R&T

Driver Fatigue and Alertness Study (DFAS). The DFAS, performed by the Essex Corporation, was the most comprehensive over-the-road study of commercial driver alertness ever conducted. It was a collaborative effort involving FHWA's OMC, Transport Canada, the Trucking Research Institute (TRI) of the American Trucking Associations (ATA) Foundation, three motor carriers, and other research and industry organizations.

The study involved real revenue runs, four different driving schedules, 80 drivers, and more than 200,000 miles of highway driving. Numerous measures were taken of the drivers' physiology, alertness, and performance during driving and of their physiology during off-duty sleep periods. They included driving task performance (e.g., lane tracking), performance on microcomputer-based vigilance tests, continuous video monitoring of the drivers' face and the road ahead, and physiological measures (e.g., "brain waves") during both driving and sleep.

The DFAS results are major scientific inputs to the current re-examination of FHWA's 60-year-old driver HOS regulations. Major findings included:

- Driver alertness and performance were more consistently related to time-of-day than to time-on-task. Driver drowsiness episodes were eight times more likely between midnight and 6:00 a.m. than during other times.

- Drivers in the study did not get enough sleep compared to their "ideal" sleep needs.

Drivers obtained an average of about two hours less sleep than their daily "ideal" requirements.

- Drivers' stated self-assessments of their levels of alertness do not correlate well with objective measures of performance. Drivers were not very good at assessing their own levels of alertness.

- There were significant individual differences among drivers in levels of alertness and performance.

The DFAS Executive Summary is available through the OMC home page (**www.fhwa.gov/omc/es-5g.html**). The 60-page Technical Summary (PB 97-129688) and the 562-page project report (PB 98-102346) are available from the National Technical Information Service (NTIS), 5285 Port Royal Road, Springfield, VA, 22161, 703-487-4828. (OMC Project Manager: Deborah Freund, 202-366-5541).

CMV Rest Areas: Making Space for Safety. This recently-completed TRI study determined what public rest area and privately owned rest stop services are needed by CMV drivers and how well the current system meets these needs/demands. The study documented a significant shortage of rest area parking for commercial vehicles and drivers. It also revealed private sector efforts to expand spaces at truck stops to meet this need. Partly in response to the study, about half of U.S. States now permit their CMV weigh stations to remain open as rest areas when they are not being used as weigh stations. The U.S. Congress added in the 1995 National Highway System Designation Act a provision for 100% Federal funding of safety rest area modification and construction, a measure expected to stimulate the construction of expanded rest areas for trucks. The final study report (PB 97-124705) is available from NTIS). (OMC PM: Bob Davis, 202-366-2981).

Multi-Trailer Combination Vehicle Stress and Fatigue. FHWA and the National Highway Traffic Safety Administration (NHTSA) co-sponsored this study, performed by Battelle, of the effect of multiple-trailer combination vehicle (MTCV) operation on driver stress and fatigue. Its goal was to determine whether there are differences in driver alertness and performance arising from driving single-trailer versus two different types of triple-trailer combinations: those employing A-dollies and those employing C-dollies. Twenty-four experienced MTCV drivers each drove 6 round trips (2 with each configuration). Total mileage per driver was about 2,700 miles. Trailer configuration was found to affect driver stress/fatigue as measured by lanekeeping, driver subjective workload, and physiological state. Task demands were greatest with triple/A-dollies, followed by triple/C-dollies and, last, single-trailers. However, stress/fatigue differences relating to trailer configuration were small compared to the individual differences among drivers. Alertness correlated much more highly with driver individual differences than with vehicle configuration. A summary report on this study was submitted to Congress in March 1996. The final report will be available in early 1998. (OMC PM: Deborah Freund, 202-366-5541).

Conference on Driver Vigilance Monitoring. In December, 1996, OMC and TRI jointly sponsored a conference focusing on technological approaches to counteracting fatigue, with emphasis on in-vehicle continuous monitoring of driver alertness/ performance and actigraphic monitoring of driver sleep/wakefulness. The conference included 20 presentations by FHWA and NHTSA officials, leading researchers, and device inventors/vendors. Topics addressed included basic research findings on driver alertness, updates on technology, and strategies for future deployment. Conference proceedings are available from Dr. Bill Rogers, TRI, (703) 838-7912. Similar future conferences are under consideration.

Conference on Managing Fatigue In Transportation. This international, multi-modal conference, jointly sponsored by FHWA, ATA, NHTSA, the Association of American Railroads, the Federal Railroad Administration, and the National Transportation Safety Board (NTSB), addressed ways to improve transportation operator alertness and lower crash risk. Speakers and panelists from the U.S., Australia, Canada, and Sweden addressed a variety of topics related to the improvement of operator fatigue management, including improving sleep, monitoring operator alertness, alternative approaches to HOS regulation, and new methods and technologies in fatigue management. Conference proceedings are available from Government Institutes, Inc.; 4 Research Place, Suite 200, Rockville, MD 20850, (301) 921-2355.

Assessment of Electronic On-Board Recorders for HOS Compliance. This project, performed by the University of Michigan Transportation Research Institute (UMTRI) through a contract with the Private Fleet Management Institute (PFMI) of the National Private Truck Council (NPTC), assessed the costs and benefits of the use of electronic on-board recorders (EOBRs) for compliance with the HOS regulations. Average EOBR acquisition and installation costs averaged approximately $2,000 per vehicle; annual operating and maintenance costs were about $200 per vehicle. EOBRs use and benefits vary widely for different segments of the motor carrier industry; almost all current use is by private fleets. The benefits associated with electronic HOS recording lie largely in the time savings for drivers in maintaining HOS logs.

These savings averaged about 20 minutes per driver per day. Also, managers of fleets using EOBRs saved an additional 20 minutes per driver per month in management review and administration time. (OMC PM: Bryan Price, 202-366-5720).

Shipper Involvement in HOS Violations. Congress directed the FHWA "to determine the scope, nature, and extent of shipper involvement in driver noncompliance with the safety regulations." This study employed focus groups to generate qualitative data about shipper demands on motor carriers and drivers. The study found that pick-up and delivery demands by shippers do lead to HOS violations, but that *all* involved parties -- receivers, shippers, brokers, schedulers, dispatchers, and drivers themselves -- contribute to the problem of HOS noncompliance. Moreover, all of these parties have a role to play in resolving the problem. This study was completed by Global Exchange, Inc. in December, 1997. (OMC PM: Elaine Riccio, 202-366-2981).

Local/Short Haul Driver Fatigue Crash Data Analysis. This small analytical study, performed by UMTRI, developed several definitions of local/short haul versus over-the-road trucks and examined the prevalence of driver fatigue as a principal factor in truck crashes. Data sources included the 1992 Truck Inventory and Use Survey and 1991-93 Trucks Involved in Fatal Accidents files. Not surprisingly, trip distance was found to have a major effect on the percentage of fatal crashes that were fatigue-related; shorter trips are associated with a much lower incidence. The risk of local/short haul truck involvement in fatigue-related fatal crashes is a fraction of that of over-the-road trucks. The project final report will be available in February, 1998. (OMC PM: Ron Knipling, 202-366-2981).

Ongoing R&T

Driver Work/Rest Cycles and Performance Modeling of HOS Alternatives. The objectives of this study are to (1) gather data on representative wake-sleep cycles of CMV drivers operating in real work settings; (2) determine quantitative relationships among driving task performance and drivers' physiological and subjective responses; and (3) extend and validate a numerical model to predict performance based on prior wake-sleep cycles, sleep quality and quantity, and circadian state for a next-generation wrist-worn activity monitor or actigraph. The study will also provide information concerning potential use of actigraphs to improve the management of CMV driver fatigue.

Both field and laboratory data collections are complete. The field study involved 25 local and

25 long-distance drivers who wore actigraphs during their normal activities for up to 21 days. The laboratory data collection consisted of 15-day lab studies of 60 drivers using computer-based driving simulators, physiological and performance monitoring, and performance tests. The drivers were allowed 3, 5, 7, or 9 hours of sleep daily. Extensive data were collected on driving performance, physiological state, psychomotor and cognitive performance, and subjective self-assessments. The numerical model has undergone extensive first-round testing and assessment. Additional refinements will be made based upon the results of the laboratory study. Project findings will be published during 1998. This project has provided unprecedented opportunities for cross-modal (FHWA, Federal Aviation Administration, Federal Railroad Administration) and cross-agency (DOT, Department of Defense, National Institutes of Health) coordination and resource sharing. (OMC PM: Deborah Freund, 202-366-5541)

CMV Driver Sleep Apnea. In response to Congressional direction, OMC is obtaining an estimate of the prevalence of sleep apnea in a population of truck drivers who may be at high risk for the disorder, and estimating the level of sleep apnea at which driving impairment becomes important. Study data may be used to raise industry awareness of the need for screening, diagnosis, and treatment of sleep apnea among drivers. The data may also form the basis for future research to identify remedial measures including new screening and detection technologies. The University of Pennsylvania Medical Center is performing this project through TRI. (OMC PM: Sandra Zywokarte, 202-366-2987)

Fitness-for-Duty Testing. Phase I of this study documented the concept and feasibility of employing in-terminal and in-vehicle testing devices for accurately determining the fitness of CMV drivers to safely operate their vehicles. Phase I results were reported in FHWA publication #FHWA-MC-95-011. The Phase II field testing, currently in progress, is integrating a continuous lane-tracking monitor with the fitness-for-duty test. This work is being performed by Evaluation Systems, Inc. through TRI. (OMC PM: Kate Hartman, 202-366-2742)

Crash Investigation Project/Crash Causation Study. OMC has contracted with UMTRI to perform a crash causation study employing a sample of fatal CMV crashes and a full taxonomy of crash causes (including drowsiness/ fatigue as well as other forms of driver inattention). Crash investigation reports are being collected from State special investigation files to determine the degree to which representative samples can be generated and crash causal factors reliably identified. (OMC PM: Ralph Craft, 202-366-0324)

Scheduling Practices and their Influences on Driver Fatigue. This TRI study will survey a representative sample of CMV drivers, carriers and shippers to determine operational scheduling requirements and recommended practices from the primary standpoint of fatigue management. Currently in progress are focus group sessions to identify key issues for the survey. A planned outcome will be a symposium addressing scheduling practices for safer trucking operations. The project subcontractors are Western Highway Institute and Iowa State University. (OMC PM: Philip J. Roke, 202-366-5884)

Effect of Loading and Unloading on Driver Fatigue. Phase I of this study involved focus groups, a driver survey and interviews to understand the loading/unloading requirements across the industry. The amount of physical labor associated with long haul driving varies, depending upon type of cargo and other factors. For bus drivers, loading/unloading luggage was not a significant fatigue factor. For many trucking segments, drivers only supervise loading/unloading. Far more complaints were heard about fatigue from lengthy waiting periods associated with loading/unloading than from physical activity *per se*. Drivers who actually perform loading/unloading were concentrated in two commodity groups: household goods movers and grocery haulers. These groups reported that they often become fatigued during and after the loading/unloading tasks. In some cases they claimed that this affects their driving alertness.

Phase II, currently in progress, is a driving simulator-based study of the effects of the physical activity of loading/unloading on subsequent driver alertness. This experiment will also measure and document drivers' "weekend" rest/recovery process over the 58-hour periods between successive weeks of simulated driving. TRI and Star Mountain, Inc. are the study contractors. (OMC PM: Bob Carroll, 202-366-9109)

Local/Short Haul Driver Fatigue Human Factors Study. The Virginia Polytechnic Institute and State University Center for Transportation Research (VPISU CTR) is conducting this 3-year study to employ focus groups and direct observation to determine the role played by drowsiness/fatigue and inattention in driver errors and incidents in local/short haul truck operations. The focus group phase of the study has been completed. The top five ranked safety issues were: problems caused by drivers of private vehicles, stress due to time pressure, inattention, problems caused by roadway/dock design, and fatigue. Fatigue was identified as a critical safety issue by about one-third of the focus groups. When questioned specifically about fatigue, drivers identified 22 issues. The top five ranked fatigue-related issues were: not enough sleep, hard/physical workday, heat/no air conditioning, waiting to unload, and irregular meal times. Instrumented vehicle data collection will begin next Spring and will include a capability for driver error/incident capturing and determination of antecedent conditions, including fatigue. (OMC PM: Bob Carroll, 202-366-9109).

Validation of Eye and Other Psychophysiological Monitors. This OMC Intelligent Transportation Systems/Commercial Vehicle Operations (ITS/C.O.)-funded effort is being managed by the NHTSA Office of Crash Avoidance Research. Under the program, the University of Pennsylvania has conducted laboratory experiments to evaluate the validity, sensitivity, and reliability of selected personal (psychophysiological) fatigue detection devices and measures, including eye closure measures such as PERCLOS, a measure of eyelid droop identified in earlier NHTSA research as being a promising index of fatigue. Preliminary results corroborate the validity of PERCLOS and related eyelid droop measures; these findings will be announced in early 1998. Follow-up field testing of promising devices and measures is planned. (PM: Paul Rau, NHTSA OCAR, 202-366-0418).

Sleeper Berths and Driver Fatigue. This three-year study, begun in May 1997, will determine the effects of sleeper berth use on driver alertness and driving performance. It will assess the quality of rest achieved while vehicles are stationary and in motion and evaluate the effects of irregular schedules and sleeper berth usage patterns on driver alertness and performance. VPISU CTR is conducting this work. Preliminary results from driver focus groups, currently in progress, indicate that sleeper berth usage is often constrained by the driver's perception of the safety and privacy of available parking. Following completion of the focus groups, this study will gather empirical data using instrumented vehicles and standard sleep monitoring methodologies. (OMC PM: Bob Davis, 202-366-2981)

Ocular Dynamics as Predictors of Driver Fatigue. This driving simulator-based study will determine whether directed eye movements (saccades) and other eye activities can be monitored as "leading indicators" of fatigue; i.e., measures obtainable *before* a driver reaches a dangerous level of fatigue. Recent advances in eye tracking make this technically-feasible. Applied Science Laboratories and the Institute for Circadian Physiology, under subcontracts to TRI, are also gathering data on napping as a fatigue countermeasure. Data collection is now underway; preliminary results are supporting the concept of early ocular indicators of fatigue. (OMC PM: Bob Carroll, 202-366-9109).

CMV Crash Rate by Time-of-Day. This small analytic study, performed by UMTRI in support of the current HOS rulemaking, will access available crash data files and information on CMV mileage exposure to determine the CMV crash involvement rate (per mile traveled) by time-of-day. This issue is relevant to current OMC deliberations on CMV HOS and may also provide useful information to help carriers schedule truck trips to minimize crash risk. (OMC PM: Ron Knipling, 202-366-2981).

Planned/Proposed R&T

Driver Compensation Practices and Safety. This analytical study will seek to determine whether and how driver pay compensation method (e.g., by-the-hour, by-the-mile) influences CMV safety as measured by such indices as crash rate and citations (e.g., HOS violations). The Phase 1 feasibility study will determine whether available data will provide definitive findings. Contract award is imminent. (OMC PM: Chuck Rombro, Analysis Division, 202-366-5615).

Intelligent Vehicle Initiative Driver Monitoring Research. Driver monitoring is a key component in the new ITS Intelligent Vehicle Initiative (IVI). Plans are underway for an operational test of continuous driver monitoring systems as well as other on-board monitors (e.g., on-board diagnostics, "black boxes") beginning in FY'99. The heavy truck portion of the IVI will be managed by the OMC ITS/C.O. division in collaboration with other DOT modal administrations. The planned operational test will foster the ability of both system suppliers and truck users to commercially deploy effective on-board technologies. (OMC PM: Kate Hartman, 202-366-2742).

Additional R&T programs planned or under consideration include development of a national data system for fatigue-related and other human factors research, a study on the effectiveness of naps in enhancing alertness, conferences on fatigue management, and additional research on technological countermeasures to fatigue. The latter study is expected to include research on individual differences in susceptibility to fatigue,

RULEMAKING

OMC applies its research results to the development of cost-effective and safety-promoting CMV regulations. An Advance Notice of Proposed Rulemaking (ANPRM) on CMV HOS was issued in late 1996 (61 Federal Register 57252); the comment period closed in mid-1997. This HOS rulemaking will likely result in the first major changes to the CMV HOS in nearly 60 years. The rulemaking addresses the potential for both conventional HOS rules and performance-based alternatives, with the latter including both fleet management and individual monitoring approaches to performance-based regulation. An advantage of performance-based approaches is the potential for synergy between these approaches and driver awareness of their alertness changes. The introduction of valid fleet safety and/or individual driving performance measures and the provision of this information as feedback to drivers will likely increase their awareness of their alertness and of sleep hygiene practices which affect alertness and performance. Of course, the effectiveness of alertness monitors or other performance-based approaches must be demonstrated before their use is permitted as an alternative to prescriptive HOS.

Although OMC has set its long-term sights on performance-based rules, the current focus is on prescriptive HOS rules. Critical HOS-related issues include:

- Maximum driving time (currently 10 hours).
- Minimum off-duty time (currently 8 hours).
- Work cycle implications of the above (current rules encourage an 18-hour cycle for maximum productivity in long-haul operations)
- Distinctions between driving and non-driving duty time; currently, the maximum continuous on-duty time is 15 hours, of which 10 hours may be driving.
- Day/night differentials to encourage day driving (currently there are none).
- Maximum cumulative on-duty hours (currently, 60 hours in 7 days or 70 hours in 8 days).
- Providing flexibility for different CMV operations and situations, while at the same time ensuring adequate daily and weekly rest for all CMV drivers.
- Special provisions; e.g., current sleeper berth provision allowing splitting of off-duty hours.

These and other issues are being addressed in the HOS rulemaking. A Notice of Proposed Rulemaking (NPRM) containing specific proposed prescriptive HOS rule changes is planned for mid-1998. In this and other regulatory initiatives, the agency seeks to achieve "win-win-win" outcomes for public safety, CMV productivity, and driver quality-of-life. The FHWA firmly believes that the current HOS rules can be improved in all three respects.

For information on the HOS rulemaking, contact David R. Miller, (202) 366-1790, FAX (202) 366-8842. Internet users may access the HOS ANPRM at the Federal Register web page at: **www.access.gpo.su_docs**.

EDUCATION/OUTREACH

Together with the ATA, NPTC, and other industry partners, the OMC has undertaken an active fatigue education/outreach effort. This outreach program uses a multimedia approach to inform a wide range of audiences -- the general public, motor carriers, professional truck driver associations, and truckers themselves -- about the hazards of driving while fatigued. A major goal is to educate all seven million commercial drivers license holders on how to recognize fatigue and about the importance of adequate rest and healthy work and lifestyle choices.

OMC's Fatigue Outreach project has provided funding to the TRI for several major information/education initiatives. This has included the development and distribution of "Awake at the Wheel" public service announcements to radio stations nationwide. In addition, nearly 1,000,000 "Awake at the Wheel" print brochures have been printed and are being distributed to truck drivers, carriers, and other organizations involved in motor carrier safety. Under the same program, TRI has produced a video to educate truckers and their families about fatigue and the importance of adequate sleep. A train-the-trainer instructional program has also been developed and is being conducted for fleet safety managers and truck driver training personnel. The OMC Fatigue Outreach manager is Dave Longo, 202-366-0456.

In addition to educating drivers on fatigue and safety, we want them to understand and appreciate how sleep, diet, exercise, and lifestyle contribute to a personal state of "wellness" and enhanced driving performance. Through the TRI and the PFMI, FHWA is funding the development of a fleet-based driver wellness program. The project will determine the critical issues and concerns relevant to driver wellness, including ways to change personal attitudes and self-perceptions of wellness, and will lead to several model wellness programs that can be adapted and integrated into company operations. Health and wellness information will also be disseminated for use by drivers and their families. The OMC PM is Albert Alvarez, 202-366-4766.

FHWA is also encouraging the installation of exercise facilities for drivers at private truck stops near Interstate highways. A new program is assessing the level of use of these facilities and, indirectly, their economic viability as a truck stop amenity to attract drivers. A major obstacle to better fitness and health for over-the-road drivers is the lack of available fitness facilities. The first two facilities have opened in at Pilot Truck Stops on I-40 in Little Rock, AR and Oklahoma City, OK. This initiative is being performed in association with TRI and the Rolling Strong Company. This innovative government-industry partnership has received widespread press and media exposure including stories by People Magazine and ABC's Good Morning America. The OMC PM is Jerry Robin, 202-366-2985.

Finally, OMC has worked cooperatively with the ATA to provide low-cost training to managers of small and new interstate fleets on how to comply with the Federal Motor Carrier Safety Regulations (FMCSRs). This includes compliance with regulations on driver qualifications and HOS/log book compliance. In 1997, 35 one-day seminars on FMSCR compliance were conducted at locations across the country. For information on these programs, contact Jai Kundu, ATA, 703-838-1852.

ENFORCEMENT/CONSULTATION

OMC, with its state partners, uses a variety of enforcement tools to ensure motor carrier and driver compliance with HOS and log book requirements as prescribed in the FMCSRs. Each year, more than 14,000 hours-of-service and log book violations are cited during compliance reviews of motor carrier operations. Carriers in compliance, but still experiencing a high crash rate, are provided with consultation regarding specific crash countermeasures and management approaches for minimizing driver fatigue.

Roadside driver/vehicle inspections performed under the Motor Carrier Safety Assistance Program annually result in more than 100,000 drivers being placed out of service for HOS or log book violations. This is about 3/4 of all driver out-of-service violations and represents about 5% of all drivers inspected annually.
More severe enforcement actions are taken against carriers or drivers with repeated or egregious violations. Usually, such cases result in fines or other civil penalties, but criminal sanctions are applied to extreme violators. In one recent case, Federal criminal charges were brought, and a conviction obtained, against a carrier that forced drivers to falsify logs and violate HOS rules. In another case, a $60-million-a-year aluminum/steel hauler was ordered by FHWA to cease interstate operations because of chronic HOS and log book violations.

An anticipated future trend, consistent with OMC's fatigue education/outreach and FMCSR compliance programs, is a greater emphasis on the consultative role of OMC Safety Inspectors in helping to prevent and correct driver fatigue and HOS-related problems before enforcement actions against carriers are necessary.

Relevant Driver Fatigue Programs Of The National Highway Traffic Safety Administration (NHTSA)

The National Highway Traffic Safety Administration (NHTSA) of the U.S. DOT has two major program areas relating to driver fatigue. In the area of Traffic Safety Programs, NHTSA is responding to a U.S. Congressional directive to develop and evaluate a drowsy driving public information & education program. In its R&D program, NHTSA is developing and testing a prototype drowsy driver detection system and related ITS countermeasures, as well as performing statistical analyses of the drowsy driver crash problem.

Public Information and Education

NHTSA's drowsy driving public information and education program is directed toward high-risk groups within the non-commercial driving population. Elements of this program include the following:

- Analyze the role of fatigue, sleep disorders, and inattention in highway crashes. This work is being performed in cooperation with the National Center on Sleep Disorders Research of NIH. A report will be available in the Spring of 1998.
- Investigate the incidence and nature of fatigue in motor vehicle operation.
- Develop a strategy for a public information campaign.
- Develop, test, and evaluate an information/ education program.
- Implement the validated information/education program on a broader scale. A grant announcement for this work is anticipated in the Spring of 1998.

For further information on the NHTSA driver fatigue public information & education program, contact Dr. Jesse Blatt of the NHTSA Office of Research & Traffic Records (202-366-5588).

Crash Avoidance Research & Development

NHTSA's R&D program on drowsy/fatigued drivers focuses on the development of a vehicle-based drowsy driver detection and warning system. NHTSA is supporting R&D on detection algorithm refinement, sensor development, and the driver interface; i.e., advisory messages and alerting stimuli. The project has field-tested a prototype system for use in trucks. Over-the-road data has been collected on system performance and useability; these data are currently under analysis and will be published during 1998.

NHTSA R&D is also sponsoring a Small Business Innovation Research (SBIR) grant to demonstrate the feasibility of an unobtrusive driver eyelid droop monitor. Eyelid droop has been found in other NHTSA-sponsored research to be a reliable and valid measure of fatigue.

The convergence of ongoing ITS R&D related to other classes of crashes will also have the potential to reduce drowsy driver crashes. ITS projects include programs on single vehicle roadway departure crashes and rear-end crashes, and the ITS concept of Automatic Collision Notification intended to speed the delivery of emergency medical response to crashes.

NHTSA's research program seeks to better assess the driver fatigue problem. Analyses of NHTSA crash databases have improved our understanding of crash characteristics and, in particular, have led to better estimates of crash problem size. Ultimately, however, direct observation of drivers using in-vehicle monitoring devices will provide the most valid data on driver drowsiness. NHTSA has developed, and is deploying, a sophisticated, unobtrusive instrumentation suite in test vehicles to obtain "real world" data on safety-related driver performance, behavior, and alertness.

Further information on the NHTSA program may be obtained from Dr. Paul Rau of the NHTSA Office of Crash Avoidance Research (202-366-0418).

CONCLUSION

CMV driver fatigue is a complex issue and can be viewed in several different ways. The commercial driver, like any transportation operator, is performing a complex sensory-motor task in a vehicle. A classic problem for human factors research is to understand operator errors and devise operational or design changes to prevent them. Physiological and behavioral studies have revealed ways that driver alertness and performance deteriorate due to fatigue, and how intelligent devices may potentially act to detect and prevent this deterioration. Driver off-duty lifestyle and sleep habits play a key role in determining alertness and performance while on-duty driving. At the level of the physical transportation environment -- such as the highway -- design changes such as rumble strips may prevent crashes due to operator fatigue. CMV drivers also perform their work against the backdrop of an operational environment including government HOS regulations, penalties for violations, and company management practices including selection, training, scheduling, and incentives for safe performance. To the greatest extent possible, the operational environment must support alert driving. At the broadest conceptual level, there is the cultural environment of public information and attitudes about the role of sleep and alertness in safety. The benefits of sleep must be valued by society just as we value the benefits of exercise and a healthy diet. FHWA OMC's driver fatigue programs reflect this multi-level conception of driver fatigue and ways that industry, government, and society can work together to improve fatigue management.

FURTHER INFORMATION

For further information on the driver fatigue R&T programs of the FHWA/ OMC, please contact Ronald R. Knipling, Ph.D., Chief, Research Division (HCS30), Office of Motor Carrier Research and Standards, FHWA, U.S. DOT, 400 Seventh Street, SW., Washington, DC 20590; Phone (202) 366-2981; FAX (202) 366-8842; e-mail **ron.knipling@fhwa.dot.gov**. The FHWA OMC is also interested in receiving copies of new research relating to driver fatigue. Please send such reports to the address above, or fax/e-mail them to OMC, attn: Ron Knipling.

FHWA OMC maintains a web site providing information on the full range of its regulatory and other activities to promote truck and bus safety. The address is: **www.fhwa.dot.gov/omc/omchome.html**.

DOT-wide activities related to transportation operator fatigue are described in the DOT's Fatigue Resource Directory at **www.dot.gov/general/human-factors/fredi**. The directory contains a wealth of information on government and non-government fatigue-related R&T and educational programs.

17

Fatigue Among Ship's Watchkeepers: A Qualitative Study Of Incident At Sea Reports

Richard Phillips, Division of Community and Rural Health, Faculty of Medicine, University of Tasmania

Introduction

The Tavistock Institute of Human Relations (Bryant 1991) identified that the human element was found to be present in over 90 per cent of collisions and in over 75 per cent of contacts, fires and explosions. The United Kingdom Mutual Steam Ship Assurance Association (Bermuda) Ltd (representing 20 to 25% of the world fleet) analysis of major claims identified that the most frequent cause of claims was that of deck officer error, encompassing 27% of all claims from 1987 to 1992 (The United Kingdom Mutual Steamship Assurance Association (Bermuda) Ltd 1993).

This study focuses on a content analysis of 100 Australian Department of Transport Marine Incident Investigation Unit Incidents at Sea Reports (currently 128 reports), looking for evidence of fatigue-induced behaviours among ship's watchkeepers. The role of the ship's watchkeeper is defined by the 1995 International Convention on Standards of Training, Certification and Watchkeeping for Seafarers (STCW) and the National Competency Standards for the Maritime Industry.

The potential for error in the navigational duties of watchkeepers has been described by Bryant (1988) and Brown (1989), where the watchkeeper is defined as a human operator in a complex technological system. Bryant (1991) observes that separation of the human and technological elements of complex systems is difficult in practice and argues that "seafaring is a way of life that cannot be separated from the physical environment and the technology that supports it". Similar views are expressed by Nitka and Dolmierski (1987), Nitka (1989 and 1990), and Dolmierski et al. (1990).

The central problem in any analysis of fatigue and accidents is *whether fatigue contributes to incidents and, if so, to what extent?* With the advent of decreased manning levels and increased automation within the maritime industry, this problem has increasing relevance. The difficulty lies in attempting to reconstruct and understand the reality of fatigue-induced incidents from retrospective analyses, given the difficulty that human actors exhibit in describing and interpreting such an indefinite concept as fatigue (McCallum et al. 1996, Brown 1990). The problem facing marine investigators is establishing what is meant by fatigue and differentiating between various manifestations of fatigue as contributing causes of accidents.

METHODOLOGY

Two methodological approaches were used in obtaining and analysing data for this study. First, a model of fatigue behaviours was developed from a review of the scientific literature (Table 1). This model was subsequently applied to the second methodological component of this study - a content analysis of Incidents at Sea Reports (IASR). Content analysis, as defined by Krippendorff (1980), is "a research technique for making replicable and valid inferences from data to their context".

The summaries and conclusions of 100 consecutively published IASRs were analysed using the non-numerical unstructured data indexing, searching and theorising software NUDIST (Version 4.0 - Qualitative Solutions and Research Pty Ltd). NUDIST is a program designed for the storage, coding, analysis and retrieval of text and is regarded by Weitzman and Miles (1995) as "one of the best thought out qualitative analysis programs around". (For a discussion on the use of computers in qualitative analysis see Richards and Richards (1994).) The summaries of the IASRs comprise a brief outline of the incident and the context of the investigation, with the conclusions identifying the different factors contributing to the incident. Summaries and conclusions of the IASRs were scanned using optical character recognition software or downloaded from the Marine Incident Investigation Unit Web site (http://www.miiu.gov.au) and imported into NUDIST. The paragraph was chosen as the unit of analysis and is termed a "text unit" by NUDIST - each conclusion of an incident forming a single paragraph in most IASRs. The conclusions component of each IASR averaged approximately 14 text units. NUDIST accommodates an indexing system of nodes (categories of related text units) and documents can be displayed, browsed and searched using Boolean, set logic, and sophisticated matrix and vector operators.

LIMITATIONS OF THE METHOD

According to the Tavistock Report on the Human Element in Shipping Casualties (Bryant 1991), "direct evidence on fatigue as a cause of collisions is hard to come by". The contribution of fatigue to incidents at sea is not established outside the anecdotal evidence detailed in reports of such incidents. Bryant (1991) notes that lack of supporting documentary evidence on the work patterns and watchkeeping system on board hampers the investigator's ability to determine the extent to which fatigue may have been a factor. There are many factors that may contribute to the phenomenon of fatigue which may present in a number of critical behaviours other than lapses of attention, awareness or vigilance, as described by the Tavistock team (Bryant 1991). Since 1991, marine incidents in Australia have been investigated by a Marine Incident Investigation Unit operating under statute. Investigators determine the circumstances of an incident and publish a report (Marine Incident Investigation Unit 1996). It is not the role of the Unit to assign fault or determine liability (Marine Incident Investigation Unit 1996).

Although the Australian Navigation Regulations provide criteria for inclusion of incidents in the Marine Incident Investigation Unit database, the possibility of excluding some incidents (through non-reporting) does exist. Such exclusions should be minimal where other statutory (e.g. coronial, environmental and Port Authority By-Laws) and insurance reporting mechanisms exist. In addition, the identification of a second party in most collisions and the highly conspicuous nature of groundings renders non-reporting less likely.

The pilot study focusing on the quantitative aspects of marine casualties, conducted by the Tavistock Institute of Human Relations (reported in 1988), showed that "valuable information on the human factor could be extracted from the Department's casualty records (Bryant 1991). Similarly, Remi Joly of Transport Canada (personal communication, 1996) states that "accident reports...have relevant data...to draw on" for the purposes of researching the effects of fatigue on the performance of ship watchkeepers. However, these views conflict with that of Brown (1989) who states that "the Marine Surveyor's files are not a reliable source of evidence in the contribution of fatigue to shipping accidents, because those files provide inadequate information on accident-involved individuals' prior working arrangements and this information usually provides the only clue to their possible fatigued state". In addition, Finkelman (1994) claims that the measurement of fatigue in the real world is difficult, due to the great variety of moderating variables.

Brown (1989) notes the concern expressed in the Tavistock Report, which expresses difficulty in investigating the effects of fatigue, given the strong commercial pressures to sail and the unlikelihood that fatigue would be proffered as a possible cause of accident in this environment. As stated in the Tavistock Report, "Information derived from any reporting scheme is likely to be systematically biased by the very process of data collection itself" (Bryant 1991).

This study, however, concentrates on fatigue behaviours present in IASRs rather than identifying fatigue as a contributing factor. The concerns expressed by Bryant (1991) relate to Type 2 errors (overlooking fatigue as contributory to an incident); such errors are less likely to apply to a broad list of behaviours determined by an objective investigation than the more specific determination that fatigue was a factor.

Literature Review - Fatigue Behaviours

The literature revealed a variety of behaviours associated with fatigue and its independent variables such as time on task, sleep loss and time of day (Table 1). These behavioural manifestations were obtained from laboratory experiments (including perceptual, motor and cognitive tests, sleep propensity, reaction time and simulations) as well as field experiments including shipboard, aviation and road transport studies. Using a neurological model, behaviours were able to be sorted into the following categories:

(a) Activation problems - attentional failures, slips and lapses
(b) Perception limitations - limiting visual and auditory sensation
(c) Information processing problems - interpretation, encoding and correlation deficits
(d) Aversion to effort - failure to act
(e) Differing effort - failure to act appropriately.

The literature review, however, also referred to non-behavioural measures of fatigue such as subjective assessments of fatigue (rating scales, sleep quality, records and logs) and physiological tests (body temperature rhythm, EEG, heart rate and clinical chemistry assessments). As IASRs rarely report such potential data, subjective assessment and physiological tests, although potentially sensitive indicators of fatigue, were not included in this study.

Table 1. Fatigue Behaviours

Fatigue Behaviour	Primary Reference
1. Activation Problems	
Decreased vigilance (during a constant task)	Mackworth 1940, Krueger 1989
Decreased alertness (to a possible problem)	Hockey 1986, Haworth 1988, Stokes and Kite 1988, Broughton 1988, Vidacek 1993, Condon et al. 1988, Folkard and Monk 1979
Gaps, lapses or blocks	Haworth 1988, Hockey 1986, Brown 1989
2. Perception (and Sensory Input) Limitations	
Reliance on visual (eyes and radar) inputs	Bryant 1991
Decreased attention to peripheral instruments	Hockey 1986
Uncertainty of observations	Bohnen and Gaillard 1994
Decreased nighttime communication	Ohashi and Morikiyo 1974, Graeber 1989, Bryant 1991
3. Information Processing Problems	
Decreased encoding/registration of recently acquired information	Hockey 1986
Failure to interpret information as part of a single, integrated system	McFarland 1971
Decreased ability to correlate dynamic processes	Luczac 1991
Decreased ability to process lower and peripheral processes	Sablowski 1989, Gaillard and Steyvers 1988
Information processing deficiencies in secondary task	Sablowski 1989
4. Aversion to effort	
Low effort, low probability of success	Hockey 1986
Easy, but risky alternatives	Hockey 1986
Response latency/decreased speed of execution	Hockey 1986
Lower standards of accuracy and performance	McFarland 1971
5. Differing effort	
Increased variability of timing of actions	Hockey 1986
Decreased performance with lower/peripheral processes	Gaillard and Steyvers 1988
General performance decrement	Grandjean 1970, Folkard and Monk 1979, Condon et al. 1988, Haworth et al 1988, Brown 1989, Stokes and Kite, 1994, How et al. 1994, Parasuraman 1986, Hockey 1986

Content Analysis Of Incident At Sea Reports

The following categories (nodes) were obtained from conclusions of IASRs using NUDIST:

(a) Documents coded by incident type (collision, grounding, fire, foundering, fatality/injury and other)

(b) Human contribution to the breakdown of complex systems (after Reason 1990)

(c) Error type (skill-based slips and lapses, rule-based mistakes and knowledge-based mistakes (after Reason 1990))

(d) Investigator's determination that fatigue was a contributory factor

(e) Reference to time of day by watch period (based on the six watch system)

(f) Reference to fatigue behaviour (from the literature review).

Both qualitative (the nature of the behaviour) and quantitative (the number of instances that a particular coding appeared at a node) analyses of the incidents were undertaken. Quantitative data was not (statistically) analysed but used to identify nodes where particular behaviours predominated.

Incident type and distribution

Of the 100 Incidents at Sea Reports analysed, there were 38 groundings and 24 collisions. The remainder comprised founderings, fires, injuries/fatalities and a small number of other incidents such as structural failures and incidents alongside wharves (Figure 1).

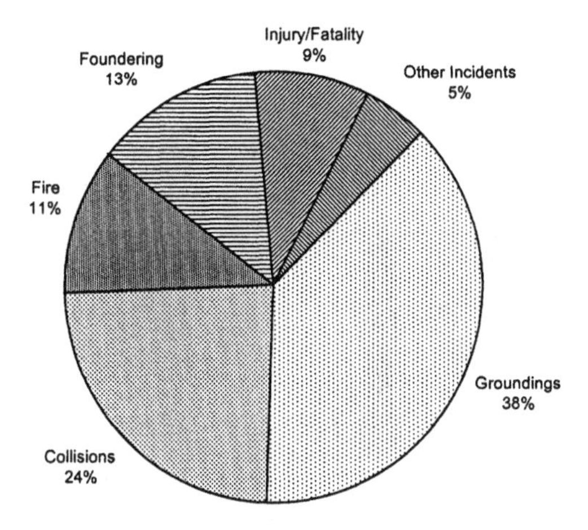

Figure 1. Distribution of Australian Incidents at Sea (from 100 consecutive reports)

Founderings (13 incidents) involved vessels lost at sea (some with all hands); however, incident time and contributing details were often undetermined. Fires (11 incidents) usually involved latent failures such as design and maintenance deficiencies, and usually involved machinery spaces. Human factors were difficult to determine in both of these categories.

The distribution of incident type by watch period varied; collisions and groundings exhibited a peak during the early morning, then decreased during the day and increased again during the afternoon. The frequency of other incident types remained reasonably static throughout the 24 hour period (Figure 2).

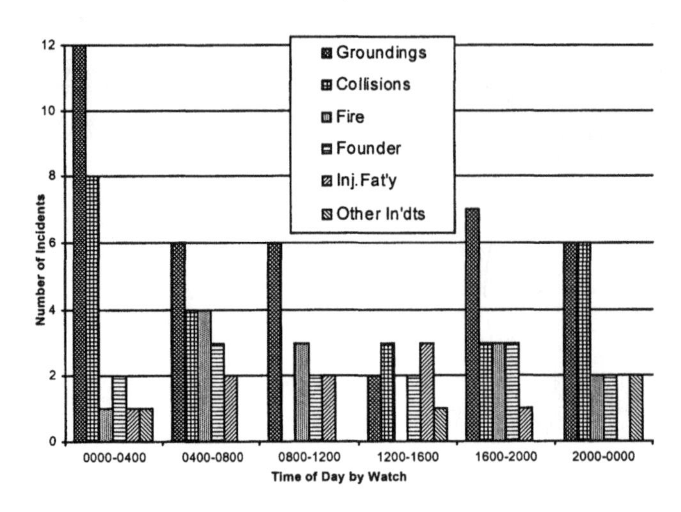

Figure 2. Incident type and time of day

Given the greater proportion of latent errors in fires, and meteorological and structural factors in founderings, these categories were excluded from time of day analyses. Similarly, the disparate nature of the incidents in the injury/fatality and "other" categories, in addition to their relatively small number, also justified their exclusion.

There appeared to be a diurnal distribution of the 62 collisions and groundings, peaking during the 0000 to 0400 watch, with a trough during the day (0800 to 1200 for collisions, 1200 to 1600 for groundings) (Figure 3). Such a distribution of accidents has been described by Filor (1996), Mitler et al. (1988), Hockey (1986), and Folkard and Monk (1979). More recently, Folkard (1996) describes a 24-hour patterning of accident risk which exists in road transport, marine and industrial situations and proposes the circadian rhythm of sleep propensity in association with time on shift effects as possible contributory factors.

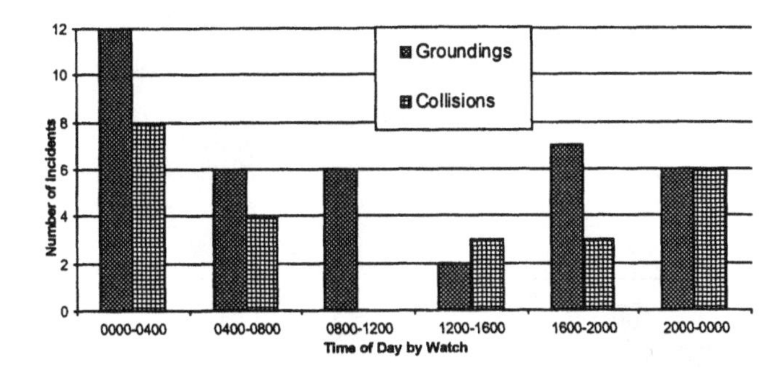

Figure 3. Collisions and groundings and time of day

Human factors in incidents at sea - error type and error form

Reason (1990) claims that there are a limited number of ways in which errors may manifest themselves, depending on the "'computational primitives' by which stored knowledge structures are selected and retrieved in response to current situational demands". Reason's (1990) Generic Error-Modelling System (GEMS) classifies human error types as skill-based slips and lapses, rule-based mistakes and knowledge-based mistakes, which reflect various processing, representational and task-related factors.

All conclusions were coded according to Reason's (1990) Generic Error-Modelling System in an attempt to locate the origins of human error types within incidents (Figure 4). All conclusions were also coded according to Reason's (1990) human contribution to the breakdown of complex systems in order to exclude latent failures from further analysis.

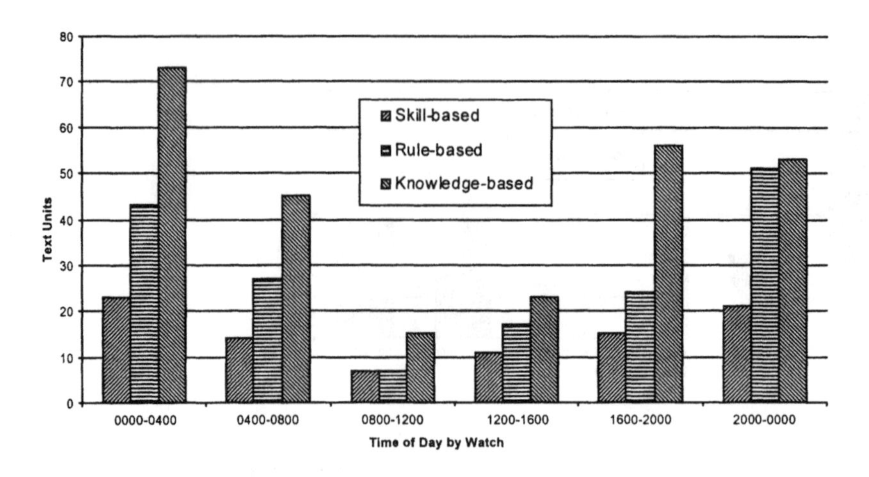

Figure 4. Time of day and error type based on the Generic Error-Modelling System of Reason (1990)

The distribution of error type (skill-based, rule-based, knowledge-based) for all conclusions appeared to follow the diurnal variation of collisions and groundings - knowledge-based errors having the highest incidence, skill-based errors the least.

Incidents where investigators determined that fatigue was a contributing factor

Investigators determined that fatigue was a factor in seven of the 100 incidents in the study. Five of the seven incidents were groundings, one a collision and one an overpressurisation. Four of the seven incidents occurred between 0000 and 0400, with two more occurring between 0000 and 0400. Only the overpressurisation incident occurred on day watches.

Factors contributing to these incidents included prior use of alcohol, sleep deprivation, failure to use all available information in navigation, failure to keep a proper lookout and failure to monitor progress. The most common unsafe act in groundings was a failure to alter course; in the collision, failure to keep clear.

Fatigue behaviours and time of day

In order to constrain behaviours to watchkeeper error, the analysis of fatigue behaviours was limited to collisions and groundings which were coded according to fatigue behaviours obtained from the literature review.

The initial analysis identified the frequency of particular fatigue behaviours in each watchkeeping period (Figure 5).

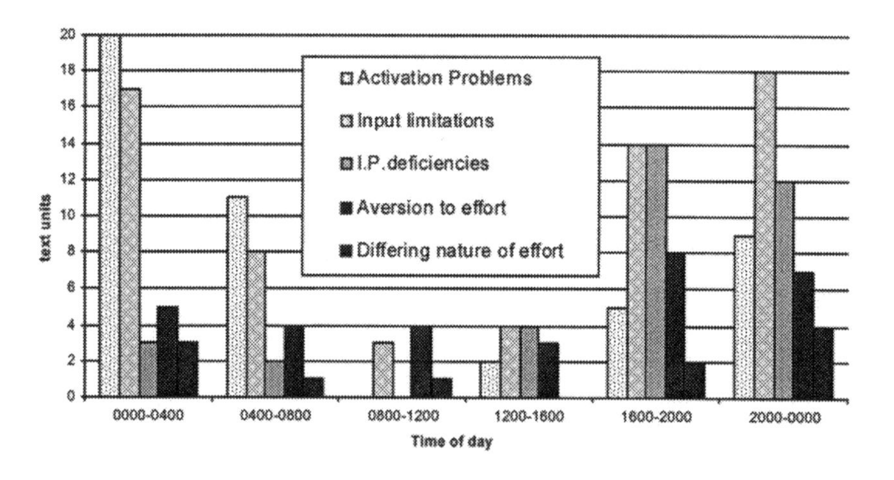

Figure 5. Fatigue behaviours and time of day.

The frequency of all fatigue behaviours was higher during the night watches and lower during the morning and afternoon watches. This distribution was most noticeable in the activation behaviours, with a ten-fold increase in the 0000 to 0400 watch compared to the 1200 to 1600 watch.

As the distribution of collisions and groundings exhibited a strong diurnal variation, fatigue behaviours were expressed as relative frequencies - the number of text units corresponding to a particular fatigue behaviour relative to the total number of conclusions in that watchkeeping period. Figure 6 shows the relative frequencies of all fatigue behaviours, which highlights watchkeeping periods where particular fatigue behaviours predominate. The 0000 to 0400 watchkeeping period was also corrected to accommodate one report comprising a disproportionately large number of text units (the *TNT Alltrans* Report being a Marine Court of Inquiry rather than a standard incident report) by attributing the average number of text units for the watchkeeping period to that report.

When the relative frequency of each category of fatigue behaviour was analysed within each watchkeeping period, the distribution revealed that activation problems and information processing deficiencies exhibit a strong diurnal variation, with the former peaking during the 0000 to 0800 period and the latter peaking during 1600 to 2000. Both activation problems and information processing fatigue behaviours were absent during the 0800 to 1200 watch; however, this was the peak of the aversion to effort and differing nature of effort fatigue behaviours. Input limitations remained fairly evenly distributed throughout most of the day but were slightly higher in the 1600 to 0000 watchkeeping periods.

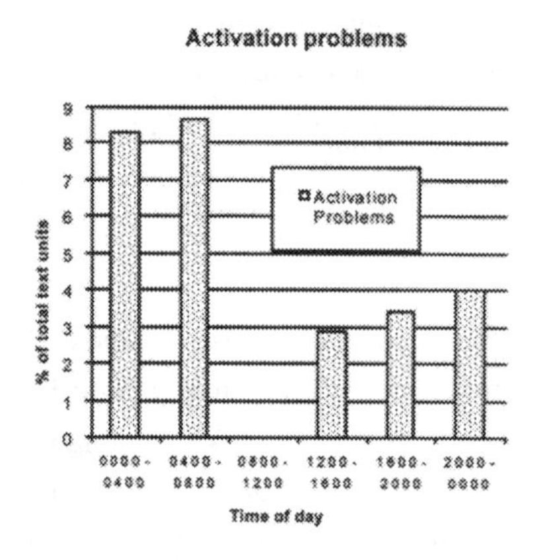

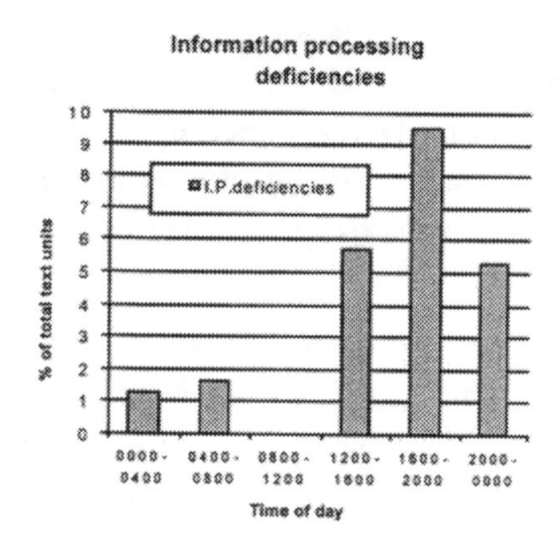

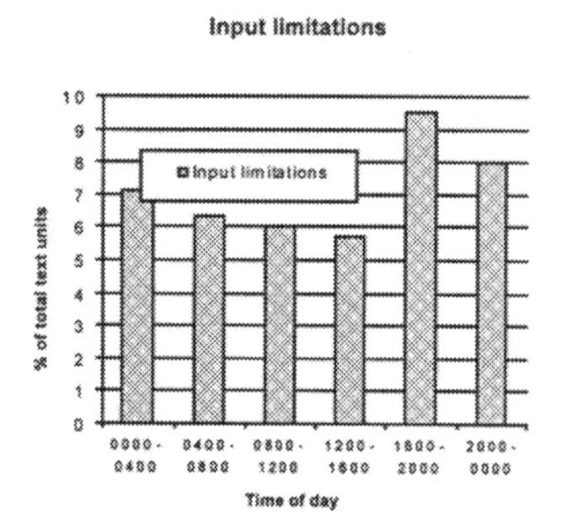

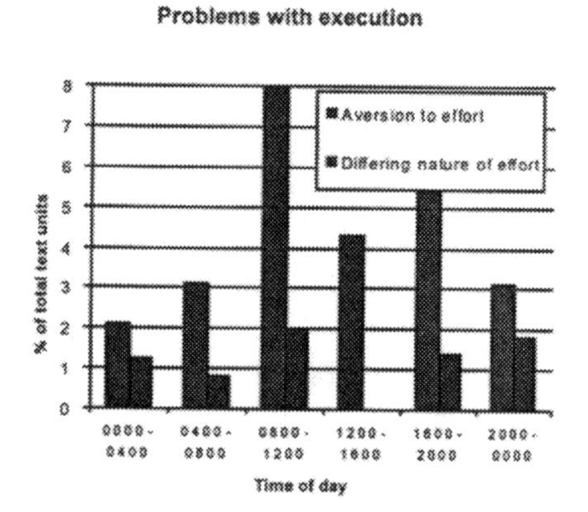

Figure 6. Relative frequency (percentage) of fatigue behaviours and watchkeeping period

DISCUSSION

Incidents where fatigue was determined to be a contributing factor

Of the 62 collisions and groundings, investigators found fatigue a contributing factor in only five (8%). Activation problems and input limitations were the only fatigue behaviours identified in these incidents; no other fatigue behaviours were evident. Activation problems were manifest in the case of the *Carola* (report 79) where investigators found that the mate was overcome by drowsiness, and in the case of the *Peacock* (report 95) the pilot lost situational awareness and fell asleep. The second mate of the *Svenborg Guardian* (report 82) left the bridge and failed to return because he fell asleep, resulting in the ship effectively out of control for almost five hours; in the case of the *TNT Alltrans* (report 8), the watchkeeper, due to alcohol and excessive tiredness, failed to alter course, resulting in the 35 000 tonne vessel, again effectively out of control, steaming onto Lady Musgrove Island at over 13 knots with those in charge of the vessel being unaware of the predicament for over 20 minutes. Input limitations commonly included lack of proper monitoring and lack of effective communication on the bridge (between master and watchkeeper and between pilot and watchkeeper).

The theme in five of the seven fatigue incidents was a decrease in watchkeeper alertness and either failing to initiate a course change or take action to keep clear of another vessel or obstruction. Due to a combination of fatigue and personal problems, the deck officer of the *Osco Star* (report 63) miscalculated the time to fill a tank, loaded the tank at an excessive rate and failed to act on an alarm, resulting in overpressurisation and structural failure of that tank. In the case of *Boa Force* (report 66), fatigue resulting from the operational program, although not explicitly stated as a contributing factor, was not ruled out.

Fatigue behaviours in the form of deficiencies in information processing or problems with execution were not identified by investigators in reports where fatigue was a contributing factor. It appears that fatigue was determined to be a factor only where the investigation process was able to demonstrate lack of activation (by decrease in vigilance or alertness) by the watchkeeper. As activation problems were generally most prevalent during the 0000 to 0800 watchkeeping periods, most incidents where fatigue has been determined to be a factor occur during these watches. This issue is pursued by Brown (1989) who questions the Tavistock team's finding of "'no direct evidence of the presence of fatigue in the Marine Surveyors' files' in the light of findings of carelessness, over-confidence and other risky behaviour, with errors of judgement and excessive speed in poor visibility being plausibly associated with the mental exhaustion associated with fatigue". It may appear that unless fatigue is explicitly presented as evidence in an inquiry, behaviours resultant from fatigue may be overlooked or attributed to some other cause.

Fatigue behaviours in collisions and groundings

The distribution of these incidents is diurnal - most occurring during the period 1600 to 0800, least during 0800 to 1600. Human factors weigh heavily in collisions and groundings, with most errors being errors of navigation. Activation problems and input limitations are the modal fatigue behaviours, which closely follow the circadian performance rhythm described by Folkard (1996).

The predominant unsafe act during the night watches was failing to keep a proper lookout, as with the watchkeepers on board the *Metal Trader* (report 25), *Jin Shan Hai* (report 31), *Khudozhnik Iaganson* (report 35), *Far East* (report 49), *Libra* (report 54), *Iron Prince* (report 81) and *Midas* (report 91). Although the possibility of fatigue existed in these incidents, there are many other explanations for such errors such low personal standards of performance, an on-board culture of poor watchkeeping practices or preoccupation with a more pressing task. For example, the Second Mate of the *Jovian Loop* (report 36) busied himself with writing up the log book and failed to notice an incorrect course setting on the autopilot. Other specific behaviours during the night included failing to alter course and to monitor progress - behaviours also prevalent in incidents where fatigue was determined a contributing factor. Activation problems during the day watches were less frequent, with the predominant behaviour being failure to keep a proper lookout.

Though activation problems can be explained in terms other than fatigue, there was a ten-fold increase in problems with activation and three times the proportion of activation behaviours per report during the midnight watch compared to the midday watch. It is problematic whether errors resulting from decreases in activation at certain times of the day can be attributed to fatigue. Just because an error occurs at one o'clock in the morning and a decrease in human performance occurs at that time doesn't automatically mean that the operator was fatigued. Fatigue behaviours result from many other contributing factors, including time on task and prior wakefulness. The United States Coast Guard (McCallum et al. 1996) have developed a fatigue index score as a potential indicator of the fatigue contribution to an incident which incorporates fatigue symptoms, hours worked and hours slept. It is noteworthy that time of day is not considered a parameter in this study despite the unequivocal evidence linking time of day with human performance. For example, Åkerstadt and Folkard's (1995) interactive computer program links time of day and hours of prior waking with alertness.

Limitations to perception were characterised by failure to utilise all available information (particularly navigational information) and failure to ensure effective communication (verbal, radio and signals). Such behaviours were fairly evenly distributed over both day and night watches as a proportion of text units. The most common input-limiting behaviour during both day and night watches was failure of the watchkeeper to make use of all available navigational equipment. Such behaviours included failure to take visual bearings but, more commonly, failure to fully utilise the radar and Automatic Radar Plotting Aids (ARPA). As supporting evidence, Bryant (1991) also notes misinterpretation of radar (despite improvements in training and equipment) and reliance on visual cues (where navigational aids are available) in an analysis of findings on collisions and groundings.

As with incidents where fatigue was determined a contributing factor, misunderstandings and failing to advise others of intentions were frequent input-limiting behaviours, such as the master of the *Jovian Loop* (report 36) who failed to advise the pilot that he was leaving the bridge. In several incidents, such as the *Jhanski Ki Rani* (report 13), the *Mobil Endeavour* (report 16), the *TNT Carpentaria* (report 37), the *Kapitan Serykh* (report 72), the *River Torrens* (report 80) and the *Maresk Tapah* (report 103), those on the bridge failed to effectively exchange information, thereby limiting the decision-making process of the watchkeeper. In support, Ohashi and Morikiyo (1974) found that the extensiveness and uncertainty of information sources increased at night-time, although communication problems predominated in daytime incidents in this study.

This issue of ship-to-shore communication problems was noted in the Tavistock Report where Vessel Traffic Services (VTS) fail to communicate the very information that would have prevented a collision (Bryant 1991). Analysis of accidents has previously shown that factors such as anxiety, boredom and fatigue influence the navigator's ability to perceive and process information and there is a reliance on sight or radar as the primary navigational equipment, discounting much of the information that is available to prevent the incident (Bryant 1991).

The distribution of information processing deficiencies differed from the previous two fatigue behaviours in that the peak occurred in the late afternoon and evening - the incidence of information processing deficiencies during the 0000 to 1200 watches was quite low. Common specific behaviours during the day watches included planning deficiencies, making assumptions and making incorrect assessments. Information processing deficiencies in association with a secondary task was also evident in many incidents, a phenomenon also reported by Sablowski (1989). For example, the captain of the *Charles H McKay* (report 7) and the skipper of the *Saltiford* (report 10) both failed to make an appraisal of a developing collision due to preoccupation with the radios. The master of the *New Noble* (report 86) was preoccupied with recovering an anchor and the third officer of the *Iron Cumberland* (report 10) was assigned to a phone call which reduced his ability to act as a lookout and contributed to the collision with the *Saltiford* (whose skipper was preoccupied with the radios!).

The ability to correlate dynamic processes is a feature of complex systems and fatigue has been documented as affecting this, particularly where a secondary task predominates. Brown and Greger (1990, in Luczac 1991) suggest that "the human ability to correlate dynamic processes is rather low, especially if the variables in these processes change their states rapidly, and if the anticipation of change is impossible because of the stochastic signal characteristics of unforeseen action to reaction cycles". Whereas activation problems predominate during the early morning, information processing problems occur mainly during the evening watchkeeping periods. While the distribution of activation problems is consistent with the circadian rhythm of subjective alertness, as described by Folkard (1983, in Monk 1989), the distribution of information processing deficiencies is more consistent with the decrease in accuracy over the working day as described by Monk and Leng (1982, in Monk 1989). This phenomenon may well be related to time on task effects in addition to circadian influences, due to the difficulty in separating on-duty time from off-duty time on a ship.

The information processing fatigue behaviours that occurred late in the day included making assumptions, incorrect assessments and inadequate planning (although aspects of the latter may be latent). Behaviours under contrived experimental conditions may not be easily transferable to analysis of real world situations. Indeed, one researcher claims that speed accuracy trade-off only occurs under strict laboratory conditions where subjects are asked to make a speeded response (Diane Williams 1996, personal communication).

Aversion to effort includes such behaviours as failing to reduce speed, failing to avoid a close quarters situation and failure to take avoiding action. This behaviour predominates during the day (late in the day in absolute frequency, but earlier in the day as a relative frequency). Examples include failure of the chief mate of the *Ruca Challenge* (report 18) to continue to plot the vessel's position, failure of the second officer of the *Han Gil* (report 4) to take avoiding action and failure of the watchkeeper of the Gumbet (report 106) to take a sufficiently wide berth when passing fishing vessels. In a number of incidents such as the *Jhanski Ki Rane* (report 13), *Yue Man* (report 7), *Great Brisbane* (report 17), *River Embly* (report 19), and *Berlin Express* (report 53), appropriate action in reducing speed was not taken. Navigation, particularly in confined waterways, could be described as a machine-paced task and Williams et al. (1959 in Kreuger 1989) noted that performance on machine-paced tasks is often affected by small amounts of sleep loss, leading to errors of omission. Similarly, it has been claimed that task-based (i.e. machine) pacing is most influenced by fatigue (Remi Joly 1996, personal communication).

As with the previous behaviour, problems with execution follow the reported phenomenon of a decrease in accurate performance during the day (Monk 1989). Other authors have described forms of aversion to effort as a consequence of time on task (Hockey 1986) but inevitably, in real world situations, a combination of partial sleep loss, time on task and circadian performance rhythm will all interplay in producing a pattern of behaviour.

As the frequency of differing nature of effort fatigue behaviours was typically less than 2% (corresponding to only 11 text units), no analysis of temporal distribution was made; however, one unsafe act predominated in this category. In eight separate instances, port instead of starboard helm was inappropriately applied, resulting in the grounding of the *River Boyne* (report 9), the *Alam Indah* (report 14), the *Berlin Express* (report 53), the *TNT Carpentaria* (report 37), and collisions involving the *Iron Cumberland* (report 10), *Yue Man* (report 7), *Longevity* (report 41) and the *Searoad Mersey* (report 64). (In most instances a course alteration to starboard is the correct action to avert a collision.)

Conclusion

The ship's watchkeeper is a fallible human operator in a complex technological system and unsafe acts by such operators may result in subsequent incidents being investigated. Collisions and groundings are two categories of incident where the fatigued watchkeeper may perform unsafe acts. Fatigue is one factor that contributes to such unsafe acts but many authors feel that its contribution is under-reported; indeed fatigue was identified as a contributing factor in only seven of 100 Incidents at Sea Reports. In this study, a model of fatigue behaviours was developed from a review of the scientific literature and subsequently applied to a content analysis of Incident at Sea Reports of collisions and groundings.

The literature revealed fatigue presenting as activation problems, sensory limitations, information processing problems, aversion to effort and differing nature of effort, and that these behaviours had characteristic distribution throughout the working day. Activation problems were most evident during the early morning watches, whereas information processing problems predominated during the afternoon and evening watches. Aversion to effort was a behaviour most evident during the morning and afternoon watches. Specific unsafe acts were associated with each category of fatigue behaviour; for example, activation problems commonly presented as failing to keep a proper lookout and monitor progress whereas information processing problems presented as planning and assessment deficiencies, frequently in association with secondary tasks.

Fatigue behaviours as reported in the scientific literature are evident in real world situations - in this study the fallible behaviours of ship's watchkeepers. Fatigue is an elusive variable; however, the increasingly automated and technological working environment of the ship's watchkeeper provides an ideal environment for further study.

References

Åkerstedt, T. and Folkard, S. (1995) Validation of the S and C Components of the Three-Process Model of Alertness Regulation. Sleep, 18 (1), 1-6

Bohnen, H.G.M. and Gaillard, A.W.K. (1994) The Effects of Sleep Loss in a Combined Tracking and Time Estimation Task. Ergonomics, 37 (6), 1021-1030

Broughton, R.J. (1988) Vigilance and Sleepiness: A Laboratory Analysis in Coblentz, A. (ed) Vigilance and Performance in Automated Systems, Kluwer, Dortrecht

Brown, I.D. (1989) Study into the Hours of Work, Fatigue and Safety at Sea, Medical Research Council, Cambridge, England

Bryant, D.T. (1991) The Human Element in Shipping Casualties. HMSO, London

Condon, R., Colquhoun, W.P., Plett, R., DeVol, D., Fletcher, N. (1988) Work at Sea: A Study of Sleep, and of Circadian Rhythms in Physiological and Psychological Functions in Watchkeepers on Merchant Vessels; Part IV - Rhythms in Performance and Alertness. International Archives of Occupational and Environmental Health, 60, 405-411

Filor, K. (1996) The Original Twenty-Four Hour Society: Issues of Fatigue and Incidents at Sea in Proceedings of the Second International Conference on Fatigue and Transportation, Fremantle, Australia, 11-16 February 1996

Finkelman, J.M. (1994) A Large Database Study of the Factors Associated with Work-Induced Fatigue. Human Factors, 36 (2), 232-243

Folkard, S. (1996) Black Times: Temporal Determinants of Transport Safety. Accident Analysis and Prevention, 29 (4), 417-430

Folkard, S. and Monk, T.H. (1979) Shiftwork and Performance. Human Factors, 21 (4), 483-492

Gaillard, A.W.K. and Steyvers, F.J.J.M. (1988) Sleep Loss and Sustained Performance in Coblentz, A. (ed) Vigilance and Performance in Automated Systems, Kluwer, Dortrecht

Graeber, R.C. (1989) Long-Range Operations in the Glass Cockpit: Vigilance, Boredom and Sleepless Nights in Coblentz, A. (ed) Vigilance and Performance in Automated Systems, Kluwer, Dortrecht

Grandjean, E.P. (1970) Fatigue. American Industrial Hygiene Association Journal, July-August, 401-411

Haworth, N.L., Triggs, T.J., Grey, E.M. (1988) Driver Fatigue: Concepts, Measurement and Crash Countermeasures, Federal Office of Road Safety, Canberra

Hockey, G. (1986) Changes in Operator Efficiency as a Function of Environmental Stress, Fatigue and Circadian Rhythms in Boff, K., Kaufman, L. and Thomas, J.P. (eds) Handbook of Perception and Human Performance, John Wiley and Sons, New York

Krippendorff, K. (1980) Content Analysis: An Introduction to its Methodology. Sage Publications, Beverly Hills

Krueger, G.P. (1989) Sustained Work, Fatigue, Sleep Loss and Performance: A Review of the Issues. Work and Stress, 3 (2), 129-141

Luczac, H. (1991) Work Under Extreme Conditions. Ergonomics, 34 (6), 687-720

Mackworth, N.H. (1950) Researches on the Measurement of Human Performance. His Majesty's Stationery Office, London

McCallum, M.C., Raby, M. and Rothblum, A.M. (1996) Procedures for Investigating and Reporting Human Factors and Fatigue Contributions to Marine Casualties. United States Coast Guard, Washington

McFarland, R.A. (1971) Understanding Fatigue in Modern Life in Hashimoto, K., Kogi, K. and Grandjean, E (eds) Methodology in Human Fatigue Assessment. Taylor and Francis Ltd, London

Marine Incident Investigation Unit. (1996) Marine Incident Investigation Unit 1991 to 1995. Department of Transport and Regional Development, Canberra

Mitler, M.M., Carskadon, M.A., Czeisler, C.A., Dement, W.C., Dignes, D.F. and Graeber, R.C. (1988) Catastrophes, Sleep and Public Policy: Consensus Report. Sleep, 11 (1), 100-109

Monk, T.H. (1989) Circadian Rhythms in Subjective Activation, Mood and Performance Efficiency in Kryger, M.H., Roth, T and Dement, W.C. (eds) Principles and Practices of Sleep Medicine. W.B. Saunders Company, Philadelphia

Nitka. J. (1989) Evaluation of the Psychical State of Deck Crew Seamen with Long Period of Service at Sea. Bulletin of the Institute of Tropical and Maritime Medicine Gdynia, 40 (1/2), 35-40

Nitka. J. (1990) Specific Character of Psychiatric Problems among Seafarers. Bulletin of the Institute of Tropical and Maritime Medicine Gdynia, 41 (1-4), 47-52

Nitka. J. and Dolmierski, R. (1987) Psychosocial Factors Causing Specificity of Work at Sea. Bulletin of the Institute of Tropical and Maritime Medicine Gdynia, 38 (3/4), 193-198

Ohashi, N. and Morikiyo, Y. (1974) Differences in Human Information Processing for a Ship Manoeuvring in the Daytime and at Night. Journal of Human Ergology, 3, 29-43

Parasuraman, R. (1986) Vigilance, Monitoring and Search in Boff, K.R., Kaufman, L. and Thomas, J.P. (eds) Handbook of Perception and Human Performance. John Wiley and Sons, New York

Reason, J. (1990) Human Error. Cambridge University Press, Cambridge

The Revised STCW Convention (1995) International Shipping Federation, London

Richards, T. and Richards, L. (1994) Using Computers in Qualitative Research in Denzin, N.K. and Lincoln, Y.S. (eds) Handbook of Qualitative Research. Sage Publications, Thousand Oaks

Sablowski, N. (1989) Effects of Bridge Automation on Mariners' Performance in Coblentz, A. (ed) Vigilance and Performance in Automated Systems. Kluwer, Dortrecht

Stokes, A. and Kite, K. (1994) Flight Stress: Stress, Fatigue, and Performance in Aviation. Avebury Aviation, Aldershot

The United Kingdom Mutual Steamship Assurance Association (Bermuda) Ltd, (1993) Analysis of Major Claims. Thomas Miller P&I, London

Vidacek, S., Radosevic-Vidacek, B., Kaliterna, L., and Prizmic, Z. (1993) Individual Differences in Circadian Rhythm Parameters and Short-Term Tolerance to Shiftwork: A Follow-up Study. Ergonomics, 36 (1-3), 117-123

Weitzman, E.A. and Miles, M.B. (1995) Computer Programs for Qualitative Analysis. Sage Publications, Thousand Oaks

18

FATIGUE IN FERRY CREWS - A PILOT STUDY

L.A. Reyner, Sleep Research Group, University of Loughborough and S.D. Baulk, Seafarers International Research Centre, University of Wales, Cardiff

INTRODUCTION

In recent years, significant changes have occurred to the occupational and social structure of the shipboard environment. These changes include increased commercial and economic pressures, new technologies, reduction of crew sizes, and crews being recruited from an increasing diversity of cultures overseas. The nature of the shipping industry as a 24 hour operation has resulted in greater pressure for crews to work longer hours, with shorter rest periods and with increased workloads, resulting in a higher risk of fatigue.

The issue of fatigue in the maritime industry as well as, indeed other modes of transportation has therefore become increasingly important. Statistics suggest that where human error is involved, fatigue has been identified as either the primary cause, or a major contributory factor in an estimated 70-80% of marine accidents (Donaldson, 1994). In some cases these accidents have resulted in loss of life and in other cases large costs have been incurred by the industry. Approved guidelines for the investigation of accidents where fatigue may have been a contributory factor (IMO/ILO 1993) are an essential means of identifying the possible causes of fatigue prior to the accident occurring through accurate reporting. In the interests of personal health and safety and economic implications, this information is vital to both the seafarer and the industry.

Fatigue and Transport

Worldwide concerns for the effects of fatigue in the transport industry have led to regulations for hours of working, and rest breaks for professional drivers and airline pilots. Research on sleepiness and accidents has been well documented (Work hours, sleepiness and accidents:

Journal of Sleep Research, Vol.4: Entire Supplement No.2, 1995). Although new work regulations for Seafarers were implemented in 1995, it would seem that these concerns have not yet filtered into the shipping industry.

Fatigue in Seafarers

Marine accidents involving both passenger and cargo ships have regularly captured the world's attention in the last 50 years. Although many of these accidents cannot be easily attributed to one factor, it is clear that incidents such as the 'Herald of Free Enterprise,' at Zeebrugge did involve sleep/fatigue related problems (Crainer, 1993).

In certain parts of the world today, new (but controversial) regulations allow one-man bridge operations. The Peacock was an example of one such ship, operating in Australian waters on the Great Barrier Reef. This was a refrigerated cargo ship which became grounded for nine days during July 1996. Investigations concluded that the pilot had fallen asleep 15 minutes before the incident, which occurred at 0155h, and therefore the ship failed to change course appropriately (MIIU Report No.95, 1996).

Investigation of the causes and effects of fatigue at sea are complicated not least because of the huge variation world-wide in ship sizes and structures, journey duration's, crew schedules and workload, and it is necessary to examine several different ship types, on different international routes, rather than one specific seafaring scenario.

A report by Brown (1989), provides an outline of the regulations of hours of work for Seafarers in different countries, and highlights the need for 'objective evidence' of fatigue at sea. This report highlights the need to identify the characteristics of fatigue related accidents, and as has been proved with certain types of accident (collisions, groundings etc., which are caused by lack of monitoring/situational awareness on the part of the watchkeeper), there is a definite need for more information on watchkeeping schedules. For example, the information below is essential after an accident, in order to determine whether the accident could be attribute in whole or in part, to fatigue:

- When a crew member last slept, and for how long
- The exact time of day of the incident
- How long the crew member had been on the bridge or on watch.

This information cannot prove either way that fatigue was wholly or partly to blame for any incident, but it can be used as an invaluable insight to what may have happened, and in terms of fatigue/sleepiness, is vitally important. Such information has been particularly useful in recent research into sleepiness related vehicle accidents (Horne & Reyner, 1995), and would help to explain many shipping accidents.

This means that some kind of log is needed of all this information, as suggested by Brown (1989), and that watchkeepers should be questioned on the information within, in the event of any accident. The information could also be monitored, so that the Master of a ship would always be aware of the sleeping habits/patterns of his crew members, particularly those keeping watch. Brown made several conclusions about what is required in the maritime industry in an attempt to abate the fatigue problem and the casualties (human, mechanical and environmental) which are too often caused.

Sanquist et al. (1997), also carried out a survey on mariners from tankers and freight ships on the US west coast, and found that the 4 hrs-on, 8 hrs-off traditional work schedule was the primary cause of fatigue in seafarers. Using subjective sleep logs, he found that they slept an average of 6.6 hours per 24 on board ship, compared to 7.9 hours while at home. Similarly, watchkeepers slept less, and their sleep was of poorer quality (due to fragmentation and inappropriate physiological timing). Tanker stewards and those taking the 0400-0800 watch on freighters had the shortest sleep duration of all. Sanquist also reported that tanker personnel worked longer days than other seafarers. These findings seem to indicate that the characteristics of the seafaring lifestyle lead to poor quality sleep. Perhaps most importantly, Sanquist et al. identified the key features of the sleep of Seafarers in general:

- An overall reduction in sleep time between working at sea and at home.
- Fragmented (and therefore poorer quality) sleep.
- Having to attempt to sleep at physiologically inappropriate times.
- Insufficient breaks for rest between shifts.
- Poor sleep quality in the main sleep period (due to environmental conditions, e.g. noise).
- Long work days.

Sanquist et al. concluded that Seafarers showed an inconsistency in alertness levels over the day, with a substantial drop in alertness between the hours 2000-0000, and a significant decline in alertness on the 0400-0800 watch. They also found that the crew had a tendency to over-estimate alertness on the 0000-0400 watch. Overall, there was no data to justify the traditional 4 hour watches.

A recent report from the National Union of Marine, Aviation and Shipping Transport Officers (NUMAST 1997: 'Give Us a Break') highlights the failure of messages such as 'tiredness can kill' to filter through to the maritime industry. It concerns a simple survey which was conducted with crews of different shipping sectors in the UK, into awareness and compliance with hours of work regulations etc., and the effect of hours of work on safety. 95% of those surveyed said that their work hours had not reduced as a result of the new hours of work regulations (1995). 90% said they worked 10 or more hours per day approximately, and 22% of this was over 13 hours. 56% said they considered that working hours present a danger to health/safety of the ship etc., and a majority of 52% said that extra manning is the best way to reduce fatigue at sea, and make ships a safer place to live/work. NUMAST's overall conclusion is that the 1995 hours of work regulations have had little impact on the shipping industry.

Aims of the study

This was a pilot study, with the main aims being (1) to investigate the quantity and quality of sleep on board ship, (2) to evaluate the extent to which poor quality sleep occurs and (3) to identify causal factors relating to poor sleep quality.

Sleep (quality and quantity) data and subjective reports of alertness and sleep quality were recorded over a two-week period from 12 crew members of varying rank of a European passenger ferry. Their normal week-on, week-off work schedule allowed comparisons to be made between a working week on board ship, and a week spent at home. A researcher was also present on board to interview participants, assist in the collection of subjective data, and to observe and photograph life on board the ship for a period of 8 days.

Differences in sleep quality and duration between the work and non-work weeks might well be greater for those crew members required to work split-shifts whilst on duty. Furthermore, they may experience greater sleep disturbance and generally shorter sleep periods than crew members working single shifts in each 24 hour period.

METHOD

Design

This investigation used a repeated measures design, where 12 subjects (two separate groups of 6) were tested both objectively and subjectively for one week at work on board ship, and (during the following week) for one week at home. The study was carried out in two phases, phase 1: February 1997, and phase 2: July 1997.

UK - Ireland	Ireland - UK
03:15 - 06:45	09:00 - 12:30
15:00 - 18:30	21:50 - 01:20

Figure 1. Ship Schedules on the Crossing

There are 4, 3½ hour crossings per 24 hours on the route, with a turnaround time of approximately 2-3 hours, sailing from the UK at 03:15h and 15:00h, and from Ireland at 09:00h and 21:50h. The evening stop in Ireland is the longest of all (3h 20 mins) and this is the period when lifeboat drills etc. are carried out.

Participants

All participants were male, the mean age was 43.17y with SD=5.34, and age range being from 31 to 50y. The participants had worked at sea for an average of 24 years, and for the company, for an average of 9 years. The crew reported working 86.92 hours per week (SD=8.2h). Six participants were tested for two weeks during each phase, one week on board ship followed by one week at home. Participants were matched (between the two ships) for task and shift schedule, to allow comparison between split-shift, and non split-shift workers. All participants were members of the crews of the two ships.

Each of the two groups comprised:

Master	The Master is in overall command of the ship and works a single daytime shift which does not change or rotate. At the end of the shift in the evening, command of the ship is handed over to the Mate/Master.
Mate/Master	The Mate/Master works through the night, relieving the Master, and taking over full responsibility of the ship. His work schedule runs 2030-0400 and 0600-1000.
2nd Officer	The Second Officer is essentially the navigational officer, in charge of watchkeeping on the bridge during crossings, and also on the car decks during loading, also during berthing at either port.
2nd Engineer	The Second Engineer answers to the Chief Engineer, and keeps watch in the engine control room etc.
3rd/4th Engineer	These engineers work split-shifts from 0000-0600, and 1130-1600. Their watchkeeping/maintenance duties are similar to the 2nd Engineer.
Hotel Repair Man (HRM)	The HRM is responsible for the maintenance and repair of domestic facilities in passenger accommodation. He works from 0700-2100 daily, and is on 24 hour call on alternate days. This position is manned by two repair men, working simultaneously to cover the 24 hour period.

Figure 2. The participants and their job descriptions

Measurements

General Questionnaire This was used to obtain general background information about participants, and (during the second phase of the study), was completed on board the ship with the researcher present.

Log Books These covered the 7 day work schedule and were completed every day to show participants' work/rest activities. Each day also featured a subjective sleepiness scale (The Karolinska Sleepiness Scale, KSS - Åkerstedt & Gillberg, 1990) which subjects completed every 2 hours. This is a subjective rating scale which ranges from 1 ('Extremely Alert') to 9

('Very Sleepy, great effort to stay awake, fighting sleep'). This scale has been used extensively in sleep research, to obtain subjective sleepiness data. These were handed in at the end of the first week (work) and participants were given a similar book to complete during the week at home.

1. Extremely Alert
2. Very Alert
3. Alert
4. Rather Alert
5. Neither Alert nor sleepy
6. Some signs of Sleepiness
7. Sleepy, but no effort to keep awake
8. Sleepy, some effort to keep awake
9. Very Sleepy, great effort to keep awake, fighting sleep

Figure 3. The Karolinska Sleepiness Scale - Åkerstedt & Gillberg, 1990

Actimeters Each subject wore an actimeter for the two-week testing period, although the data from the first week (work) was downloaded before the second week (at home) began. The actimeters are worn on the wrist like a watch, and measure movement in 30 second epochs. There is a small circular 'event' button on the top of each actimeter which clicks audibly when pressed. This is to indicate sleep periods. i.e. subjects press it once when about to go to sleep, and then again upon waking up. The information obtained from the actimeters is used to determine the quality and quantity of subject's sleep.

Procedure Data was collected in two phases: Phase 1 took place in February 1997. Six crew members of varying rank wore actimeters for one week at sea, and one week at home (schedules running Tuesday-Tuesday). They also completed background information questionnaires regarding their sleeping habits, and use of stimulants such as caffeine etc. Phase 2 took place in July 1997. The second group of 6 crew members again wore actimeters for a week at sea and a week at home. In addition to the data collected in phase 1, these participants completed subjective logbooks incorporating the KSS which was completed every two hours. For phase 2, a researcher remained on board the ship with participants for the working week, in order to assist in the completion of questionnaires, collection of subjective and objective data, and also to photograph the facilities for crew on board the ship, and to get a clearer picture of what life on board is like.

During phase 2, the Subjective Logbooks were distributed to participants on the first day of the working week, and collected before they left the ship to go home, when they were also given a second logbook to cover the week at home. Similarly, the Actimeters were explained and given to participants as soon as they arrived on board the ship, and these were all set-up to begin recording at 1500 hours on the Tuesday, 15th July (with the exception of the Master, who arrived one day later).

RESULTS

Subjective Log Book and KSS Ratings

The graph in Figure 4. below shows the mean subjective sleepiness of a typical subject across the 24 hour period, both at work and at home. As the graph shows, during the working week, he remains awake longer, although waking at a similar time, and his alertness deteriorates towards the end of the day more markedly than during the week at home. Ideally, while at work, participants should not be sleepy, rating themselves at 5 or less.

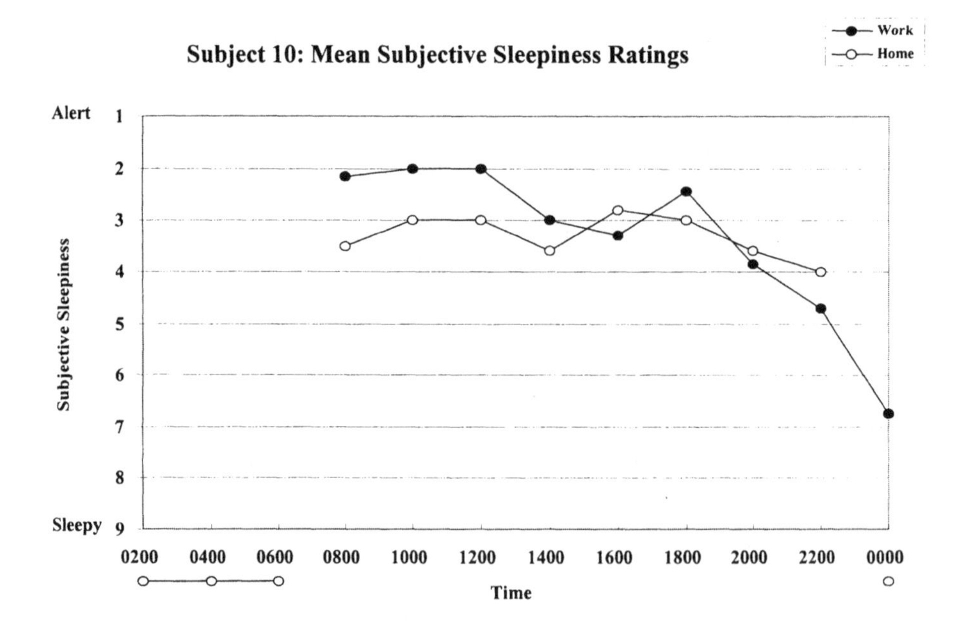

Figure 4. Subject 10, Mean KSS Ratings

Objective Actigraph Data

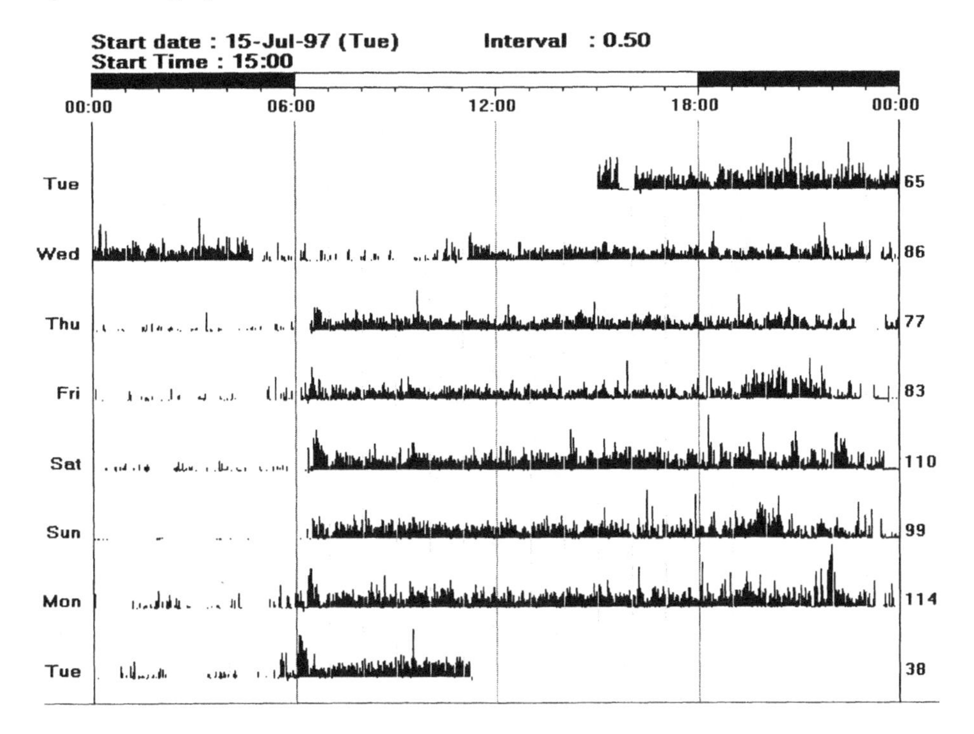

Figure 5. Example of Actigraph data

The actigraph data shows the periods of activity and inactivity throughout the 24 hour period (an example is given in fig.5). From this we can estimate the onset and offset of sleep, the duration of each sleep period, and the percentage of this sleep which is disturbed. The latter is known as the Sleep Disturbance Index, or SDI, and can then be plotted (see Fig.6) against sleep length. Participants can experience a lengthy sleep, but it can be greatly disturbed, and therefore less beneficial. The SDI can be compared with the KSS subjective sleepiness data. For example, when a subject experiences a short sleep with a high level of disturbance, there is increased subjective sleepiness during the following day (and vice-versa), which is seen in the sleep profile graph (see fig.7), displaying sleep length against SDI for one subject both at work (black) and at home (white). As the graph shows, this subject works a split-shift, and therefore has two sleep periods per day, compared to one per day at home. It is clear from this graph that 1) sleep length is greater and more stable at home than at work, and 2) that SDI is similarly constant at home, and less predictable at sea.

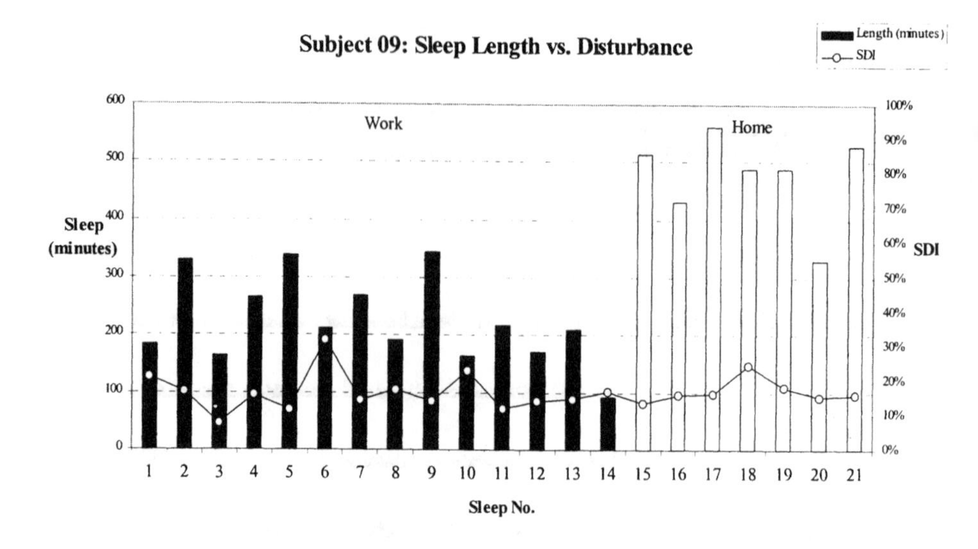

Figure 6. Subject 09 Sleep Length vs. SDI

This can be related to the sleep profile graph (fig.7, below) showing the mean KSS ratings for every period of wakefulness (upper black horizontal bars) the length of each sleep period (lower black horizontal bars) and the percentage of each sleep which is disturbed (SDI - sleep disturbance index - vertical black bars). As this graph (of the week at home) shows, although Sleep length and SDI remain reasonably constant throughout the week, the mean KSS at the beginning of the week is below 6 ('Some signs of sleepiness'), and this slowly increases as the week goes on. This demonstrates that it took this participant about 4 days to recover from the working week, and return to a reasonable level of subjective alertness (i.e. above KSS 5).

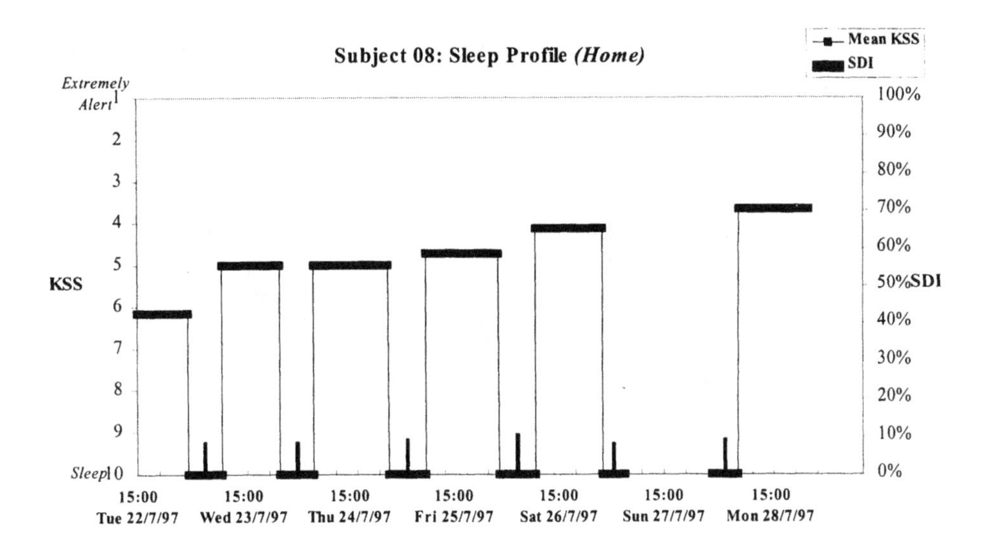

Figure 7. Subject 08 Sleep Profile (Home)

The two graphs below show the mean sleep length and SDI per 24 hours for each subject at work (fig. 8) and at home (fig. 9). Participants 01-06 were tested during phase 1, while subjects 07-12 were tested in phase 2. The vertical black bars show the mean sleep duration per 24 hours for each subject, at work and at home, while the smaller grey bars show the SDI or sleep disturbance index for the same period. It is immediately clear from the two graphs that firstly there seems to be a difference in sleep length between the home and work periods, and secondly that there is a similar difference for SDI.

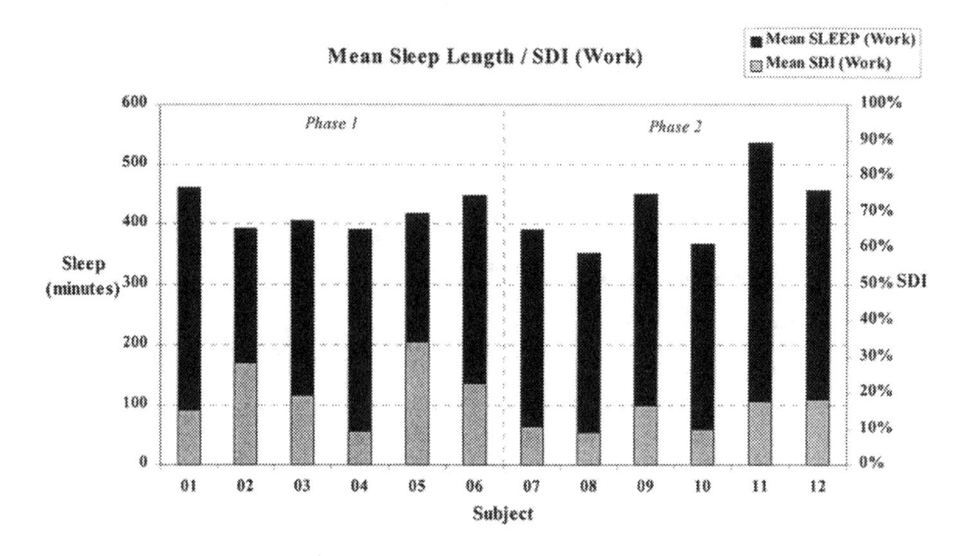

Figure 8. Mean Sleep Length vs. SDI (Work)

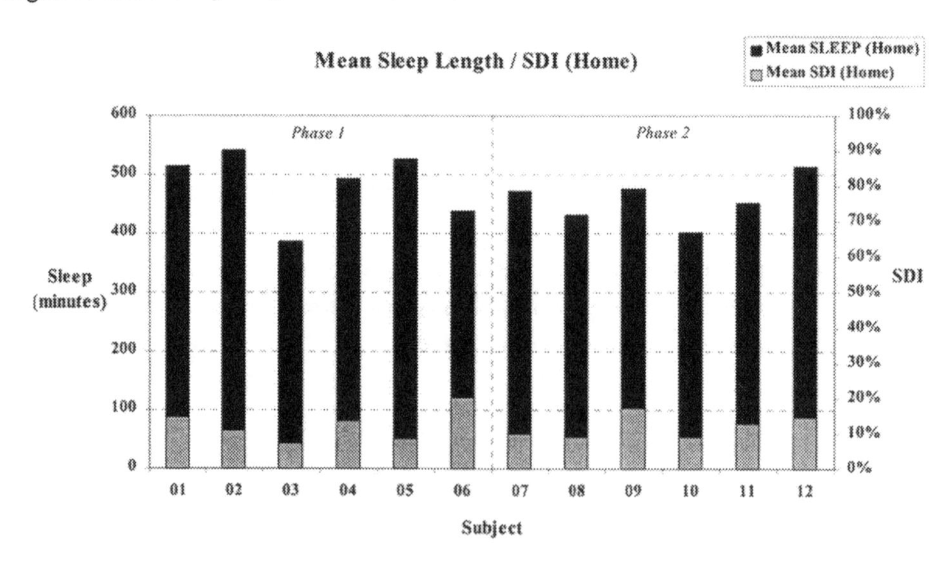

Figure 9. Mean Sleep Length vs. SDI (Home)

Statistical Analysis

Sleep Length

For the group of 12 participants, tested with a paired samples t-test, the examination of work vs. home produced a significant result (t=2.65, df.=11, p=0.023) for sleep length, i.e. subjects slept for significantly longer per 24 hour period at home than at work.

Sleep Disturbance Index

Similar analysis of the work/home data for SDI highlighted a near significant trend towards a difference in sleep disturbance index (t=-2.09, df.=11, p=0.061) i.e. that participants tended to experience more sleep disturbance while at sea than at home.

Discussion

Overall, there appeared to be no serious sleepiness problems on these two ships. Although the crew had significantly less sleep at sea than at home, and there was a trend towards similar differences in the sleep disturbance index (i.e. participant's sleep was generally more disturbed at sea than at home), their daytime sleepiness measurements did not reveal any serious difficulties. However, as the KSS data from the subjective logbooks shows, there is evidence to support an increase of subjective sleepiness over the week at work. This in turn would suggest that performance would deteriorate. Hence there is a need for measures of performance in order to ascertain the true impact of any fatigue problem on different aspects of the seafarer's job.

This study probably represents 'the best case scenario,' and there is much need for further research into more problem areas of the industry. Therefore this study is important in terms of its use as a marker for a relatively 'good ship' in terms of fatigue etc. Investigations should be made into other areas of the shipping industry, using some of the methods developed in this study. This and related research could be applied to the controversial use of one-man bridge operations, as used in Australian waters along the Great Barrier Reef, exemplified by the grounding of the Peacock in July 1996.

However, there are some problems with the study design. Because of the small sample size, it was not possible to reach any conclusions about differences between the sleep of the shiftworkers and the non-shiftworkers.

It was useful to have on board the experimenter/observer during phase two of the study. The participants became familiar with the experimenter and were able to provide him with information. One subject aboard the Beatrix mentioned the completion of work logbooks: 'that it takes longer as the shift goes on.' Another subject noted that there was no time between sleep and work to wake up, that 'work starts straight away, when we are not at our best, mentally.....it takes 12-15 minutes to wake up mentally.' Other (split-shift) participants claimed that the 'by the time your mind is working at it full rate, it is time to relax.' This causes other distractions to be made worse. Participants would like to work for longer, and then have more time off. They also claim that the asymmetry in alertness between those coming onto-, and those going off-watch, causes problems.

Although this study produced a good response rate from participants regarding subjective data, this may be attributed to the presence of the researcher on board ship during phase 2 of the study, which leads us to suggest having reward systems, with incentives to improve response rates. Also this could be used to help in the collection of objective (actimeter) data, as participants can sometimes forget to replace actimeters after bathing etc., and this leaves 'gaps' in the data.

CONCLUSIONS

This investigation indicates that the methodologies employed within are effective, and suitable for testing seafarers while at work (at sea) and at home. There is no real evidence to suggest that there is a severe problem with fatigue aboard such passenger ferries. This study can contribute towards a standard with which to measure fatigue etc. aboard other types of ship, with more demanding schedules.

There is further need to conduct parallel research, on anecdotal evidence, and fatigue-related marine accident reports. Clearly there is strong support for Brown's (1989) suggestion that more information relevant to circadian factors is required in the investigation of accidents, and that this information should be logged for all watchkeepers, and available for monitoring by the Master of the ship.

The full extent of the fatigue problem in the maritime industry is not yet clear, due to the lack of vital information relevant to circadian biological rhythms and also a lack of any real objective evidence on the problem. More detailed investigations such as this one, investigating fatigue on different ship types, and in different areas of the world are necessary to obtain a true picture.

References

Åkerstedt, T. & Gillberg, M. (1990) Subjective and Objective sleepiness in the active individual. International Journal of Neuroscience, 52, pp. 29-37.

Åkerstedt, T. (1995) Work hours, sleepiness and the underlying mechanisms. Journal of Sleep Research 4, Suppl. 2, 15-22.

Åkerstedt, T. (1995) Work hours, sleepiness and accidents: Introduction and Summary. Journal of Sleep Research 4, Suppl. 2, 1-3.

Brown, I.D. (1989) Study into Hours of Work, Fatigue, and Safety at Sea. Medical Research Council.

Colquhoun, W.P. (1985) Hours of work at sea, watchkeeping schedules, circadian rhythms and efficiency. Ergonomics 28, 637-653.

Colquhoun, W.P. (1987) A shipboard study of a four crew rotating watchkeeping system. Ergonomics 30, 1341-1352.

Crainer, S. (1993) Zeebrugge: Learning from Disaster, Lessons in Corporate Responsibility. Herald Families Association.

Dinges, D.F. (1995) An Overview of Sleepiness and Accidents. Journal of Sleep Research 4, Suppl. 2, 4-14.

Donaldson (1994) Safer Ships, Cleaner Seas. Report of Lord Donaldson's Enquiry into the Prevention of Pollution from Merchant Shipping.

Folkard, S. & Monk, T.H. (eds. 1985) Hours of Work: temporal Factors in Work Scheduling. John Wiley & Sons.

Harma, M. (1995) Sleepiness and Shiftwork: Individual differences. Journal of Sleep Research 4, Suppl. 2, 57-61.

Horne, J.A. & Reyner, L.A. (1995) Sleep Related Vehicle Accidents. British Medical Journal, Vol. 310, pp 565-567.

Kecklund, G. & Åkerstedt, T. (1995) Effects of timing of shifts on sleepiness and sleep duration. Journal of Sleep Research 4, Suppl. 2, 47-50.

Knauth, P. (1995) Speed and direction of shift rotation. Journal of Sleep Research 4, Suppl. 2, 41-46.

MIIU: Marine Incident Investigation Unit (Australia). Report No. 95 (1996).

Monk, T.H. & Folkard, S. (1992) Making Shift Work Tolerable. Taylor & Francis.

Reyner, L.A. (1995) Sleep, Sleep Disturbance and Daytime Sleepiness in Normal Subjects. Loughborough University Doctoral Thesis.

Rosa, R. (1995) Extended workshifts and excessive fatigue. Journal of Sleep research 4, Suppl. 2 51-56.

Rutenfrantz, J., Knauth, P. & Colquhoun, W.P. (1976) Hours of work and shiftwork. Ergonomics 19, 331-340.

Sanquist, T.F. et al. (1997) Work Hours, sleep patterns and fatigue among merchant marine personnel. Journal of Sleep Research, 6, 245-251

Torsvall, L. et al. (1987) Sleep at sea: A diary study of the effects of unattended machinery space watch duty. Ergonomics 30, 1335-1340.

Part IV.

New Approaches to Managing Fatigue

19

THREE FATIGUE MANAGEMENT REVOLUTIONS FOR THE 21ST CENTURY

Ronald R. Knipling, Ph.D., Chief, Research Division, Office of Motor Carrier Research and Standards, Federal Highway Administration, U.S. Department of Transportation, Washington, D.C.

INTRODUCTION

The late 1990s and early 2000s will be remembered as an exciting, transformational time in the history of commercial motor vehicle (CMV) driver fatigue management. The next few years will likely see *three* successive revolutions in the way that CMV driver fatigue and alertness are managed in the United States and elsewhere in the industrialized world (see Figure 1). In the U.S., the U.S. Department of Transportation Federal Highway Administration (FHWA) and its sister agency, the National Highway Traffic Safety Administration (NHTSA), are working to ensure that each of these revolutions results in safety and economic benefits for drivers, industry, and the U.S. public (FHWA, 1998; Knipling, 1997a). This paper describes these revolutions and some of the major scientific, policy, and practical issues surrounding them.

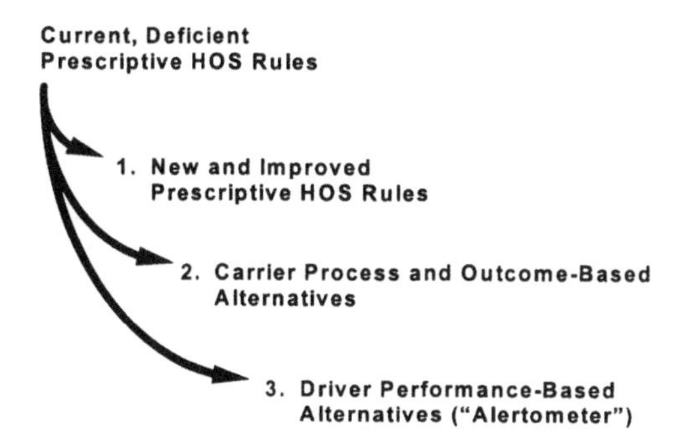

Figure 1

Revolution #1: Reform Of Prescriptive Hours-Of-Service

The first revolution in U.S. CMV driver fatigue management is happening already; i.e., reform of the 60-year-old prescriptive HOS rules. An Advance Notice of Proposed Rulemaking (ANPRM) on CMV driver HOS was issued in November, 1996 (61 Federal Register 57252) and was open for public comment through June 30, 1997. FHWA is currently reviewing the ANPRM docket and related scientific research to formulate a Notice of Proposed Rulemaking (NPRM) to be issued in mid-1998. The NPRM will likely focus on specific elements of the U.S. prescriptive rules such as the following:

- Maximum driving time (currently 10 hours).
- Minimum off-duty time (currently 8 hours).
- Work cycle implications of the above (current rules encourage an 18-hour cycle for maximum productivity in long-haul operations)
- Distinctions between driving and non-driving duty time; currently, the maximum continuous on-duty time is 15 hours, of which 10 hours may be driving.
- Day/night differentials to encourage day driving (currently there are none).
- Maximum cumulative on-duty hours (currently, 60 hours in 7 days or 70 hours in 8 days).
- Providing flexibility for different CMV operations and situations, while at the same time ensuring adequate daily and weekly rest for all CMV drivers.
- Special provisions; e.g., current sleeper berth provision allowing splitting of off-duty hours.

From a safety perspective, there are at least two major concerns regarding the current HOS rules. The first of these is the minimum continuous off-duty period of 8 hours. Since 7-8 hours sleep is a nightly requirement for most people (Rosekind, 1995; Belenky, 1997), an off-duty period of 8 hours does not appear to provide sufficient time for sleep for most CMV drivers. In the Driver Fatigue and Alertness Study (Freund and Vespa, 1997; Wylie *et al*,

1996), 8 hours off-duty were associated with less than 5 hours of nightly sleep, which was about 2 hours less than the drivers' stated ideal sleep time of about 7 hours.

A second concern is the duration of the work/rest cycle that is encouraged by the current regulations; for truckload operations, in particular, the most productive work/rest schedule is successive cycles of 10 hours driving followed by 8 hours off-duty. This 18 hour cycle is 6 hours disparate from an optimal circadian cycle of 24 hours. Moreover, the "rotation" over successive days is backwards -- drivers are being asked to accelerate their natural circadian cycle by six hours every cycle. "Backward" rotation is especially difficult and unnatural for humans and other animals (Rosekind, 1995).

In formulating a new prescriptive HOS regiment, FHWA is attempting to address the above two concerns while, at the same time, sustaining or even improving the productivity and operational flexibility of industry. There are quality-of-life issues in the HOS rulemaking as well; in particular, the need to provide drivers with sufficient work hours to make a living but also as much off-duty time as possible *at home* to promote normal family lives (Hill *et al*, 1998, McCartt *et al*, 1997).

In this and other regulatory initiatives, FHWA seeks to achieve "win-win-win" outcomes for public safety, CMV productivity, and driver quality-of-life. The FHWA firmly believes that the current HOS rules can be improved in all three respects and that reform of the U.S. prescriptive HOS rules in the dawn of the new century will constitute the first revolution in CMV driver fatigue management.

Revolution #2: Carrier Process And Outcome-Based Fatigue Management

Prescriptive HOS rules, no matter how good, encourage the management of *hours of work* as opposed to the direct management of *fatigue and alertness*. The second revolution in fatigue management will be regulations that focus on carrier fatigue management practices and processes, and related safety outcomes (Arnold and Hartley, 1997; Mahan, 1997; Hartley, 1997; Hartley, 1996). These are referred to here as "process and outcome-based" approaches. These approaches are not *purely* performance-based (as compared to Revolution #3, below), but they do nevertheless go beyond prescriptive HOS regulations. For example, the state of Queensland in Australia is conducting a pilot study of a Fatigue Management Program (FMP; Mahan, 1997) which focuses primarily on scheduling and other fleet management practices of participating fleets. These fleets are granted relief from prescriptive HOS limits but they must design and implement a comprehensive management program that prevents driver fatigue through enlightened scheduling practices, education of drivers and managers on sleep hygiene practices, health and fitness screening of drivers, and a zero-tolerance policy regarding fatigue-related crashes. This program emphasizes self-regulation but allows government verification/enforcement through periodic audits. The Western Australia Department of Transport has proposed a Code of Practice to encourage similar fleet management practices (Arnold and Hartley, 1997; Hartley, 1997) and the Canadian Trucking Association has proposed a FMP-like pilot study for the province of Alberta (Cooper, 1997). The U.S. Transportation Research Board of the National Research Council (NRC) has recognized the potential advantages of this approach and has proposed an NRC policy study of it (Morris, 1997).

While primarily non-technological, such management process approaches to fatigue management could be enhanced through the use of technologies such as the actigraph and related "expert systems" for assisting and/or assessing driver scheduling practices. These technologies do not directly *monitor* individual driving performance and/or driver alertness but they nevertheless represent potential enhancements over compliance with prescriptive HOS. The actigraph is a wrist-worn accelerometer and processor (Belenky, 1996). It measures and records arm movements which are highly correlated with sleep-wakefulness. The device scores sleep quantity and quality and predicts performance based on a sleep/performance model which factors in the amount and timing of sleep, time-of-day (as influenced by circadian rhythms), and number of hours awake. The use of actigraphs and related expert systems to manage driver alertness could represent a distinct improvement over conventional HOS because it would, in essence, allocate work hours based on actual *sleep* history as opposed to prescribing a rigid and generic "time on-time off" regimen that is insensitive to the actual amount of sleep obtained by a driver. In a sense, the actigraph would be prescriptive, but it would prescribe *sleep* requirements as opposed to work/non-work requirements. The actigraph could also greatly facilitate individual drivers' efforts to improve their sleep hygiene since it would provide daily feedback on amount of sleep and a continuous assessment of *predicted* alertness based on measured sleep/wakefulness status. It would encourage drivers to take greater responsibility for their own sleep hygiene practices and resulting levels of alertness.

FHWA is interested in pursuing process and outcome-based alternatives to prescriptive HOS. They are a step in the direction of performance-based regulation and focus fatigue management and regulation on the key cause of fatigue: lack of sufficient sleep. However, there are practical challenges faced by process and outcome-based alternatives. The first challenge is to achieve sufficient objectivity and verifiability in the specification and assessment of required management processes; for example, requiring that drivers have certain specified minimum and average *predicted* alertness levels based on actigraph and expert system predictions. Another challenge is the need for simplicity, both for fleets and government auditors. Like any other successful regulatory reform, fatigue management reform should reduce paperwork and management time, not increase it. This may become problematic if the required processes and outcomes are more complex than the current prescriptive HOS rules. Customization of rules to address individual carriers' routes and delivery requirements could be a "double-edged sword;" it could increase the relevance of the rules to those individual fleets but could add to the bureaucratic burden of the regulation, both from the perspective of the carrier and the government. A third challenge is the need for reliable, practical safety outcome measures. The most obvious outcome measure is frequency of fatigue-related crashes. However, such crashes are relatively rare events, even for poorly-managed fleets. The attribution of the crash to fatigue is problematic (Knipling and Wang, 1994) and is a potential source of contention among drivers, carriers, and regulators. Only the very largest of fleets would have sufficient vehicles and drivers for fatigue-related crash rate to be a reliable outcome measure of carrier fatigue management performance. And, even if fleets were sufficiently large for such measures to be reliable, crash rates could never be used to assess individual drivers except for those with extreme alertness problems. The above challenges would need to be addressed for a process and outcome-based approach to fatigue management to be acceptable and successful.

A potential method to help bridge the gap between prescriptive HOS and pure performance-based fatigue management, and to minimize potential shortcomings, would be a *hybrid* approach which requires certain elements of the prescriptive HOS (e.g., daily minimum time off) but which provides performance-based or management process and outcome-related alternatives to other prescriptive HOS elements (e.g., cumulative on-duty hours over the course of a work week). Such a hybrid approach might provide many of the benefits of non-prescriptive approaches while ensuring compliance with the most fundamental and inescapable prescriptive requirement: sufficient daily sleep.

REVOLUTION #3: DRIVER PERFORMANCE-BASED FATIGUE MANAGEMENT

The third, and most dramatic, revolution in fatigue management will be regulations that directly specify a criterion level of *alertness and driving performance* for individual drivers. Such an approach to fatigue management requires valid and practical *measures* of alertness and performance. Research over the past decade has demonstrated "proof-of-concept" of such measures (Wierwille *et al*, 1994, 1996a,b,c). **Figure 2** (from Knipling and Wierwille, 1994) illustrates this "proof of concept;" it shows results for one typical sleep-deprived subject on the Virginia Tech driving simulator. The solid line shows physiological drowsiness as measured by PERCLOS, the percent of time that the eyelids are closed 80% or more. The dots show aggregated driving performance (comprising measures of lane tracking and steering) derived through multiple regression analysis using group data from 12 subjects and cross-validated on 12 other subjects. Note the high correlation and the relative slowness of the trend toward drowsiness. High correlations between psychophysiological and behavioral deterioration imply that either, or both, processes could be used for detection. That is, deterioration in either domain indicates deterioration in both. The relative slowness of the trend implies that *gradations* of drowsiness are measurable (and thus recordable) and that there is the opportunity for feedback (and perhaps intervention) well before the driver actually reaches a dangerously impaired state. The fact that *group*-derived algorithms can be used to accurately assess *individual* drivers means that these phenomena and relationships are

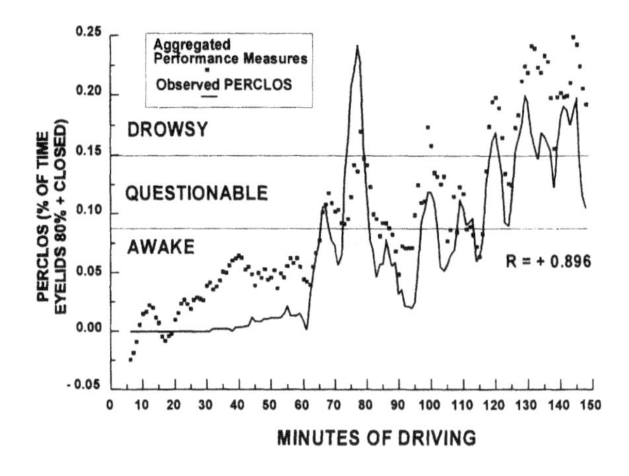

Figure 2

largely universal and theoretically may be applied "off the shelf" to most or all drivers.

Development efforts are underway to produce practical devices for providing such **continuous in-vehicle driver monitoring** in a real truck on a real road; (Grace, 1996; Rau, 1997, Knipling *et al*, 1996, FHWA, 1998). Many of these efforts are now subsumed under the U.S. DOT Intelligent Vehicle Initiative (IVI), a broad Intelligent Transportation System (ITS) program to improve the safety and efficiency of motor vehicles. The continued development, validation, operational testing, and deployment of in-vehicle driver monitoring is a key element in the overall IVI program, and will be pursued first in commercial vehicles.

The monitoring of individual driver alertness -- accomplished through technological devices to measure driving performance and/or the driver's psychophysiological state of alertness -- would constitute a direct assurance of a specified level of driving safety for individual drivers and, in the aggregate, carrier fleets and the trucking industry in general. Moreover, performance-based regulations employing monitoring technologies would generally be preferable to prescriptive HOS for both carriers and drivers because, in addition to the safety benefits, they would provide more operational flexibility and, potentially, greater productivity.

The potential advantages of continuous in-vehicle driver monitoring over other fatigue-related technologies are apparent from its name; i.e., the driving task and driver are themselves monitored and this monitoring is continuous. Of course, the monitoring of performance and/or alertness does not *in itself* enhance safety. Safety is enhanced through the use of decision algorithms to translate raw performance/ psychophysiological data into meaningful metrics of alertness and continuous feedback to drivers regarding their performance/alertness state. This feedback would prompt drivers to consider whether they need to stop for a nap or terminate their driving day. Most importantly, driver alertness data would also be recorded for post-trip review by drivers themselves and fleet managers. **Figure 3** presents a schematic of this system concept. If the use of such devices were permitted by federal motor carrier regulations, then State and Federal safety inspectors would also be able to directly review alertness records.

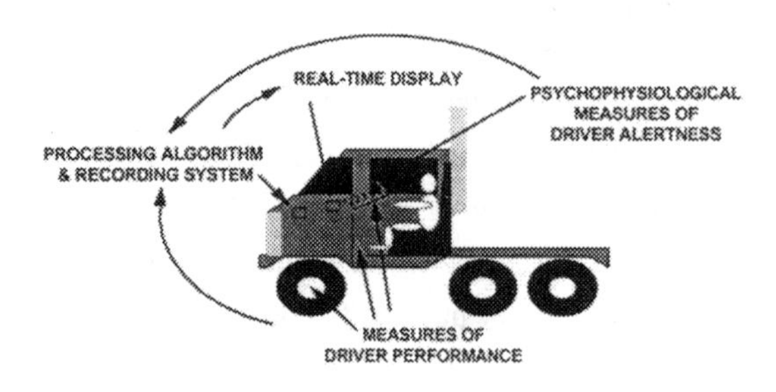

Figure 3

The design of the system driver interface is the subject of considerable research (Wierwille *et al*, 1996b) and a matter of potential debate. The interface could consist of a simple continuous indicator display of driver alertness level ("alertometer"), advisory messages, warning signals, or even alerting stimuli such as blasts of cool air. For commercial drivers, it is envisioned by FHWA that continuous in-vehicle monitors would be used in a manner analogous to current use of speedometers (standard equipment on all large trucks) combined with on-board speed recorders (used on many large trucks). On-board speed recorders enable post-trip review and management action as necessary to help ensure the development of a "habit of compliance" by drivers with speed limits and company policies. The FHWA deployment concept emphasizes measurement, display/feedback, recording, and establishment/enforcement of performance standards. Reliance on warning signals and alerting stimuli for fatigued drivers is de-emphasized; indeed, these may not be necessary or desirable as system components for *commercial* vehicle applications. This is a question for future FHWA and NHTSA research, in particular *in-situ* operational field tests employing prototype systems and instrumented vehicles to provide in-vehicle observations of driver behavior and performance.

Challenges -- and Reasons for Optimism

True performance-based fatigue management must function at the level of the individual driver; that is, each driver must be responsible for maintaining a specified level of alertness and performance as measured by *validated* on-board monitors. Validity is the most important requirement for alertness/performance monitors -- they must measure alertness accurately. Detection of fatigued states must also be *anticipatory*; that is, it must occur before a driver becomes so fatigued that there is an immediate danger of a crash. In addition, such devices must be *reliable* (i.e., durable in operational environments), *acceptable* to drivers and managers (e.g., not obtrusive or perceived as invasive of privacy), *affordable* (both marketable and cost-beneficial), and *user-friendly*. Such devices should reduce, not increase, paperwork and other regulatory compliance costs.

In spite of the above daunting challenges, there are several reasons to be confident that technology can be applied to provide performance-based fatigue management. As described above, landmark driving simulation studies by Walter Wierwille and others have provided scientific "proof-of-concept" (Wierwille *et al*, 1994, 1996a,b,c). These studies have demonstrated that both driving performance and human physiology deteriorate in predictable ways as drivers are overcome by fatigue, and that this process is gradual enough to permit measurement, display, and recording of various levels of fatigue. In particular, this research has demonstrated that eyelid droop measures of fatigue such as PERCLOS may be considered "gold standards" for the in-vehicle measurement of alertness and fatigue. Recent FHWA/NHTSA-sponsored laboratory validation studies conducted by David Dinges and his colleagues at the University of Pennsylvania have further demonstrated the validity of PERCLOS as a fundamental measure of alertness and resulting vigilance performance (Dinges, 1998).

The recently-completed FHWA-sponsored Driver Fatigue and Alertness Study (DFAS; Freund and Vespa, 1997; Wylie *et al*, 1996) provided additional proof-of-concept of alertness measurement. The DFAS found that drivers' subjective self-assessments of fatigue level are not very accurate. Objective performance-based or psychophysiological measures are likely to be more accurate than driver self-assessments. In addition, the DFAS found huge individual differences in driver susceptibility to fatigue. Fourteen percent of the drivers accounted for 54% of video-observed drowsiness episodes. Extrapolating these percentages to a detection system implies that a system capable of diagnosing the worst 14% of drivers at any given time would reduce the overall incidence of extreme drowsiness episodes by half!

Another reason for optimism is the potential *synergy* of multiple alertness/performance sensors. Numerous sensor types (e.g., steering movements, lane tracking, eyelid droops, head motion) have shown promise; systems which combine and integrate information from multiple sensor types are likely to be more accurate than single-sensor systems. Extremely high levels of detection accuracy seem theoretically possible.

A Psychological Revolution

The capability to continuously monitor, measure, and record driver alertness will certainly constitute a technological revolution, but will it also result in a psychological, behavioral revolution? FHWA research and regulatory planning are predicated on the conviction that continuous in-vehicle alertness monitoring technology will create the conditions for dramatic, positive, and long-term behavioral change.

Three elements of in-vehicle alertness monitoring and planned deployment strategies may combine synergistically to facilitate long-term positive changes in CMV driver lifestyle and behavior. The three elements are:

Performance Measurement and Feedback. In-vehicle alertness monitors will function as "alertometers" that provide continuous performance measurement and feedback to drivers. Alertness measurements may be used to generate summary statistics regarding overall driver alertness for a particular trip, week of driving, or even over a driving career. A fundamental principle of human psychology is that such feedback facilitates learning and long-term behavior change (Holland, 1975; Behavioral Science Technology, Inc., 1997). In group settings such as CMV fleets or, more broadly, the entire population of long-haul CMV drivers, such feedback can be used to establish group norms regarding expected or admirable levels of performance (Allen, 1987; Geller *et al*, 1989), further increasing their motivational effects. Most people strive to improve their performance levels when given continuous or frequent objective feedback regarding their performance. Technology makes such feedback possible.

Incentives. An even more fundamental principle of behavior is the Law of Effect; i.e., the principle that behavior is controlled by rewards and punishments (Holland, 1975; Geller *et al*, 1989). Behavior that is rewarded increases in frequency and over the long term can become firmly established as a behavioral "habit." Deployment of in-vehicle alertness monitors will be most successful if supported by regulatory incentives offered by the government to industry (e.g., full or partial relief from prescriptive HOS) and/or by incentives offered by industry to drivers (e.g., monetary or other rewards). Incentives would be tied not just to the use of these devices but rather to the achievement of a criterion level of driving alertness/safety performance.

Education and Training. People cannot change their lifestyles and behavior unless they know *how* to change. Understanding *why* certain behaviors affect sleep and alertness positively or negatively also increases the probability of change. Efforts are already underway to educate the entire CMV industry regarding the risks of fatigue and ways to enhance sleep and alertness (e.g., American Trucking Associations Foundation [ATAF], 1996, 1997; Knipling, 1998). Continuing such efforts will be essential for supporting driver behavior changes in the future.

The above three elements form a "motivational triangle" for long-term behavioral change as shown schematically in **Figure 4**. FHWA OMC's long-term program involves R&T, rulemaking, and education/outreach to build this motivational triangle as the foundation for CMV driver fatigue management for the 21st century.

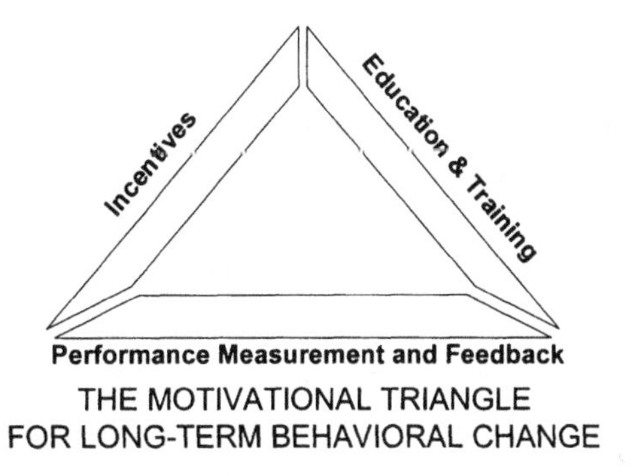

Figure 4

CRASH PROBLEM SIZE AND COUNTERMEASURE "BUDGET"

To be successful, fatigue countermeasures must be economically viable; their cost must be justified by crash-reduction and/or productivity enhancement benefits. Furthermore, unless they are mandated, they need to be marketable to the industry. A first step toward assessing the economic viability of safety devices is to estimate the monetary costs associated with their target crash problem. **Table 1** (excerpted from Knipling *et al*, 1996) provides a monetary perspective on the U.S. CMV drowsy driver crash problem. Two levels of monetary value for these crashes are shown:

- "Economic" (E) values based on narrow economic loss criteria; i.e., the cost of the crash in terms of actual monetary loss including medical care, legal services, vehicle repair/replacement and, significantly, lost productivity.

- "Comprehensive" (C) values representing a higher level of monetary valuation; i.e., incorporating both economic losses (as described above) *and* a valuation of less tangible human consequences such as "pain and suffering" and loss of life. Comprehensive value includes not only the monetary value *of* crash consequences but also the additional monetary value society *places upon* crash consequences such as loss of life or disability.

Table 1 shows these two levels of monetary value for four different monetary metrics for three vehicle type categories: all vehicle types (combined), combination-unit trucks, and single-unit trucks. The four metrics are total annual U.S. monetary cost, per-police-reported crash cost, per registered vehicle annual cost, and per vehicle operational life cycle cost. The crash statistics forming the basis of these economic estimates were from the NHTSA General Estimates System (1989-93 average), but with a 50% correction for missed drowsy/fatigue driver cases (per the findings of Knipling and Wang, 1995 and Wang *et al*, 1996a). In other words, it was assumed that the "real" number of drowsy/fatigue driver crashes was 50% greater than GES statistics indicate. This is considered a reasonable but *conservative* assumption based on available research. It is conservative because it considers only estimates of fatigue as the *principal* cause of crashes; crashes where fatigue plays a *contributory* role in causing driver attentional lapses or other mental errors are not included because their number is completely unknown.

Table 1 shows that combination-unit trucks represent a relatively small portion of the overall U.S. national picture (e.g., $280M of $3.8B, or 7%), but that their per-vehicle costs, both annually and over a vehicle life cycle, are many times that of other vehicles. The high per-vehicle-produced monetary costs for combination-unit trucks ($2,060 [E] and $5,600 [C]) mean that these vehicles are by far the most promising platforms for cost-effective applications of vehicle-based fatigue countermeasures. A positive benefit/cost ratio will be much easier to achieve for combination-unit trucks than for any other vehicle type. An overall countermeasure deployment strategy points to combination-unit trucks as the first target for device implementation, followed by other vehicle types as devices are refined to bring costs come down and effectiveness up.

In contrast, single-unit trucks are a relatively unpromising platform for the application of vehicle-based fatigue technologies. Their per-vehicle-life cycle crash costs are less than half than those of the vehicle fleet in general and less than 1/20 of those of combination-unit trucks. These statistics reflect the combined effects of low fatigue-related crash rate per mile traveled (related to their primarily daytime, local/short haul usage patterns), relatively small mileage exposures, and the lower severity of their crashes (compared to combination-unit trucks) (Massie *et al*, 1997; Knipling *et al*, 1996).

Table 1.
Monetary Estimates of the U.S. Drowsy/Fatigued Driver Crash Problem (excerpted and adapted from Knipling *et al*, 1996)

Vehicle Type Category: / Monetary Statistical Metric:		All Vehicles	Combo-Unit Trucks	Single-Unit Trucks
Total annual U.S. monetary cost*	E	$3.8B	$280M	$32M
	C	$12.5B	$765M	$87M
Percent of all annual crash costs for this vehicle type**	E	2.5%	3.2%	0.64%
	C	3.1%	3.8%	0.81%
Per-police-reported crash cost	E	$34K	$87K	$48K
	C	$120K	$234K	$127K
Crash costs per registered vehicle annually*	E	$20	$170	$10
	C	$68	$470	$20
Crash costs per vehicle operational life cycle*	E	$220	$2,060	$90
	C	$730	$5,600	$240

* Inflated by 50% for undercounting in GES per the narrative discussion; ** Calculated using denominator (same years, same monetary algorithms) from Wang *et al*, 1996b

E: "Economic" value; **C**: "Comprehensive" value; **B**: Billion, **M**: Million.

The relation of drowsiness/fatigue crashes to the overall crash picture for each respective vehicle type category (as quantified in Wang *et al*, 1996b using the same years of GES data and the same monetary algorithms) varies somewhat among the three vehicle types shown in Table 1. The combination-unit truck percentages (e.g., 3.2%, E) are higher than those for all vehicles (2.5%, E), but the single-unit truck percentages are dramatically lower (e.g., 0.64%, E), reflective of their primarily daytime, short-haul operational use patterns.

Based on the statistics in Table 1 and various assumptions regarding device effectiveness, one may estimate the maximum "break-even" costs (annual or vehicle life cycle) of fatigue countermeasures. For combination-unit trucks, the average vehicle life cycle fatigue-related economic (E) crash costs are estimated to be $2,060. Accordingly, a device preventing 100% of these crashes would be worth exactly $2,060 in life cycle costs. This means that a device preventing 50% of fatigue crashes -- a more realistic assumption -- would be worth $1,030. Based on these calculations and assumptions, $1,000 per vehicle (life cycle cost; not just purchase price) seems to be a reasonable target price for these devices.

The above $1,000 "break-even" cost estimate should be regarded as very conservative. As noted above, it is based on the more conservative of the two monetary estimates (E rather than C) and it is based on a conservative count of target crashes. Even more importantly, it does not consider possible productivity gains resulting from technology-enabled performance-based fatigue management. A 5-10% increase in per-vehicle productivity would easily amount to several thousand dollars in economic benefits *annually* for the life of each vehicle.

RESEARCH & TECHNOLOGY CHALLENGES

The numerous research and technology (R&T) and deployment questions and challenges relating to CMV driver monitoring may be summarized under six headings: target problem, sensor suite, processing unit, driver interface, system deployment, and future applications (Knipling, 1997b). A few of the most important and intriguing of these are stated below. FHWA and its R&T partners have a number of programs underway addressing these issues. In particular, NHTSA is conducting research on many of these issues, as discussed in Rau (1997).

Target Problem

- CMV driver drowsiness/fatigue is recognized as a priority concern (FHWA, 1998), but the crash problem has not been well-quantified. Exactly how big is the problem, and what does this tell us about potential crash-reduction benefits from countermeasures?

- In addition to its role as a primary cause of some crashes, fatigue plays a *contributing* role to some percentage of the many crashes caused by attentional and other mental lapses. How many such crashes are causally related to drowsiness/fatigue (Knipling *et al*, 1996)?

Sensor Suite

- Can a one-sensor driver monitoring system (e.g., steering, eye activity monitoring) ever be accurate enough?

- If more than one sensor is needed, what are the best combinations? In other words, what are the most promising sensor synergies (Wierwille *et al*, 1994, Wierwille *et al*, 1996a,b,c)?

- Actigraphs *predict* current alertness but they do not *measure* it. As such, can they be the primary technological basis for alertness management as envisioned in Revolution #2?

- When will there be low-cost, reliable hardware for lane-tracking? There have been a number of recent and current development efforts in this area but no product appears ready for marketing and operational use (ATAF, 1996).

- When will there be low-cost, reliable, valid, and acceptable hardware for eye-monitoring? Past and recent validation studies of PERCLOS reinforce the tremendous need for such devices and the potential payoffs to vendors who successfully develop such devices.

- What are the best "leading indicators" of drowsiness/fatigue that can be captured well before there is risk of falling asleep and/or crashing? Are PERCLOS and correlated performance decrements valid and predictive at the very earliest stages of degradation?

Processing Unit

- How is independent information from two or more sensors best combined; e.g., AND versus OR decision functions (Wierwille *et al*, 1996)?

- After they are deployed, can "off the shelf" processing units be designed to further customize the system and its algorithm(s) to individual performance and behavior (Rau, 1997)?

- *If* warning alarms are used, what is the optimal system threshold given the trade-off between detection accuracy and false alarm rate (Knipling and Wierwille, 1994)? This issue is strongly related to the human factors of the driver interface (see below); i.e., how high a false alarm rate will drivers accept?

- Relating to the above, should algorithms be driver-adjustable; i.e., high or low drowsiness threshold with attendant accuracy and false alarm rates?

Driver Interface

- Should there be a simple continuous indicator of driver alertness level -- an "alertometer?" Assuming an affirmative answer, what is the best display design?

- When drowsiness/fatigue is detected, should a warning alarm (with alerting stimuli) be activated (as opposed to a simple advisory message along with the indicator)? Studies of the effectiveness of alerting stimuli (e.g., noise, cold air, vibration) on sleep-deprived subjects have generally indicated that their arousing effects are short-lived (Åkerstedt, 1997; Dinges, 1998). Perhaps the negative ramifications of the use of such stimuli (e.g., drivers having a false sense of security and continuing to drive when they should stop; in effect, compensatory risktaking) would outweigh the positive ramifications (temporarily forestalling loss-of-alertness).

- *If* warning/alerting stimuli are needed, what stimuli should be used (Wierwille *et al*, 1996b)? Should drivers be allowed to choose and/or adjust their own warning stimuli?

System Deployment

- Will at-risk driving *decrease* or *increase* after the installation of a drowsiness monitoring system? In other words, will behavioral adaptation be positive (i.e., feedback-supported lifestyle change and performance enhancement) or negative (e.g., compensatory risktaking) (Tijerina, 1995)? As discussed above, FHWA believes that behavioral adaptation will be positive (i.e., that at-risk driving will decrease) -- but this must be demonstrated in operational tests.

- Will CMV drivers and their unions accept driver monitoring systems, which some may perceive as "big brother" surveillance (McCartt *et al*, 1997)? The FHWA program assumes that drivers will accept these systems if they understand them and if there are benefits from their use.

- For commercial drivers, will there be "motivational synergy" among a triad of factors -- performance measurement/feedback, incentives, and education/training -- which leads to long-term behavior change in the direction of better lifestyles and sleep hygiene? As discussed above, this is the hypothesis of the FHWA program.

- If performance-based fatigue management programs are successful, will it be because they "chop off the tail" of the current driver alertness level distribution (thereby preventing asleep-at-the-wheel crashes) or because they move the whole curve (thereby, presumably, reducing the risk of all crash types)? Or both?

Future Applications

- After further development on the individual systems, can driver monitoring data (e.g.., steering, lane-tracking, psychophysiological) be integrated with data from other crash avoidance systems (e.g., headway monitoring) to enhance both systems?

- Can sensors and algorithms developed for drowsiness/fatigue detection generalize to other forms of driver inattention (distraction, "daydreaming") or impairment?

CONCLUSION: FATIGUE MANAGEMENT, ALERTNESS, AND THE REDUCTION OF CRASHES

FHWA's short-term focus is on Revolution #1; the current U.S. prescriptive HOS rules have significant deficiencies and must be reformed. Successful implementation of these reforms over the next few years is essential but, ironically, may actually reduce government and industry incentives to pursue the more dramatic changes and opportunities possible through Revolutions #2 and #3. Nevertheless, there are inherent limitations to prescriptive HOS regulations, no matter how good (Hartley, 1996, 1997). The long-term FHWA strategic program for improving CMV driver fatigue management will pursue performance-based alternatives to prescriptive HOS rules. FHWA assumes that in-vehicle driver monitoring technology will be ready, that incentives for improved fatigue management will be available and effective, and that drivers will respond in a responsible and positive manner when given accurate, continuous feedback on their alertness/performance and education on how to obtain sufficient sleep and make other lifestyle changes to ensure safe driving. If successful, Revolution #3 will be a technological, regulatory, *and* behavioral revolution.

FHWA's "win-win-win" goals for all of these initiatives are enhanced safety, productivity, and quality-of-life. Enhanced productivity and quality-of-life will be ensured by the greater flexibility possible through performance-based regulations. What about the primary goal, enhanced safety?

In spite of the intense interest in motor vehicle driver drowsiness and fatigue over the past decade, it should be remembered that driver drowsiness/fatigue ("asleep-at-the-wheel") is *not* known to be a leading *principal* cause of motor vehicle crashes, including large truck crashes (Massie *et al*, 1997; Wang *et al*, 1996a, Knipling *et al*, 1996, Najm et al, 1995; NHTSA, 1995; Knipling and Wang, 1994; Treat *et al*, 1979). Reduction in the number of asleep-at-the-wheel crashes probably will not in itself *dramatically* change the overall large truck crash picture. Based on the monetary statistics presented earlier in Table 1, a 50% reduction in asleep-at-the-wheel crashes would reduce the overall combination-unit truck crash problem size by about 2%. This is significant and probably cost-beneficial (depending on the cost of the interventions), but not *dramatic*. It should be noted that certain *types* of combination-unit truck crashes would be dramatically reduced; most notably, fatal-to-the-truck driver crashes would be reduced by at least 5% (based on NHTSA, 1995) and possibly as much as 15% (based on NTSB, 1990).

The above hypothesized ~2% reduction in the overall combination-unit truck crash problem size (based on a 50% reduction in asleep-at-the-wheel crashes) would be significant, but it is hypothesized here that the biggest benefits of improved fatigue management would come from "across-the-board" enhancements to CMV driver alertness and driving performance. Although the term "alertness" is often used as the simple converse of "fatigue," it really relates to much more than just the avoidance of asleep-at-the-wheel crashes. The largest known cause of traffic crashes is driver *attentional lapses*, including distraction, "daydreaming," and non-descript states of inattention (Wang *et al*, 1996a, Najm et al, 1995; Knipling and Wang, 1994; Treat *et al*, 1979). The contributory role of fatigue to driver inattention *in the real driving world* is not known, and is the subject of current FHWA and NHTSA instrumented vehicle research (Knipling *et al*, 1996; Wang *et al*, 1996a). But we do know that the alert, effective driver is not just awake; he or she is *attentive*. Alertness monitoring or other improved fatigue management approaches will support both "awakeness" and, in all likelihood, "attention." Defensive driving is also a key element in CMV safety; more than two-thirds of truck-car crashes are not related primarily to unsafe actions or errors of the truck driver at all but rather to those of the car driver (NHTSA, 1996). The alert driver avoids attentional lapses of his or her own and also is much more likely to be a defensive driver who anticipates and compensates for the mistakes of others. The safety benefits of improved CMV driver fatigue management will include the avoidance of asleep-at-the-wheel crashes as well as, and more importantly, a general optimization of driver behavior, attention, and performance.

REFERENCES

Åkerstedt, T. (1997) Readily available countermeasures against operator fatigue. *Proceedings of the International Conference on Managing Fatigue in Transportation*. American Trucking Associations Foundation, Inc. (ATAF), Pp. 105-122, April.

Allen, R.F. (1987) Group norms: their influence on training. *Training and Development Handbook, 3rd Edition*, Edited by Robert L. Craig, American Society for Training and Development, Pp.180-194.

American Trucking Associations Foundation, Inc. (ATAF). (1996) *Awake at the Wheel.* (Sleep education brochure, available from ATAF, (703) 838-1966, FAX (703) 838-0291). December.

Arnold, P.K. and Hartley, L.R. (1997) Hours of service for truck drivers: recent research on fatigue and a proposed approach to dealing with the problem. *Proceedings of the International Large Truck Safety Symposium.* University of Tennessee Transportation Center, Knoxville, Pp. 189-198, October.

ATAF. *Proceedings of the International Conference on Managing Fatigue in Transportation.* (1997) (Available from Government Institutes, Inc., 4 Research Place, Suite 200, Rockville, MD 20850), Tampa, April.

ATAF. *Proceedings of the Technical Conference for Enhancing Commercial Motor Vehicle Driver Vigilance.* (1996) (Available from Dr. William Rogers, ATAF, (703) 838-1966, FAX (703) 838-0291). Tysons Corner, VA, December.

Behavioral Science Technology, Inc. (1997) Ojai, California. *BST Safety Seminar: The First Step to Behavior-Based Safety; TQM Principles Applied to Safety.*

Belenky, G. (1996) Truck driver sleep deprivation study. *Proceedings of the Technical Conference for Enhancing Commercial Motor Vehicle Driver Vigilance.* Published by the ATAF in conjunction with FHWA and NHTSA. Pp. 53-63. December.

Belenky, G. (1997) Sustaining performance during continuous operations; the U.S. Army's Sleep Management System. *Proceedings of the International Conference on Managing Fatigue in Transportation.* ATAF. Pp. 95-104. April.

Cooper, G. (1997) Plan for Alberta pilot study of fatigue management program. Presentation to fatigue management working group of the Canadian Council of Motor Transit Administrators, Toronto, June.

Dinges, D. (1998) Personal communication based on FHWA/NHTSA-sponsored validation study of psychophysiological indices of fatigue. Univ. of Pennsylvania.

Freund, D.M. and Vespa, S. (1997) The Driver Fatigue and Alertness Study: from research results to safety practices. *Proceedings of the International Large Truck Safety Symposium.* University of Tennessee Transportation Center, Knoxville, Pp. 179-188, October.

FHWA OMC. (1998) Update on the U.S. FHWA Commercial Driver Fatigue Research & Technology, Rulemaking, Education/Outreach, and Enforcement Program. *Third*

International Conference on Fatigue and Transportation, Fremantle, Australia, February.

Geller, E.S., Lehman, G.R., and Kalsher, M.J. (1989) *Behavior Analysis Training for Occupational Safety*. Make-A.Difference, Inc., Newport, VA.

Grace, R. (1996) Field testing a prototype drowsy driver detection and warning system. *Proceedings of the Technical Conference for Enhancing Commercial Motor Vehicle Driver Vigilance.* ATAF. Pp. 27-30. December.

Hartley, L.R. (1997) Beyond one size fits all hours of service regulations. *International Conference Proceedings: Managing Fatigue in Transportation*, ATAF, Pp. 9-28, Tampa, April.

Hartley, L.R. (Ed.). (1996) Recommendations of the Second International Conference on Fatigue. Institute for Research in Safety and Transport, Murdoch University, Perth, Western Australia.

Hill, M.J., Hudson, N.W., and Lantz, B.M. (1998) Commercial Vehicle Driver Family Issues Assessment. Paper # 980098. Transportation Research Board 77th Annual Meeting, Washington, DC, January 11-15.

Holland, M.K. (1975) *Using Psychology: Principles of Behavior and Your Life*, Little, Brown, and Company, Boston.

Knipling, R.R. (1997a) The technologies, economics, and psychology of commercial motor vehicle driver fatigue monitoring. *Proceedings of the International Large Truck Safety Symposium*. University of Tennessee Transportation Center, Knoxville, Pp. 215-224, October.

Knipling, R.R. (1997b) Supplement on driver monitoring: FHWA OMC R&T programs and scientific challenges. *Proceedings of the Transportation Research Board Conference on Intelligent Transportation Systems, Highway Safety, and Human Factors*, Washington, DC, March.

Knipling, R.R. (1996) The promise of technology for fatigue management: The Federal Highway Administration Perspective. *Proceedings of the Technical Conference for*

Enhancing Commercial Motor Vehicle Driver Vigilance. Published by ATAF in conjunction with FHWA and NHTSA. Pp. 8-13. December.

Knipling, R.R. and Wang, J.S. (1995) Revised estimates of the U.S. drowsy driver crash problem size based on General Estimates System case reviews. *39th Annual Proceedings, Association for the Advancement of Automotive Medicine*, Chicago, October.

Knipling, R.R., and Wang, J.S. (1994) *Crashes and Fatalities Related to Driver Drowsiness/Fatigue.* National Highway Traffic Safety Administration Research Note, U.S. Department of Transportation, November.

Knipling, R.R., Wang, J.S, and Kanianthra, J.N., (1996) Current NHTSA drowsy driver R&D. *Proceedings of the 15th International Technical Conference on Enhanced Safety of Vehicles (ESV)*, Melbourne, May.

Knipling, R.R. and Wierwille, W.W., (1994) Vehicle-based drowsy driver detection: current status and future prospects. *Proceedings of the IVHS America 1994 Annual Meeting,* Pp. 245-256, Atlanta, April 17-20.

Mahan, G. (1997) New approaches to fatigue management: a regulator's perspective. *International Conference Proceedings: Managing Fatigue in Transportation*, ATAF, Pp. 145-154, Tampa, April.

Massie, D.L., Campbell, K.L., and Blower, D. (1997) Short-Haul Trucks and Driver Fatigue. University of Michigan Center for National Truck Statistics. Report prepared for the FHWA Office of Motor Carriers, In press. September.

McCartt, A.T., Hammer, M.C., and Fuller, S.Z. (1997) Understanding and managing fatigued driving; a study of long-distance truck drivers in New York state. *Proceedings of the International Large Truck Safety Symposium.* University of Tennessee Transportation Center, Knoxville, Pp. 199-213, October.

Morris, J. (1997) Summary of National Research Council planning meeting on commercial operator fatigue and transportation safety. Transportation Research Board of the National Research Council. July.

Najm, W.G., Mironer, M., Koziol, J.S.Jr., Wang, J.S., and Knipling, R.R. (1995) *Synthesis Report: Examination of Target Vehicular Crashes and Potential ITS Countermeasures.* Report for Volpe National Transportation Systems Center, DOT HS 808 263, DOT-VNTSC-NHTSA-95-4, June.

National Highway Traffic Safety Administration (NHTSA) Office of Crash Avoidance Research. (1995) *Analysis of Combination-Unit Truck Fatal Crashes by Location of Fatality.* NHTSA Research Note, September.

NHTSA (1996) National Center for Statistics and Analysis. *Traffic Safety Facts: Large Trucks.*

National Transportation Safety Board (NTSB). (1990) *Safety Study: Fatigue, Alcohol, Other Drugs, and Medical Factors in Fatal-to-the-Driver Heavy Truck Crashes.* PB90-917002, NTSB/SS-90/01, February.

Rau, P.S. (1997) Heavy vehicle drowsy driver detection and warning systems: technical challenges and scientific issues. *Proceedings of the 1997 International Large Truck Safety Symposium.* Knoxville, Pp.173-178, October.

Rosekind, M.R. (1995) Physiological considerations of fatigue. *Fatigue Symposium Proceedings.* National Transportation Safety Board and NASA Ames Research Center. November.

Tijerina, L. (1995) Key human factors research needs in intelligent vehicle-highway system crash avoidance. *Human Performance and Safety in Highway, Traffic, and ITS Systems*, Transportation Research Board Record No. 1485, Transportation Research Board, National Research Council, National Academy Press.

Treat, J.R., Tumbas, N.S., McDonald, S.T., Shinar, D., Hume, R.D., Mayer, R.E., Stansifer, R.L., & Catellan, N.J. (1979) *Tri-Level Study of the Causes of Traffic Accidents: Final Report Volume I: Causal Factor Tabulations and Assessments*, Institute for Research in Public Safety, Indiana University, DOT Publication No. DOT HS-805 085.

Wang, J.S., Knipling, R.R., and Goodman, J.R. (1996a) The role of driver inattention in crashes; new statistics from the 1995 Crashworthiness Data System. *40th Annual Proceedings, Association for the Advancement of Automotive Medicine*, Vancouver,

October.

Wang, J.S., Knipling, R.R., and Blincoe, L.J. (1996b) Motor vehicle crash involvements: a multi-dimensional problem size assessment. *Proceedings of the Intelligent Transportation Society of America Sixth Annual Meeting.* Houston, TX, April.

Wierwille, W.W. M.G. Lewin, and R.J. Fairbanks, III. (1996a) *Final Report: Research on Vehicle-Based Driver Status/Performance Monitoring; Part I.* Vehicle Analysis and Simulation Laboratory, Virginia Polytechnic Institute and State University, Publication No. DOT HS 808 638, September.

Wierwille, W.W. M.G. Lewin, and R.J. Fairbanks, III. (1996b) *Final Report: Research on Vehicle-Based Driver Status/Performance Monitoring; Part II.* Vehicle Analysis and Simulation Laboratory, Virginia Polytechnic Institute and State University, Publication No. DOT HS 808 638, September.

Wierwille, W.W. M.G. Lewin, and R.J. Fairbanks, III. (1996c) *Final Report: Research on Vehicle-Based Driver Status/Performance Monitoring; Part III.* Vehicle Analysis and Simulation Laboratory, Virginia Polytechnic Institute and State University, Publication No. DOT HS 808 638, September.

Wierwille, W.W., Wreggit, S.S., Kirn, C.L., Ellsworth, L.A., and Fairbanks, R.J. (1994) *Research on Vehicle-Based Driver Status/Performance Monitoring; Development, Validation, and Refinement of Algorithms for Detection of Driver Drowsiness.* Vehicle Analysis and Simulation Laboratory, Virginia Polytechnic Institute and State University, VPISU Report No. ISE 94-04, NHTSA Report No. DOT HS 808 247, December.

Wylie, C.D., Shultz, T., Miller, J.C., Mitler, M.M., and Mackie, R.R. *Commercial Motor Vehicle Driver Fatigue and Alertness Study: Technical Summary.* Essex Corporation. FHWA Report No. FHWA-MC-97-001, November, 1996.

20

New York State's Comprehensive Approach To Addressing Drowsy Driving

Anne T. McCartt, Mark C. Hammer, Sandra Z. Fuller, Institute for Traffic Safety Management and Research, University at Albany, State University of New York

Introduction

In the U.S., scarce resources in highway safety are allocated largely in accordance with the perceived effects of specific problems on highway crashes. Thus, the fact that sleepiness-related driving[1] has not been a major focus of highway safety initiatives aimed at the general motoring public may result, at least in part, from analyses of crash data that suggest that sleepiness-related driving represents a relatively low crash risk, when compared with other risky behaviors such as alcohol-impaired driving or speeding. Researchers with the U.S. National Highway Traffic Safety Administration estimate that drowsiness or fatigue is a causal factor in 1.2 - 1.6% of all police-reported crashes and 3.6% of fatal crashes (Knipling and Wang, 1994, 1995). However, police crash reports are an incomplete source of information on the role of sleepiness in crashes due to the lack of standardized reporting by states on the role of fatigue/drowsiness in crashes, the difficulty of police officers in detecting the involvement of drowsiness in a crash, and the fact that other contributing factors (e.g., driver intoxication) may be more easily identified. The limitations of police-reported crash data have led to the use of additional research methods to study drowsy driving. These methods have included, for example, self-report studies, laboratory studies, and *in situ* over-the-road studies.

Over the past few years, there has been a growing interest in the U.S. in the role of sleepiness and other forms of inattention in highway crashes. In addition, for many years, government agencies, transportation professionals, researchers, the motor carrier industry, safety advocates, and others have been concerned with the extent and causes of sleepiness-related driving among long-distance truck drivers. As other papers at this conference will discuss in some detail, a primary component of the government response to this concern has been the promulgation and enforcement of federal hours-of-service regulations.

---

[1] The terms "fatigued driving," "drowsy driving," and "sleepiness-related driving," are used synonymously.

In the U.S., the responsibility for highway safety is shared among the various levels of government. With some governmental functions, such as regulations related to vehicle safety, the federal government has the primary responsibility. With other functions, such as the enactment of traffic laws, the states have the primary responsibility. The responsibility for still other functions, such as enforcement and education, is primarily shared between state and local governments. To further complicate the situation, at each level of government the responsibility for highway safety is shared among various government agencies, and in recent years the non-profit and private sectors have emerged as important partners.

This paper focuses on the role of the states in combating drowsy driving and describes an innovative state program. Through the enforcement of federal regulations, the enactment and enforcement of traffic laws and regulations, the licensing of drivers, the registration of vehicles, public information and education efforts, and other functions, the states share the responsibility for highway safety with the federal government and localities. The recent "devolution" of power to the states has resulted in an even stronger role for the states in establishing highway safety priorities. Although federally-sponsored research and grant programs influence state policies and programs, highway safety priorities and programs vary widely among the states. Under the leadership of the National Highway Traffic Safety Administration, state and local programs have adopted a comprehensive approach that encompasses education, enforcement, legislation, and engineering improvements, and that involves both private and public sector partners. This comprehensive approach partly reflects a growing realization among the highway safety community that tougher regulations and laws and stricter enforcement are most effective when coupled with education and other preventative efforts.

New York has taken the lead among U.S. states in conducting research on drowsy driving, developing programs to address sleepiness-related driving among the general driving population, and identifying and targeting drivers at high risk for sleepiness-related driving, including commercial vehicle drivers. This paper reports the progress of New York's efforts to date. The purpose is two-fold. First, research findings from New York pertaining to sleepiness-related driving among the general population and high-risk groups are presented. Second, New York's comprehensive and innovative approach to addressing drowsy driving is outlined.

New York State's Program To Address Drowsy Driving

From its inception, New York's program in drowsy driving has been characterized by three important features. *First*, a high premium has been placed on conducting research to 1) identify the scope and nature of the drowsy driving problem, 2) identify high-risk groups of drivers, and 3) evaluate countermeasures. To a great extent, the work of the Task Force has been guided by the results of this research. *Second*, a broad, comprehensive set of countermeasures, encompassing the various components of the total highway safety system, has been developed to "manage drowsy driving." *Third*, a wide range of state and federal government agencies and private sector organizations have been involved in the development and implementation of countermeasures. Although all these organizations have donated time and resources to the Task Force, the Governor's Traffic Safety Committee has been the primary sponsor of the Task Force.

New York's foray into the area of drowsy driving began in Fall 1993, when a forum on fatigued driving was sponsored by our Institute, the Governor's Traffic Safety Committee, the National Highway Traffic Safety Administration, and the National Sleep Foundation (Institute for Traffic Safety Management and Research, 1994). In conjunction with the forum, the state and the National Sleep Foundation launched a comprehensive statewide public information campaign, "Drive Alert... Arrive Alive," which represented the nation's first comprehensive campaign on drowsy driving.

The forum generated widespread interest in drowsy driving among the state's highway safety community and government officials, and in January 1994 the Governor established a Task Force on the Impact of Fatigue on Driving. The Task Force was charged with examining the scope and causes of drowsy driving and developing recommendations. Eighty-two individuals representing a number of public and private organizations and varied areas of expertise served on the Task Force. To accomplish its mission, the Task Force was organized into teams to address the following eight areas: commercial drivers, crash reporting, curriculum, public information and education, roadside rest areas, shoulder rumble strips, legal sanctions, and research (NYS Task Force, 1994b).

The Task Force, in its final report submitted in December 1994, recommended the following initiatives, determined to have the highest priority because of their potential for reducing drowsy driver crashes (NYS Task Force, 1994a):

- development of public awareness and informational campaigns directed at the general public and at high-risk drivers

- development of a curriculum on the risks and prevention of drowsy driving for integration into driver education programs and health courses
- modification of the motor vehicle crash report form to include a more accurate and reliable description of crashes related to drowsy driving
- increased installation of roadway shoulder rumble strips
- improvements to increase the security and adequacy of roadside rest area facilities
- development of a training program for police officers to increase awareness of drowsy driving and improve identification and reporting of drowsy driving as a crash factor
- education for commercial drivers and their employers on the dangers and financial liability of drowsy driving and on countermeasures to reduce drowsy driving
- conduct of continuing research on the nature and scope of drowsy driving among the general population and high-risk sub-groups

An ongoing Task Force has monitored the implementation of the recommendations, evaluated progress, and devised new countermeasures. The research conducted and the accomplishments of the Task Force are summarized below. Since commercial vehicle safety has emerged as a primary focus over the past three years, particular attention is devoted to describing the research and activities undertaken in this area.

Research On The Scope And Nature Of Drowsy Driving Among The General Population

Either under the auspices of the Task Force or with the support of individual member agencies, a number of research studies on drowsy driving have been conducted to support the development of countermeasures. Together, these efforts have provided convincing evidence that drowsy driving is a frequent occurrence among New York motorists and that crashes and near-misses due to drowsy driving represent a serious traffic safety problem. These efforts have also provided evidence that certain types of drivers and certain types of driving situations represent an elevated risk. Due to the problematic reporting of drowsiness as a crash factor, other research methods in addition to analyses of crash data were employed. These efforts have included focus group research and surveys of drivers. This section of the paper briefly summarizes research designed to estimate the scope and nature of drowsy driving. Other research conducted to support the work of a particular aspect of the problem, e.g., commercial vehicle drivers, is described in other sections of the paper.

Crash Data

Analyses of crash data for 1990-1994, obtained from the state Department of Motor Vehicles, were undertaken by the Institute (NYS Task Force, 1996). The characteristics of these crashes were similar over this five-year period. Based on those crashes for which the police report indicated a factor of "fell asleep," there were 2,517 crashes in 1994, representing about one percent of the total crashes on the state's roadways. There were 46 fatal fall-asleep crashes in 1994, representing approximately three percent of all fatal crashes in that year.

While these statistics likely underestimate the scope of the problem in New York, the crash data are useful in identifying the characteristics of fall-asleep crashes. When compared to the characteristics of all police-reported crashes in 1994, the characteristics of fall-asleep crashes differ in the following respects:

- Fall-asleep crashes were more likely to occur on expressways or state roadways, and less likely to occur on municipal streets.
- 70.2% of fall-asleep crashes involved a single vehicle, compared to 26.8% of all crashes.
- 65.1% of fall-asleep crashes involved a collision with a fixed object, compared to 11.5% of all crashes.
- 45.4% of fall-asleep crashes occurred between 1:00 a.m. and 7:00 a.m., compared to 9.6% of all crashes; fall-asleep crashes were more likely to occur on Saturday or Sunday.
- Fall-asleep crashes were more likely to occur on dry roadways and were less likely to occur on straight and level roads.
- Alcohol was more likely to be a police-reported crash factor for the drivers who fell asleep than for the drivers in other crashes.
- Male drivers represented 75.4% of the drivers who fell asleep in 1994, compared to 66.4% of the drivers in all crashes.
- 35.4% of the drivers who fell asleep were 18-24 years of age, compared to 18.7% of all drivers involved in crashes.

Telephone Survey of Licensed Drivers

In Fall 1994, a statewide telephone survey was conducted of 1,000 randomly selected licensed drivers (Fact Finders, Inc. and Institute for Traffic Safety Management and Research Management and Research, 1994; McCartt et al., 1996). Telephone numbers were generated by random-digit dialing. The statistical sampling error associated with the overall findings ranged from +/- 1.8 - 3.1 percentage points. The sample was highly representative of the total population of New York State licensed drivers in terms of age, gender, and region of the state.

The survey indicated that not only do drivers consider drowsy driving to be a serious highway safety concern, but many drivers also frequently engage in this risky behavior.

- With "drowsy" defined as "so tired you could easily fall asleep," 54.7% of the survey respondents reported that being drowsy greatly affected their ability to drive safely. Respondents indicated that drowsiness had a greater effect on their ability to drive safely than either adverse weather or having two drinks of wine, beer, or liquor.

- 54.6% had experienced driving while drowsy (on at least rare occasions) in the last year, and 2.5% drove drowsy "very often."

As shown in Table 1, in terms of their lifetime driving experiences, 2.8% of the respondents reported that they had fallen asleep at the wheel and crashed, 1.9% had been drowsy while driving and crashed, and 22.6% had fallen asleep at the wheel without crashing. In all, 24.7% of the respondents had fallen asleep at the wheel at some point in their driving career, including incidents resulting in a crash and incidents in which no crash occurred.

Table 1. Reported Lifetime Drowsy and Fall-Asleep Incidents

Incidents	**(N=1000)**
Fell asleep at the wheel and did not crash	22.6%
Fell asleep at the wheel and crashed	2.8%
Had a crash when driving while drowsy	1.9%
None of the above	74.7%

Note: First 3 categories are *not* mutually exclusive: 24.7% have fallen asleep, including crash and no-crash incidents.

The respondents who crashed due to falling asleep or drowsiness (n=40) or who fell asleep without crashing (n=152) were queried about the circumstances of the most recent incident. Although there was a relatively small number of drivers involved in fall-asleep/drowsy incidents, the results were of great interest.

- 82.5% of the fall-asleep/drowsy crashes involved the driver alone in the vehicle, 60.0% occurred 11 p.m. - 7 a.m., 47.5% were drive-off-road crashes, and 40.0% occurred on a highway or expressway.

- 42.5% of crash-involved drivers reported that they had worked a night shift or overtime hours during the previous week, 35.0% had consumed alcohol before driving, and 10.0% had taken medication. Drivers had been awake an average of 13.5 hours and had been driving 4.0 hours, on average, before the crash occurred.

- When compared to the fall-asleep/drowsy crashes, the fall-asleep/no-crash incidents were less likely to occur between 11 p.m. and 7 a.m. or to involve driving off the road, and more likely to occur on a highway or expressway. On average, the drivers in the fall-asleep/no crash group had been driving fewer hours and awake fewer hours prior to the crash; they were less likely to have been alone in the vehicle, to have been working a night shift or a lot of overtime, and to have consumed alcohol or medication prior to the incident. Although the samples are small, these differences between the crash and no-crash incidents suggest that there may be factors that mitigate the severity of an incident if a driver falls asleep.

Stepwise multiple regression analysis was used to examine the relative importance of driver characteristics and behaviors in predicting the frequency of driving drowsy in the last year (McCartt et al., 1996). The dependent variable was the reported frequency of driving drowsy in the last year (very often, sometimes, rarely, never). Four sets of hypothesized driver correlates included demographics, and driving, work, and sleep/wake patterns. In the first phase of the analysis, a regression model was built for each of the four sets of predictor variables. Then a full model was built, including all variables from each of the four sets of factors.

The set of predictors in the combined stepwise regression model was statistically significant ($R^2 = 0.177$). Six driver characteristics were most highly associated with more frequent drowsy driving ($p<0.01$): more frequent trouble staying awake during the day, a greater number of miles driven annually, a younger age group, fewer hours of sleep per night, more education, and fewer hours that can be driven before becoming drowsy. Of lesser importance ($p<0.05$) were driver gender (with male drivers more likely to be drowsy), more frequent driving for work, and working rotating shifts. However, the relatively modest predictive power of the models suggests that additional research studies are warranted.

Crash Reporting

The goal of the Task Force team on crash reporting was to improve the accuracy and reliability of the police reporting of drowsiness or fatigue as a contributory factor in crashes. Toward this end, two major initiatives were undertaken. First, the police crash report was revised in July 1996 to incorporate a new code that provides for drowsiness or fatigue, in addition to the existing code for falling asleep. Second, the New York State Police integrated education on drowsy driving into its training courses, and the state Division for Criminal Justice Services integrated education on drowsy driving into training programs for local enforcement personnel. As part of this training, officers are encouraged to consider that they may be at risk for sleepiness-related crashes, since many work rotating or night shifts, work long hours, and have considerable driving exposure. They may also have schedules that make it difficult to live a healthy life style and experience high levels of stress, which may inhibit the ability to get adequate restorative sleep and contribute to daytime sleepiness.

Officers are encouraged to watch for the following indicators that falling asleep or drowsiness may be involved in a crash: single-vehicle run-off-road crash, lack of evidence of evasive maneuvers (e.g., skid marks), driver alone, occurring late at night or mid-afternoon, lack of evidence of other causation, and the driver's recent sleep/work history. When drowsiness is suspected, officers are advised to ask the driver whether drowsiness was involved in the crash, about his/her sleep and work history in the past 24 hours, how long he/she had been awake, and how long he/she had been driving.

Education

The curriculum team fulfilled its goal to develop a standardized, medically accurate curriculum on the risk and prevention of drowsy driving. Through the efforts of the Department of Motor Vehicles, this curriculum has been incorporated into the major educational programs for drivers. As a result, since 1995 millions of new and experienced drivers have been exposed to education on drowsy driving. More recently, the state Department of Health is developing a model community program on drowsy driving for youth, which will include a major educational component. As part of the development of this program, the department has conducted focus groups of youth across the state.

RUMBLE STRIPS

An important and highly successful countermeasure has been the installation of continuous roadway shoulder rumble strips; these are raised or grooved patterns on the roadway shoulder to alert a driver that he/she is drifting off the road. In 1990 the New York State Thruway Authority began to test the use of shoulder rumble strips as a countermeasure for drift-off-road crashes, and the Authority has continued to expand the STAR (Shoulder Treatment for Accident Reduction) program. By 1997, rumble strips had been installed on all rural segments on the Thruway, a major 641-mile interstate highway traversing the state. The reduction in cost per foot from $1.89 to less than $.30 has been a significant factor in the number of miles treated.

Analyses conducted by the Thruway Authority, using crash data through June 1997, indicate a 69% reduction in fall-asleep crashes on the roadway segments treated with rumble strips. According to the Thruway Authority, between 1991 and 1995, fall-asleep fatal crashes averaged 12 crashes per year, representing 35% of total fatal crashes; in 1996, there were only 2 fall-asleep fatal crashes, accounting for less than 10% of the total.

In February 1995 the New York State Department of Transportation issued a policy for the siting and installation of rumble strips on the interstate highways and parkways under the department's jurisdiction. In addition, the department undertook an aggressive program to install "safe strips" on interstate highways with the greatest potential for sleepiness-related crashes. The treated shoulder-miles increased from 92 miles in October 1993 to 3,150 miles in December 1997.

Surveys of drivers support the perceived efficacy of rumble strips. In the Institute's 1994 survey of licensed drivers, 58.7% of the survey respondents reported that they had driven over rumble strips, and 93.4% of these drivers believed that the strips were helpful in keeping drivers "alert and on the road." In a Spring 1997 survey of long-distance truck drivers, described in more detail under the section on Commercial Drivers, 77.2% of drivers said that they believed rumble strips are very effective, and 15.1% believed that they are somewhat effective, in preventing run-off-road crashes due to drowsiness or falling asleep (McCartt et al., 1997b). More than half (55.9%) of the truck drivers said that driving over rumble strips had alerted them that they were driving off the road due to drowsiness.

Legal Sanctions

No issue examined by the Task Force has been more difficult than the issue of a driver's legal responsibility in causing a crash due to drowsiness or falling asleep. Commercial drivers are subject to federal and state hours-of-service regulations, which are enforced through periodic road checks. State law prohibits the operation of a bus by a driver whose "ability or alertness is so impaired or likely to become impaired through fatigue, illness or any other causes, as to make it unsafe...." For drivers in general, state law allows for the discretionary suspension or revocation of a license "for gross negligence in the operation of a motor vehicle or motorcycle or operating a motor vehicle in a manner showing a reckless disregard for life or property of others." The law also prohibits reckless driving, defined as driving in a manner which unreasonably interferes with the free and/or proper use of the public highway or unreasonably endangers highway users. The statutes for drivers of passenger vehicles are rarely applied.

The majority of the members of the 1994 Task Force concluded that special sanctions for drowsy driving should not be imposed until the level of awareness about the problem had been heightened among the general population and among the enforcement and judicial communities. Despite this conclusion, however, the issue of sanctions has continued to be debated by the Task Force. In Spring 1997, the majority of a newly formed subcommittee on sanctions re-affirmed that the institution of criminal sanctions would be premature, given the difficulty in proving that a crash was caused by a driver's drowsiness or falling asleep, and given that drowsy driving is not well under-stood by most of the public, that further research is necessary, and that there are existing laws that may be used in extreme cases. Rather, the subcommittee suggested that current efforts be directed at providing a more in-depth administrative review of crashes involving a reported drowsy or fall-asleep driver, and at education for drivers, police officers, and the judiciary.

It is likely that New York and other states will continue to wrestle with issues related to the liability of drivers who cause a crash due to drowsiness or falling asleep, as well as other legal issues, such as whether drivers with diagnosed sleep disorders should have limitations placed on their driving privileges. In developing the appropriate strategies, the relevant issues include not only what the responsibility of the driver should be, but also how to craft laws that could be enforced and upheld in court. Another concern is that legal sanctions may discourage drivers who may have a sleep disorder from seeking diagnosis and treatment. Some issues related to driver liability may eventually be resolved through civil litigation resulting from crashes involving drowsy or fatigued drivers and/or drivers with untreated sleep disorders.

PUBLIC INFORMATION

In the 1994 survey of licensed drivers, about one-third of drivers had recently seen, read, or heard something related to driving while drowsy or falling asleep at the wheel (Fact Finders, Inc., and Institute for Traffic Safety Management and Research, 1994). Perhaps the strongest recommendation of the 1994 Task Force was to increase this level of awareness, and New York has become a leader in developing public information programs in drowsy driving.

Under the auspices of the 1994 Task Force, focus group research was undertaken with the following groups of drivers believed to be at high risk for drowsy driving: suburban/urban younger drivers (age 18-29), suburban/rural drivers (age 18-29), persons employed full-time in permanently rotating shift work (age 30-59), and older drivers (age 65-79) (James P. Murphy, 1994). Participants were queried about their experiences with drowsy driving. They were also asked for their reactions to examples of public information materials on drowsy driving and were asked about their suggestions for effective public information strategies. These suggestions included, for example, the use of humor, the use of situations familiar to most motorists, the use of words that are attention-grabbing, the separation of drowsy driving messages from other safety issues such as alcohol-impaired driving, and the presentation of messages during the time of day when drivers are likely to be drowsy. The results of the focus group research were used to develop subsequent public information campaigns.

Informational materials on drowsy driving and New York's programs have been widely disseminated, and numerous presentations on the program have been made at local, state, and national events. New York's comprehensive and innovative public information campaigns have included the highly successful "Break for Safety" campaign. This campaign was coordinated by the Governor's Traffic Safety Committee and the National Sleep Foundation and involved a wide array of state and federal government agencies and private sector organizations. More recently, drowsy driving announcements have been included in New York's "Choice Is Yours" public information campaign, developed by the Governor's Traffic Safety Committee in partnership with the National Highway Traffic Safety Administration. The campaign used a humorous character to convey messages about several highway safety issues, including drowsy driving.

ROADSIDE REST AREAS

Like most states, New York is concerned with improving the security and adequacy of its roadside rest area system. In the 1994 survey of licensed drivers, 45.2% of respondents said that in the last year they had stopped at a roadside rest area when they felt drowsy (Fact Finders, Inc., and Institute for Traffic Safety Management and Research, 1994). However, 29.9% also indicated that they needed or wanted to stop at a roadside rest area within the past year when one was not available. Furthermore, 68.2% of respondents indicated that they would be very likely to stop at a rest area if they felt drowsy while driving, but only 29.8% of all respondents, and 16.8% of female respondents, said that they would do so if they were driving alone at night.

To assist the state Department of Transportation in making improvements in limited-service rest areas on two major highways, in Summer 1997 our Institute conducted interviews with 303 long-distance truck drivers at rest areas. Nine in ten drivers on each road said that more commercial vehicle parking is needed (Hammer et al., 1997).

In the Spring 1997 survey of long-distance truck drivers, a large majority of the respondents (80.2%) indicated that they always/often want to stop at a public rest area at night but find that all the parking places for commercial vehicles are full (McCartt et al., 1997b). Even more troubling was that the frequency of finding full rest areas was associated with more frequent drowsy driving and with more frequent violations of the hours-of-service regulations. When asked what, if anything, discouraged them from using public rest areas for napping or sleeping, 50.8% of drivers mentioned inadequate parking; 28.2% mentioned enforcement of a statutory two-hour parking. The next most common responses were solicitation/prostitution (15.7%), lack of security (15.7%), and poor/expensive food (14.0%).

The goal established by the rest area team has been the provision of an adequate number of safe and well-maintained rest areas where motorists and their families and commercial drivers will feel comfortable stopping to rest. The challenge in meeting this goal is to secure the considerable resources needed. Partly due to the Task Force's efforts, New York has undertaken major programs of improvement to public roadside rest areas. Since 1990, the Thruway Authority has refurbished its service areas through public/private partnerships. The Department of Transportation has a number of projects planned or underway to provide new rest area facilities and to revitalize existing facilities. To assist in these efforts, the New York State Police has developed a Rest Area Policy; one component is the location at rest areas of State Police satellite stations or offices of convenience.

COMMERCIAL VEHICLE DRIVERS

Fatigued driving among commercial vehicle drivers is a concern not only for the federal government but also for the states, which enforce federal regulations covering interstate carriers, promulgate and enforce regulations for intrastate carriers, and enforce traffic laws. Efforts to reduce drowsy driving among commercial vehicle drivers have involved an active coalition of federal agencies (Federal Highway Administration, the National Highway Traffic Safety Administration), state agencies (e.g., Governor's Traffic Safety Committee, Department of Transportation, Department of Motor Vehicles, State Police), private sector groups (New York State Motor Truck Association, Bus Association of New York State Inc.), and representations of the research and medical communities. Most countermeasures described elsewhere in this report, e.g., enhanced rest area capacity and security and the installation of rumble strips, are components of a comprehensive program to manage fatigue among commercial vehicle drivers.

In addition, the Institute and other partners of the Task Force have been involved in the ongoing examination of the hours-of-service regulations of commercial vehicle drivers. The Task Force submitted a petition to the U.S. Federal Highway Administration (FHWA) requesting that rulemaking be opened with regard to the regulation of commercial drivers with untreated sleep disorders. The petition emphasized the need for further investigation regarding such issues as the method for identifying a driver with a potential sleep disorder, the degree of disorder that would be considered disqualifying, the appropriate treatment, and who would pay for treatment. FHWA is in the process of evaluating the request.

Survey of Fatigue-Related Driving Among Long-Distance Truck Drivers

A primary activity over the past year has been our Institute's conduct of the *Study of Fatigue-Related Driving among Long-Distance Truck Drivers in New York State*, undertaken under a grant from the Department of Transportation with monies from the Federal Highway Administration, and with the support of many of the Task Force partners (McCartt et al., 1997a, 1997b). One component of the study was an analysis of available crash data. A second and primary component involved a survey of long-distance truck drivers; selected highlights of the survey are presented below.

In the survey of truck drivers, conducted in Spring 1997, interviews were conducted with a representative sample of 593 long-distance truck drivers on New York's interstate highways.

The sample included 192 drivers (32.4%) interviewed at public full-service and limited-service rest areas, 233 drivers (39.3%) at private full-service rest areas (i.e. truck stops), and 168 drivers (28.3%) who had been waved through at routine truck safety inspections at public limited-service rest areas. A random sampling strategy was employed at all sites. Participating drivers had driven a tractor-trailer for at least six months, made overnight trips, and drove at least 50,000 miles/year for work. The overall participation rate was 74.9%; the participation rates were 62.1% at the public rest areas, 70.6% at the inspection sites, and 91.4% at the private truck stops.

The survey instrument gathered a wealth of information on drivers' crash involvement, and their sleepiness-related driving experiences, job characteristics, typical work schedules and sleep/rest patterns, general daytime sleepiness, compliance with the current hours-of-service regulations, and experiences and attitudes related to public rest areas and roadway shoulder rumble strips. Drivers were also questioned about their attitudes toward the current hours-of-service regulations and proposed countermeasures, including proposed regulatory changes.

Respondent Characteristics. Almost all drivers interviewed were men (98.8%). The age distribution was as follows: 34 years or younger - 22.5%; 35-44 years - 36.1%; 45-54 years - 28.4%; 55 years or older - 13.1%. Less than one-quarter of the drivers (22.3%) were licensed by New York State; 61.2% were licensed by other states and 16.5% were licensed by a Canadian province. More than three-quarters of the drivers (78.0%) had been driving a commercial vehicle for 5 years or longer, and 58.9% had more than 10 years of experience.

Selected Job Characteristics. The majority of drivers (61.3%) logged more than 100,000 miles each year for work, and nearly one-quarter (22.4%) drove more than 125,000 miles/year. Most drivers worked for a private fleet (37.7%), a company that owns trucks and employs drivers, or a for-hire fleet (35.3%); the remaining 27.0% were "owner-operators." Most drivers were paid by the mile (70.6%), did not have an on-board computer or other device to record the driving time (74.0%), and loaded or unloaded their truck at least occasionally (72.0%).

Typical Work/Sleep Schedule. More than one-third (34.4%) of the respondents drove more than 60 hours, and worked more than 70 hours (36.4%), in a typical seven-day week. About three-quarters usually made trips of four or more days duration (79.6%), had an irregular work/rest schedule (70.4%), did not usually split their 8-hour off-duty period (78.8%), and had a driving schedule that included midnight-dawn hours (79.2%). While on the road, 55.6% of drivers always took their longest sleep during nighttime hours; 10.3% usually slept

5 hours or less during the total off-duty period, including split shifts. About a quarter (23.9%) usually slept 5 or fewer hours during their longest sleep period.

Reported Violations of Hours-of-Service Regulations. As indicated in Table 2, about two-thirds of the drivers reported that on at least rare occasions, they drove longer than the 10 hours allowed by regulation; took fewer than the required 8 hours off-duty, including split shifts; and drove more hours than recorded in the log book. Only about one-third reported that they never have a schedule that does not allow delivery without speeding or violation the hours-of-service regulations; 15.3% stated that they often or always have such a schedule.

Table 2. Survey of Long-Distance Truck Drivers
Reported Violations of Hours-of-Service Regulations

	Drive > 10 Hr.	Rest < 8 Hr.	Drive > than Record in Log Book
Often/Always	19.5%	18.9%	21.4%
Sometimes	24.5%	22.5%	21.6%
Rarely	21.9%	25.3%	20.3%
Never	34.1%	33.3%	36.7%

Bivariate analyses, based on the chi-square statistic, identified a number of associations between various job characteristics and work and sleep/rest patterns and reported hours-of-service violations. For example, more frequently driving more than 10 consecutive hours and more frequently taking fewer than 8 hours off-duty were related to a method of payment other than an hourly wage, a more frequent tight delivery schedule, more frequent loading/unloading cargo, driving at night, an irregular schedule, fewer hours slept while on the road, more hours driven and worked in a typical 7-day period, and more frequently finding public rest areas full at night.

Sleepiness-Related Driving Experiences. Of primary concern was the extent to which drivers reported driving while drowsy or falling asleep at the wheel (Table 3). Nearly half (47.1%) the drivers reported that they had fallen asleep at the wheel of their truck, with or without crashing, on at least one occasion. Approximately two-thirds (65.8%) said that they had been drowsy at the wheel of their truck during the past month; 4.7% said that they drove while drowsy either almost every day or every day. Of the 21.9% of the drivers who reported at least one highway crash in the past 5 years, 7.0% said that drowsiness contributed to the cause of their most recent crash.

Table 3. Survey of Long-Distance Truck Drivers Experiences with Drowsiness-Related Driving	
Ever Fell Asleep While Driving a Truck	47.1%
Times Fell Asleep While Driving Truck in Past Year	
0	74.6%
1	7.8%
2	8.3%
3	3.0%
≥ 4	6.3%
Frequency of Driving While Drowsy in Past Month	
Every day	1.3%
Almost every day	3.4%
Sometimes	16.5%
Occasionally	44.5%
Never	34.2%

When drivers were asked when they were most likely to be drowsy while driving, even if they do not usually drive during that time, the most common response was between midnight and dawn (52.5%), followed by 2-4 p.m. (17.6%).

A wide range of factors was cited by drivers as the main contributor to their driving while drowsy. The most common factors were boredom or "highway hypnosis" (14.7%), followed by driving in rain or fog (11.0%), in a warm truck (10.8%), or after a big meal (10.3%). Less than one-tenth of the drivers cited either insufficient or poor sleep (9.4%) or driving long hours (8.9%) as the primary cause of drowsiness. Time of day was mentioned by 8.6% of the drivers.

Bivariate analyses, using the chi-square statistic, examined the association between the hypothesized work and sleep/rest correlates and the reported frequency of drowsy driving in the past month (daily/almost daily/sometimes/occasionally/never), and between these correlates and falling asleep at the wheel during the past year. The results for the outcome variable of falling asleep at the wheel are summarized in Table 4.

The frequency of a tight delivery schedule was strongly associated with more frequent drowsy driving and with falling asleep at the wheel during the past year. Among the other factors associated with more frequent drowsy driving and falling asleep were a greater number of hours driven and hours worked in a typical 7-day week, irregular daily schedules, driving at night, driving in the evening, and more frequent hours-of-service violations.

Table 4. Correlates with Falling Asleep in Past Year Based on the Chi-Square Statistic

p<.01	p<.05	p<.10
Tight delivery schedule	Type of carrier	Method of pay
Hours drive in 7-day week	Split off-duty	Load/unload truck
Hours work in 7-day week	Drive in evening	Usual duration of trip
Drive > 10 hours		Irregular schedule
Off-duty < 8 hours		Drive at night
Drive > record in log book		
Nap on road		
Epworth Daytime Sleepiness Scale		

While the bivariate associations indicate a number of significant relationships between various sleep/rest/work variables and falling asleep at the wheel and frequency of drowsy driving, the large number of hypothesized variables and the fact that many of the predictors are related to each other argues or a multi variate approach. The Institute is currently concluding multivariate analyses of the data set to identify the most important predictors of sleepiness-related driving. The results will be reported shortly.

Attitudes toward Current and Proposed Regulations. Drivers were almost evenly divided in their general attitude toward the regulations; about half believed the regulations primarily interfere with doing their job, and about half believed the regulations primarily serve to protect their safety. The proposals viewed as having the most potential for reducing drowsy driving were requiring realistic shipping schedules (90.5%); educating drivers to recognize when they are too drowsy to drive (84.8%); providing drivers with scheduling information in advance (82.3%); adopting a weekly "restart" rule after a driver is off-duty for a defined number of hours (76.5%); requiring rest breaks after a certain number of hours (61.9%);

allowing drivers to drive more than 10 hours at a time without increasing the weekly maximum (61.3%); penalizing the carrier as well as the driver for violations of regulations (60.2%). Drivers expressed the least support for limiting driving in the middle of the night (16.8%); requiring more than 8 hours off-duty (23.1%); replacing the regulations with testing for alertness (26.6%); requiring on-board computers (28.9%); allowing drivers to drive for 12 hours, followed by 12 hours off-duty (34.3%); paying drivers by the hour (45.7%).

Discussion

While one intent of this paper was to describe a model comprehensive state program to address drowsy driving, the more immediate effect may rather be to emphasize the highly complex nature of the highway safety community in the U.S. Despite this complexity, however, we in New York State believe our Task Force approach has succeeded in developing a thoughtful, coherent, and coordinated program in drowsy driving. Any successes that we have had are due to the efforts of the Task Force member organizations and individuals.

An important feature of New York's approach is reliance, whenever possible, on research results. In a 1994 survey of licensed drivers, convincing evidence was provided that drowsy driving is perceived by drivers to be a serious issue of concern. The survey also indicated that driving while drowsy and falling asleep at the wheel are common experiences. Furthermore, analyses of the survey results suggest that certain drivers may be at higher risk for these behaviors. The analyses of the survey results, as well as analyses of crash data, also suggest that certain driving situations may place a driver at higher risk for a drowsiness-related crash or near miss.

Many of the current discussions about fatigue-related driving in the U.S. are focusing on long-distance truck drivers, and New York is keenly interested in this high-risk group as well. A Spring 1997 survey of long-distance truck drivers in New York documented the high prevalence of drowsy driving and fall-asleep incidents among these drivers. About half the drivers reported falling asleep at the wheel of a truck on at least one occasion, and nearly one-quarter had done so in the past year.

As anticipated, many truck drivers reported that they work demanding schedules. A number of job factors and scheduling practices were associated with sleepiness-related driving experiences. Of particular concern was that two-thirds of drivers said that they violate, rarely or more frequently, the following hours-of-service regulations: driving more than 10 consecutive hours, taking less than the required 8 hours off-duty, and driving more hours than they record in the log book. It is also of concern that more frequent violations of these regulations were associated with sleepiness-related driving.

Especially given that we do not have full scientific knowledge about fatigue and its effects on driving, it appears that an effective approach to managing fatigue cannot rely solely on regulation or enforcement, but must include educational programs and other types of countermeasures. In considering a "win-win" approach to regulation, effective strategies to manage fatigue should consider drivers' and carriers' understanding of fatigue and how it affects performance and safety. Drivers' practices on the road and their motivations for these practices, including economic needs and scheduling pressures, should also be considered.

Lessons that may be learned from New York include the following: First, continued research at all levels of government is critical to gain a fuller understanding of the causes and consequences of drowsy driving. The significant gaps in our knowledge not only limit our ability to develop effective countermeasures, but also pose difficulties in securing governmental funds, which are allocated on the basis of documented crash risks. Second, the involvement of a wide range of public agencies and private sector organizations, as well as scientific sleep researchers and physicians, is critical. Third, a comprehensive set of programs is needed. And finally, the causes and risks of drowsiness-related driving need to be more widely publicized among the general driving population and among those sub-groups at highest risk.

ACKNOWLEDGMENT

The studies reported in this paper were supported by the New York State Governor's Traffic Safety Committee, with funding from the National Highway Traffic Safety Administration, and the NYS Department of Transportation, with funding from the U.S. Department of Transportation, Motor Carrier Safety Assistance Program. The authors would like to acknowledge the work of the other participants in the Task Force on Drowsy Driving.

REFERENCES

Fact Finders, Inc., and the Institute for Traffic Safety Management and Research. (1994) 1994 Survey on Drowsy Driving. Albany, NY.

Hammer, M.C. et al. (1997). Study of Use of Limited-Service Rest Areas by Commercial Vehicle Drivers in New York State. Institute for Traffic Safety Management and Research: Albany, NY.

Institute for Traffic Safety Management and Research. (1994) Proceedings: Highway Safety Forum on Fatigue, Sleep Disorders and Traffic Safety, Albany, New York, December 1, 1993. Albany, NY.

James P. Murphy & Co. (1994) Driver Fatigue Focus Group Study: Final Report (Abridged).

Knipling, R.R., and Wang, J.-S. (1994) Crashes and Fatalities Related to Driver Drowsiness/ Fatigue. Research Note. National Highway Traffic Safety Administration: Washington, DC.

Knipling, R.R. and Wang, J.-S. (1995). Revised Estimates of the U.S. Drowsy Driver Crash Problem Size Based on General Estimates System Case Reviews in 39th Annual Proceedings, Association for the Advancement of Automotive Medicine: 377-392.

McCartt, A.T. et al. (1996) The Scope and Nature of the Drowsy Driving Problem in New York State, Accident Analysis and Prevention, 28 (6), 709-719.

McCartt, A.T. et al. (1997a). Study of Fatigue-Related Driving Among Long-Distance Truck Drivers in New York State, Vol 1: Survey of Long-Distance Truck Drivers; Vol 2: Analyses of Crash Data. Albany, NY.

McCartt, A.T. et al. (1997b) Work and Sleep/Rest Factors Associated with Driving while Drowsy Experiences among Long-Distance Truck Drivers in 41st Annual Proceedings, Association for the Advancement of Automotive Medicine: 95-108.

New York State Task Force on Drowsy Driving. (1996) Status Report. Institute for Traffic Safety Management and Research: Albany, NY.

New York State Task Force on Drowsy Driving. (1994a) [Final Report.] Albany, NY.

New York State Task Force on the Impact of Fatigue on Driving. (1994b) Team Reports. Institute for Traffic Safety Management and Research: Albany, NY.

21

An Integrated Fatigue Management Programme For Tanker Drivers

Assoc. Prof. Philippa Gander[1], Dr David Waite[1], Mr Alister Mckay[2], Mr Trevor Seal[2], & Ms Michelle Millar[1]. [1] Otago University at Wellington School of Medicine, [2]BP Oil New Zealand Limited

Introduction

This paper describes the elements of a comprehensive, non-proscriptive fatigue management programme that we are currently implementing and assessing in tanker trucking operations. The programme is a response to three major concerns.

Fatigue is a Safety Issue

First, there is a growing body of evidence that fatigue is an important safety issue in trucking operations. Reviewing its accident investigation data, the US National Transportation Safety Board (NTSB) has found that fatigue is the most common probable cause of fatal-to-the-driver trucking accidents, being cited in 31% of cases (NTSB, 1990). The Board has also concluded that driver fatigue is implicated in 30-40% of all heavy trucking accidents (NTSB, 1995). A major truck and bus safety summit meeting in the USA in 1995 identified fatigue as the number one truck safety issue (Federal Highway Administration, 1995). In its 1996 report to the New Zealand Parliament, the Transport Committee's Inquiry into Truck Crashes (NZ House of Representatives, 1996) concluded:

> "...driver fatigue is a largely unrecognised problem, especially in New Zealand, and may well rate with alcohol and excessive speed as a significant contributor to crashes."

Fatigue-related motor vehicle accidents tend to have more severe consequences, which may be associated with the fatigued driver's reduced ability to take evasive or corrective action (NTSB, 1995; NZ House of Representatives, 1996; Maycock, 1995; Pack et al., 1995).

Fatigue is Underestimated

The second major concern is that the real contribution of fatigue to trucking safety incidents and accidents is likely to be underestimated. Fatigue is seldom addressed adequately, and often overlooked entirely, when safety events are investigated. Inadequate incident/accident investigation in general, is an important safety concern in trucking. The US Truck and Bus Safety Summit (Federal Highway Administration, 1995) rated the lack of data on truck and bus crashes as the second most important safety issue, after driver fatigue. According to the Transport Committee's Inquiry into Truck Crashes (NZ House of Representatives, 1996), in New Zealand

> "there is little quality accident investigation information available and investigation of truck crashes is sporadic and lacking in detail".

There are a number of reasons why fatigue in particular has not been investigated adequately in the past (Maycock, 1995), including: failure to recognise or acknowledge fatigue on the part of drivers involved in incidents or accidents; the fact that the symptoms of fatigue may not be evident to police or witnesses at the scene; and investigators having insufficient understanding of fatigue to know what to look for, or what questions to ask.

Fatigue is Complex

The third major concern is that, historically, fatigue has tended to be viewed as a simple consequence of the amount of time spent working (NTSB, 1995). This has been the rationale behind proscriptive approaches to fatigue management based on hours-of-work regulations (US Congress, Office of Technology Assessment, 1991). However, this focus overlooks two key types of physiological disruption that underlie many fatigue symptoms, namely sleep loss and disruption to the circadian biological clock (NTSB, 1995;US Congress, Office of Technology Assessment, 1991; Monk, 1994a, Rosekind et al., 1996ab). It also fails to address the fact that fatigue may be influenced as much (or more) by the activities of a driver outside of work, as by what takes place during scheduled work hours.

A Fatigue Management Programme

The present programme draws from experience in the aviation industry, and is an evolution of the approach to fatigue management developed by the NASA Fatigue Countermeasures

Program (Rosekind et al., 1996ab). The first step was a survey to assess sleepiness levels and the prevalence of several other accident risk factors among tanker drivers in the participating companies, and to see whether these were related systematically to rostering practices. A programme of interdependent fatigue management strategies is now being developed and implemented. These include:

1. a method for assessing the role of fatigue in incidents and accidents;
2. education and training for drivers and managers;
3. follow-up on fatigue-related issues through company medical examinations; and
4. the development of a set of rostering guidelines.

This is work in progress. The aim of this paper is to describe the rationale behind our approach, and some of the challenges already encountered, with a view to stimulating debate about the principles and practice of non-proscriptive fatigue management in trucking.

Survey Of Fatigue And Rostering Practices

All tanker drivers working for six Australasian trucking companies received a 2-page questionnaire that addressed sleepiness levels and the prevalence of other accident risk factors, and possible effects of shift characteristics. Participation was entirely voluntary and anonymous, and completed questionnaires were received from 163 drivers (70% response rate). Based on Epworth sleepiness scores (ESS), these drivers were sleepier (mean ESS = 7.0, with 2.5% of drivers scoring ≥ 16) than a sample of British heavy transport drivers (mean ESS = 5.7, with 0.3% scoring ≥ 16; Maycock, 1995)[1]. Twenty-one percent of drivers reported snoring every night, and one third had a neck size of at least 42 cm (a size "large" collar). In the British survey, accident rates were 29% higher among those who snored every night, when compared to those who snored less often or not at all. Accident rates were also 42% higher among drivers with notably large necks, compared to drivers with smaller necks.

Among the Australasian tanker drivers surveyed, 63% worked rotating day/night shifts, 27% were permanent day workers, 6% were permanent night workers, and 4% described their work patterns as "other". For drivers working rotating day/night shifts, higher sleepiness levels were associated with longer average shift lengths, and longer maximum duration of the night shift. A number of the findings from the survey influenced the development of the rostering guidelines described below.

[1] Epworth sleepiness scores ≥ 16 indicate a high level of daytime sleepiness, and have been found to be characteristic for people with severe sleep disorders, including obstructive sleep apnoea (Johns, 1991; Johns, 1993).

At the end of the study, all tanker drivers working for the participating companies received a letter of thanks from the research team and a 2-page summary of the findings, together with a new copy of the first page of the questionnaire so that they could reassess their own risk factors. The aim was to encourage individuals who were experiencing problems to seek help, either independently or through company channels. They also had ready access to the full study report through local management, if they so wished.

INCIDENT/ACCIDENT INVESTIGATION

Assessing the role of fatigue in an incident or accident is not simple. There is no single measure of fatigue that can be taken (comparable to a blood alcohol level) to indicate the level of fatigue-related impairment that a person is experiencing The approach that we are proposing is based on a method pioneered by the National Transportation Safety Board in an aircraft accident investigation in 1994 (NTSB, 1994). It relies on collecting information relevant to the physiological factors affecting fatigue, namely:

- duration of continuous wakefulness;
- acute sleep loss;
- cumulative sleep debt;
- presence of a sleep disorder;
- time of day of the accident.

The reasons for the inclusion of each of these factors are summarised below.

Duration of Continuous Wakefulness

Laboratory studies consistently show that the longer a person stays awake, the sleepier they become, and the more slowly and inaccurately they perform any type of work[2] (Akerstedt, 1991; Dinges and Kribbs, 1991; Monk, 1994a). Studies of alertness (Williamson et al., 1996) and fatigue-related accidents among truck drivers (NTSB, 1995) generally show that the number of hours a driver has been awake is not as important as the number of hours he has driven since he had last slept, or the number of hours that he has been on duty since he last slept. These measures reflect the combined effects of time since sleep and driving-induced fatigue. Changes observed with increasing driving time in trucks include: increasing fatigue,

[2] The decline in performance associated with increasing time awake is superimposed on the rises and falls in performance associated with the cycle of the circadian biological clock.

measured subjectively or objectively, decreasing physiological alertness, and increasing error rates and accident risk (Akerstedt, 1991; Keckland and Akerstedt, 1993; Williamson et al., 1996). Interestingly, these changes do not appear to be affected markedly by the pattern of breaks across the drive (Keckland and Akerstedt, 1993; Williamson et al., 1996).

Acute Sleep Loss

To be alert and able to function well, each person requires a specific amount of nightly sleep. The average for an adult is about 7-8 hours, but there are people who require more or less than this average. If this individual "sleep need" is not met, the consequences are increased sleepiness and impaired performance (Carskadon and Roth, 1991; Dinges and Kribbs, 1991). For most people, getting two hours less sleep than they need on one night (an acute sleep loss of two hours) is enough to consistently impair their performance and alertness the next day. The reduction in performance capacity is particularly marked if less than about 5 hours sleep is obtained (Carskadon and Roth, 1991; Horne, 1991).

Cumulative Sleep Debt

The effects of several nights of reduced sleep accumulate into a "sleep debt", with sleepiness and performance becoming progressively worse (Roth et al., 1994). Recovery sleep after an accumulated sleep debt is usually deeper and more efficient, and the lost hours of sleep do not need to be recovered hour-for-hour. It typically takes two good nights for sleep to return to normal after sleep loss (Carskadon and Dement, 1994). Reduced sleep is common during shift work, particularly among night workers (Akerstedt, 1991; Gander et al., 1996; Monk, 1994b; US Congress, Office of Technology Assessment, 1991).

Time of Day

There is clear evidence, from laboratory studies, workplace studies, and incidents and accidents in a variety of industries, that people are most prone to making errors, and to falling asleep inadvertently, in the early hours of the morning and again in mid-afternoon (Akerstedt, 1991; Dinges and Kribbs, 1991; Mitler et al., 1988; Monk, 1990, 1994a). This pattern is reflected in the timing of the 107 single-vehicle heavy trucking accidents reviewed by the National Transportation Safety Board in their 1995 study (NTSB, 1995). In this sample, 53% of accidents occurred between 2:00 am and 8:00 am. Furthermore, 75% of fatigue-related accidents occurred in this time interval. A similar pattern has been reported for fatigue-related motor vehicle accidents in general (Mitler et al., 1988).

Sleep Disorders

Not only the amount of sleep, but also the quality of sleep can have important effect on wake-time functioning (Roth et al., 1994). Sleep that is restless and fragmented by frequent awakenings also leaves a person sleepy and at increased risk of making errors. Sleep can be disrupted by a wide variety of factors, including physical sleep disorders and other health problems, changing work/rest schedules, poor sleep habits, and ill-informed attitudes about increasing wake-time activities by cutting back on sleep.

The most common sleep disorder, which is of particular concern in trucking, is obstructive sleep apnoea. Studies generally suggest that it affects up to 4% of the adult male population (Partinen, 1994). However, one recent US study of 156 commercial truck drivers found that 46% suffered from moderate-to-severe sleep apnoea (National Commission on Sleep Disorders Research, 1993). Another recent US study of 90 commercial long-haul truck drivers indicated that drivers with sleep-related breathing disorders (including obstructive sleep apnoea) had twice the accident rate per mile of drivers without sleep-related breathing disorders. The increase in accident rates was independent of the severity of the breathing disorder (Stohs et al., 1994). Risk factors for sleep apnoea include regular snoring, large neck size, and excessive daytime sleepiness (Partinen, 1994). In a survey study of 996 British heavy transport drivers, these three factors, together with age, explained 54% of the variability between drivers in their accident rate over the preceding three years (Maycock, 1995).

An Investigation Method for Trucking

A 2-page form has been designed that complements the existing incident/accident investigation forms used by the main participating company. A short manual has also been developed that explains the rationale behind the method and how to use it. For each individual directly involved, a duty history, and where possible a sleep history, are collected for 3 days prior to the day of the event, together with information on the amount of sleep normally needed to feel well rested. When a sleep history is available, this information permits calculation of the duration of continuous wakefulness at the time of the event, and the amount of sleep loss (acute and cumulative) that an individual was operating under. When it is impossible to obtain a sleep history (for example after a fatal accident), some idea of sleep opportunities can be gained by looking at the duty history. However, this does not give information on whether a person actually slept during those opportunities.

In addition, information is sought on: sleepiness risk factors (snoring, sleep disorders, Epworth score); the driver's self-assessment of his/her status at the time of the event; the type of incident/accident; and the type of human errors or failures contributing to the safety event. It is then possible to build an argument for the likely role of fatigue in an incident or accident.

The next stage of this project is the development of a centralised, de-identified database that will eventually permit more comprehensive analysis of the role of fatigue in safety incidents and accidents. This information will provide guidance for tailored fatigue management strategies for particular areas of risk.

Driver Education

The driver education package is based on a 2-hour live presentation accompanied by a handout covering all the presented materials. It aims to:

- explain the current state of knowledge about the physiological mechanisms that underlie fatigue, and how trucking operations affect them;
- demonstrate how this knowledge can be applied to improve driver sleep, alertness, and performance; and
- recommend scientifically-validated alertness management strategies, including strategies to assist drivers to arrive at work in the best possible (least fatigued) condition, and strategies to help maintain their alertness once they are at work.

An initial version of the presentation was "field-tested" with a group of 8 experienced drivers and 4 managers who provided invaluable input on both content and style. The group rated the overall presentation highly (average rating 4.3 on a scale from 1=poor to 5=excellent). They felt that the information would help drivers cope better with shift work (average rating 4.2 on a scale from 1=no help at all to 5=extremely helpful). Half the group indicated that they were quite likely to make changes to improve their alertness, based on the presentation, and 40% said that they would definitely make changes. Most (11/12) participants indicated that there should be recurrent training on alertness management every 1-3 years.

To disseminate the driver education package, a 2-day train-the-trainers workshop has been developed. The first day of the workshop focuses on the causes and effects of inadequate sleep, and on the effects of work demands, particularly shift work, on the circadian biological clock. The second day addresses fatigue-related performance impairment in trucking, incident/accident investigation, and strategies for managing fatigue. It also includes a presentation of the 2-hour training module, the provision and review of training materials, and the discussion of implementation issues.

Two such workshops have been held so far, and attendees have included managers, health and safety personnel, and representatives from a professional driver training organisation and from the Land Transport Safety Authority (the New Zealand regulatory organisation). Trainers are asked to have all participants complete brief, anonymous questionnaires before and after each training session, and to provide copies to the Sleep/Wake Research Centre. These will be used to assess the short-term effectiveness of knowledge transfer, and to provide guidance for the development of subsequent refresher training. Driver training is due to commence in the next few months.

Driver education is seen as a fundamental part of the overall programme, since critical aspects of fatigue management remain the responsibility of the individual driver. Education is perhaps the only legitimate way of addressing the potential impact of non work-related activities on fatigue, through an appeal to enlightened self-interest. Enabling drivers to develop better coping strategies, and in particular improved sleep, can be expected to have positive effects not only on safety and productivity at work, but also on overall health and well-being outside the workplace.

Better education can help counter the widely-held view that fatigue is somehow an indication of personal inadequacy, rather than a normal consequence of certain work demands on human physiology. Unless drivers are comfortable discussing fatigue-related issues openly, it is very difficult to develop a systematic approach to fatigue management. By way of illustration, a driver who was recently killed in an accident in which he is believed to have fallen asleep at the wheel, had been found by other drivers on three previous occasions asleep at the wheel. On one occasion, he had not even moved the truck to the side of the roadway. However, this information was not reported to the company until after the fatal accident. This incident also illustrates the responsibility of managers to foster an environment of trust in which drivers can be confident that their interests will be fairly addressed in such cases. It is also essential that drivers have a sound understanding of the causes and consequences of fatigue, if they are to have a major role in roster design (see below).

MANAGEMENT EDUCATION

Managers at all levels also need to have a sound understanding of the causes and consequences of fatigue if they are to design, implement, and maintain successful fatigue management strategies. A common knowledge base is also fundamental if fatigue-related issues are to be dealt with in co-operation with the workforce, rather than in an adversarial context, which is often the case at present (for example, in contract negotiations or in attribution of culpability for accidents).

To address this need, a 1-day intensive fatigue management workshop is being developed specifically for managers. It will cover:

- the current state of knowledge about the physiological mechanisms that underlie fatigue, and how trucking operations affect them;
- fatigue-related performance impairment in trucking;
- incident/accident investigation, and
- strategies for monitoring and managing fatigue.

Participants will also receive a comprehensive information package for future reference. This workshop is expected to be available by the middle of 1998.

COMPANY MEDICAL EXAMINATIONS

Drivers in the main participating company are required to undergo annual medical examinations to confirm their fitness for driving. Until now, these examinations have not included any information about, or discussion of, fatigue-related issues. As part of the fatigue management programme, they will now include preliminary screening for excessive daytime sleepiness (the Epworth Sleepiness Scale), as well as risk factors for sleep apnoea (snoring every night, upper body obesity and large neck size), and an opportunity for drivers to discuss their fatigue-related concerns. Drivers will be familiarised with the measures, and the reasons for collecting this new information, through the driver education package.

This strategy raises a number of issues concerning the assessment of fitness for driving and suggestions or requirements for treatment of chronic fatigue-related problems. Debate on these issues is ongoing. One area of particular concern is that sleep disorders medicine is in its infancy in New Zealand, and the waiting time for admission to public sleep disorders clinics can be as long as 3 years. If it is agreed that a suspected major sleep disorder is of sufficient concern to warrant (temporary) suspension from driving, then a procedure must be established to expedite assessment and treatment in a reasonable time frame. One possibility currently under consideration is that the company would pay a premium rate to assure access for drivers to assessment in a private clinic within two weeks. As a general principle, it is envisaged that sleep disorders or other chronic fatigue-related problems would be handled, from an occupational health perspective, in a comparable manner to other health issues such as diabetes or cardiovascular illness. This is a new area of occupational medicine, and the challenges are many. However, with the implementation of education programmes designed to heighten driver awareness of these issues, it is essential that fair and effective strategies are developed to handle possible occupational health outcomes.

ROSTERING GUIDELINES

Two different considerations prompted an initiative to develop a set of rostering guidelines. First, there is support within the main participating company for giving drivers greater autonomy in the design and manning of rosters. Guidelines would provide outer limits for this exercise. Second, some of the participating companies use sub-contractors and owner drivers. It was felt that rostering guidelines, as part of contracts, would help standardise working conditions across the workforce and ensure adherence to acceptable rostering practices.

It is accepted that there are no perfect rostering solutions to cover 24 h 7-day a week operations, and that individual drivers will respond differently to the same roster, because of their different individual characteristics and life circumstances (Folkhard, 1996). The following initial recommendations are currently being debated.

- The longest shift should not exceed 12 h, and the longest night shift should not exceed 10h.

  In the survey, among drivers working days (fixed or rotating), 34% reported that their longest day shift lasted 14 h. Among drivers working nights (fixed or rotating), 59% reported that their longest night shift lasted 12 h, and 14% reported working night shifts of 14-14.5 h. For drivers working rotating day/night shifts (63% of the sample), higher sleepiness levels were associated with longer average shift lengths, and longer maximum duration of the night shift

  Restricting shift durations is expected to limit the accumulation of time-on-task fatigue, and ensure that rest periods between shifts are of a reasonable length, given that times of shift changeover are more-or-less fixed. Drivers on night shift are particularly vulnerable to the effects of fatigue because they are most likely to be suffering from sleep loss, and are trying to work when the circadian drive for sleep, and error vulnerability, are maximal (Akerstedt, 1991; Dinges and Kribbs, 1991; US Congress, Office of Technology Assessment, 1991; Monk, 1990, 1994a). In this context, it is important to note that the National Transportation Board study comparing fatigue-related and non fatigue-related single vehicle trucking accidents (NTSB 1995) found that driving at night with a sleep deficit was far more critical than simply driving at night.

A concern raised by some logistics managers in response to these proposals was that the combination of a 12 h day shift and a 10 h night shift means that expensive assets (trucks) are not used to their full potential (24 h/day) and overall costs will rise.

- No driver shall work more than 5 consecutive shifts without a break that allows 2 nights of unrestricted sleep.

In the survey, only 43% of drivers had 2 full days off every 4-6 days, and 25% did not get 2 consecutive days off in less than 14 days. Some reported that the frequency of 2-day breaks varied seasonally, and three drivers reported having a 2-day break only when on annual leave.

The rationale in restricting the number of consecutive shifts is that it limits the accumulation of sleep debt. Two nights of unrestricted sleep are normally required for sleep to return to normal after sleep loss (Carskadon and Dement, 1994). Without an adequate opportunity for recovery, the effects of sleep loss (increasing sleepiness and performance degradation) presumably continue to accumulate from one shift cycle to the next. Ultimately, sleepiness will become overwhelming, leading to the possibility of falling asleep inadvertently at the wheel (Akerstedt,1991; Keckland and Akerstedt, 1993).

- The timing of shift changes should be considered carefully.

In the survey, a common pattern was to have 12 h day shifts and 12 night shifts, with shift changes at 3-4 am and 3-4 pm. On the one hand, this gives the night shift a longer sleep opportunity in the morning, before the circadian drive for wakefulness reaches its maximum (Strogatz, 1986; Gander et al., 1996). On the other hand it could cause considerable sleep restriction for the "day" shift, who would be expected to have difficulty going to sleep earlier in the evening, due to the evening wake maintenance zone (Strogatz, 1986; Gander et al., 1994).

A number of logistical considerations also have an important influence on the preferred timing of shift changes, including availability of loading facilities and local traffic patterns, particularly in major cities.

- The restrictions on shift durations and the number of consecutive shifts lead to a maximum work week of 60 h, with an average of 55 h for drivers working rotating shifts.

For many drivers in the survey, this would represent a reduction in work hours and in salary. It could also represent an increased cost for some operations, where extra drivers would need to be added.

- Permanent night shift should be avoided.

There is no strong evidence that permanent night shift increases circadian adaptation to night work, or health and safety on the night shift (Monk, 1990, 1994; Gander et al., 1996). On the other hand, there is considerable evidence that sleep loss, and subjective and objective measures of sleepiness and fatigue are greatest on the night shift (Akerstedt, 1991; US Congress, Office of Technology Assessment, 1991; Gander et al., 1996). Permanent night work also minimises drivers' contact with managers, who are mostly day workers, and complicates access to training opportunities. It reduces opportunities for regular contact with people who are not night workers, potentially impoverishing the family and social support networks of permanent night workers, which may be important in coping with shift work (Monk, 1994; Knauth and Costa, 1996).

- Rosters should be regular and predictable. This facilitates regular sleep patterns for different types of shifts, and planning for family and social activities.

- There should be a fair distribution of free weekends, since this is preferred time off for most people.

From discussions to date, it is clear that it will be necessary to have a mechanism to negotiate temporary exceptions, particularly because there is marked seasonal variation in demand in some operations. Thus, for example, at times of peak demand, there may be strong logistical arguments for some 14-h day shifts. In this case, one could argue for working fewer than 5 shifts before a 2-day break. On very long duty days, the distribution of workload may be particularly important in minimising the effects of fatigue. The use of breaks, and organising the order of deliveries (keeping shorter driving periods to the end of the run) are possibilities. The aim is to enable flexibility without compromising safety.

CONCLUSIONS

There are no simple solutions to fatigue management in trucking operations. Fatigue has multiple causes, different types of operations impose different demands, and individuals react to those demands differently. The approach to fatigue management described here relies on co-ordinated action and co-operation throughout a company. It is predicated on a relationship of trust and open communication between drivers and management, which is not easy to obtain or maintain.

It is reasonable to expect that better fatigue management can bring benefits (safety, health, and economic) for all parties, and that those benefits can extend to other areas of life outside the workplace. However, fatigue management interventions of the type proposed here have not often been implemented and even less often assessed for their effectiveness. These are the current challenges.

REFERENCES

Akerstedt, T. (1991) Sleepiness at work: Effects of irregular work hours. In Monk, T.H. (ed), Sleep, Sleepiness and Performance. John Wiley and Sons Ltd: West Sussex. pp 129-152

Carskadon, M. A., & Dement, W. C. (1994) Normal human sleep: An overview. In: Kryger M.H., Roth, T., and Dement, W.C. (eds.), Principles and Practice of Sleep Medicine . W. B. Saunders Company: Philadelphia. pp 16 - 25

Carskadon, M. A., & Roth, T. (1991) Sleep restriction. In Monk, T.H. (ed), Sleep, Sleepiness and Performance. John Wiley and Sons Ltd: West Sussex. pp 155-167

Dinges, D. F., & Kribbs, N. B. (1991) Performing while sleepy: Effects of experimentally-induced sleepiness. In Monk, T.H. (ed), Sleep, Sleepiness and Performance. John Wiley and Sons Ltd: West Sussex. pp 97-128

Federal Highway Administration (1995) 1995 Truck and Bus Safety Summit: Report of Proceedings. U.S. Department of Transportation: Washington DC

Folkhard, S. (1996) Effects on performance efficiency. In: Shiftwork: Problems and Solutions. Peter Lang: Frankfurt. 65-87

Gander, P. H., Gregory, K. B., Connell, L. J., Miller, D. L., Graeber, R. C., & Rosekind, M. R. (1996) Crew Factors in Flight Operations, VII: Psychophysiological Responses to

Overnight Cargo Operations (NASA TM 110380). NASA Ames Research Center: Moffett Field, CA

Gander, P. H., Graeber, R. C, Foushee, H.C., Lauber, J.K., and Connell, L. J. (1994) Crew Factors in Flight Operations, II: Psychophysiological Responses to Short-Haul Air Transport Operations (NASA TM 108856). NASA Ames Research Center: Moffett Field, CA

Horne, J.A. (1991) Dimensions to sleepiness. In Monk, T.H. (ed), Sleep, Sleepiness and Performance. John Wiley and Sons Ltd: West Sussex. pp 169-196

Johns, M. W. (1991) A new method for measuring daytime sleepiness: The Epworth Sleepiness Scale. Sleep, 14, 540 - 545

Johns, M. W. (1993) Daytime sleepiness, snoring and obstructive sleep apnoea: The Epworth Sleepiness Scale. Chest, 103, 30 -36

Keckland, G., & Akerstedt, T. (1993) Sleepiness in long-distance truck driving: An ambulatory EEG study of night driving. Ergonomics, 36, 1007 - 1017

Knauth, P., and Costa, G. (1996) Psychosocial effects. In: Shiftwork: Problems and Solutions. Peter Lang: Frankfurt. pp 89-112

Maycock, G. (1995) Driver sleepiness as a factor in car and HGV accidents. (TRL Report 169). Transport Research Laboratory, Department of Transport: Crowthorne, Berkshire, UK

Mitler, M. M., Carskadon, M. A., Czeisler, C. A., Dement, W. C., Dinges, D. F., & Graeber, R. C. (1988) Catastrophes, sleep and public policy: Consensus report. Sleep, 11, 100-109

Monk, T. H. (1990) Shiftworker performance. In A. J. Scott (ed.), Shiftwork. Occupational Medicine: State of the Art Reviews (Vol. 5) Hanley and Belfus Inc: Philadephia. pp. 183-198

Monk, T.H. (1994a) Circadian rhythms in subjective activation, mood, and performance efficiency. In: Kryger M.H., Roth, T., and Dement, W.C. (eds.), Principles and Practice of Sleep Medicine . W. B. Saunders Company: Philadelphia. pp 321-33

Monk, T. H. (1994b) Shiftwork. In: Kryger M.H., Roth, T., and Dement, W.C. (eds.), Principles and Practice of Sleep Medicine . W. B. Saunders Company: Philadelphia. pp 471-476

National Commission on Sleep Disorders Research (1993) Wakeup America: A National Sleep Alert. Report submitted to the US Congress and the US Department of Health and Human Services

National Transportation Safety Board (1990) Fatigue, alcohol, other drugs, and medical factors in fatal-to-the-driver truck crashes. (Safety Study NTSB/SS-90/01) National Transportation Board: Washington DC

National Transportation Safety Board (1994) Uncontrolled collision with terrain. American International Airways Flight 808. Aircraft Accident Report 94/04. National Transportation Board: Washington DC

National Transportation Board (1995) Factors that affect fatigue in heavy truck accidents, Volume 1: Analysis (Safety Study NTSB/SS-95/01) National Transportation Board: Washington DC

New Zealand House of Representatives (1996) Report of the Transport Committee on the Inquiry into Truck Crashes. Government Printing Office: Wellington

Pack, A. I., Pack, A. M., Rodgman, E., Cucchiara, A., Dinges, D., & Schwab, C. W. (1995) Characteristics of crashes attributed to the driver having fallen asleep. Accident Analysis and Prevention, 27, 769 - 775

Partinen, M. (1994) Epidemiology of sleep disorders. In: Kryger M.H., Roth, T., and Dement, W.C. (eds.), Principles and Practice of Sleep Medicine . W. B. Saunders Company: Philadelphia. pp 437-452

Rosekind, M. R., Gander, P. H., Gregory, K.B., Smith, R.M., Miller, D.L., Oyung, R., Webbon, L.L., and Johnson, J.M. (1996a) Managing fatigue in operational settings 1: physiological considerations and countermeasures. Behavioral Medicine, 21, 157-165

Rosekind, M. R., Gander, P. H., Gregory, K.B., Smith, R.M., Miller, D.L., Oyung, R., Webbon, L.L., and Johnson, J.M. (1996b) Managing fatigue in operational settings 2: an integrated approach. Behavioral Medicine, 21, 166-170

Roth, T., Roehrs, T. A., Carskadon, M. A., & Dement, W. C. (1994) Daytime sleepiness and alertness. In: Kryger M.H., Roth, T., and Dement, W.C. (eds.), Principles and Practice of Sleep Medicine . W. B. Saunders Company: Philadelphia. pp 40-49

Stohs, R. A., Guilleminault, C., Itoi, A., & Dement, W. A. (1994) Traffic accidents in commercial long-haul truck drivers: The influence of sleep-disordered breathing and obesity. Sleep, 17, 619 - 623

Strogatz, S.H. (1986) The Mathematical Structure of the Human Sleep-Wake Cycle. Springer-Verlag: Berlin, Heidelberg

Summala, H., & Mikkola, T. (1994) Fatigue accidents among car and truck drivers: Effects of fatigue, age, and alcohol consumption. Human Factors, 36, 315 - 326

U. S. Congress, Office of Technology Assessment (1991) Biological Rhythms: Implications for the Worker (OTA-BA-463). Government Printing Office: Washington, DC.

Williamson, A.M., Feyer, A.-M., and Friswell, R. (1996) The impact of work practices on fatigue in long distance truck drivers. Accident Analysis and Prevention, 28, 709-719.

22

THE QUEENSLAND APPROACH : THE FATIGUE MANAGEMENT PROGRAM

Gary L. Mahon, MCIT, Queensland Transport, Brisbane

INTRODUCTION

It has been widely acknowledged for some time that fatigue is one of the major causes of heavy vehicle accidents, causing up to 30% of all crashes of heavy vehicles in Australia. Indeed, many experts believe this figure to be too low. In an effort to reduce the trauma and cost to the community caused by these accidents, governments and industry all over the world are trying to find a practical and effective solution to this problem.

The last time I spoke to this forum, I discussed with you the fatigue problem itself, the ineffectiveness of the prescriptive approach, the cultural change that is required within governments, industry and the community, and the approach that Queensland has been taking to the issue.

The Prescriptive Approach

Until recently, the only method used by governments to address the fatigue issue has been to regulate to restrict driving hours. Unfortunately, the use of prescriptive hours and logbooks only attempts to manage driving hours and not the fatigue of the driver. And doesn't do this very well. In fact, it could be said that any similarity between the hours that are recorded in a driver's logbook and that driver's actual activities is purely coincidental.

Queensland's Approach

Queensland has devoted a substantial amount of its resources to the development of an alternative approach to fatigue management that has the potential to be many times more effective than the traditional approach. The Fatigue Management Program, or FMP as it has come to be known, is the result of our labours in this area. Today, I would like to briefly explain the FMP for those of you that are unfamiliar with the project, and to provide you with an update of the enhancements that have been made to the FMP model and the progress of the FMP pilot operators.

What Is FMP?

The FMP is a performance-based approach to managing fatigue that places the onus on the operator to take responsibility for and manage the fatigue of their drivers. The program is consistent with alternative compliance and quality assurance methodologies and provides a mechanism that will ensure compliance with a set of standards and a high level of safety.

The FMP is designed to identify all the factors that cause and increase the risk of fatigue and influence behaviour to improve road safety. As the FMP takes into account more than just the number of hours spent driving and holds operators to a challenging set of standards, the result is a much higher standard of safety than can be provided by the existing hours and logbook regime.

Pilot Project

The FMP pilot project is being conducted in two phases. Phase One involved several operators who assisted in the design and development of the FMP model. Three operators have been operating successfully under this phase for over 18 months. Phase Two of the pilot is designed to evaluate the effectiveness of the program and includes 16 operators, with over 1000 drivers, selected to provide a wide coverage of the country and a large variety in transport task.

There are two FMP Evaluation Projects currently being conducted by independent consultants. The first is designed to evaluate the effect of the FMP on driver fatigue and business efficiency, through a series of surveys being conducted before, during and after the completion of Phase Two of the pilot which is to run for 12 months. This evaluation is being performed by the Institute of Workplace Training and Development Consortium based in Brisbane. The second evaluation project, conducted by Doctors Anne-Marie Feyer and Anne Williamson, is aimed at developing a performance test that will assist in gauging the fatigue of individual drivers using both laboratory and on-road tests. Both of these projects have been primarily funded by the Federal Office of Road Safety (FORS).

The pilot project is being overseen by the FMP Project Team which is composed of representatives from road and traffic authorities, enforcement agencies and industry from all parts of Australia. The FMP Project Team has also benefited greatly from the support and assistance provided by FORS and our research consultant, Dr Anne-Marie Feyer.

Main Components

The main components of the FMP model are the accreditation agreement with its accompanying terms and conditions, the FMP standards and the performance management model.

The FMP operators enter into the accreditation agreement with Queensland Transport. this agreement sets out all the terms and conditions of their operation under the program. The agreement is legally binding and allows the operator to operate outside the regulations. In signing the agreement terms with its accompanying conditions the operators will be binding themselves to a comprehensive management arrangement to meet fatigue based performance outcomes.

FMP Standards

The FMP Standards identify and attempt to assist the operator with managing all of the factors the cause fatigue and increase the risk of fatigue. The standards form the basis of all FMP audits. In order to operate under the FMP, the operator must develop and implement management systems and procedures that will allow them to meet the standards and to achieve and maintain the level of performance that is required. A brief description of the FMP standards is provided below.

Scheduling. Scheduling of all trips must be planned and incorporate fatigue management measures required to undertake the transport task and provide drivers with the flexibility to reschedule driving and rest periods.

Rostering. Rostering systems must be in place to incorporate fatigue management measures and assign drivers to tasks in accordance with their recent work history, welfare and preference, where appropriate.

Time Working. Drivers must have the ability and opportunity to effectively manage their driving and non-driving work time in a way that allows them to combat the effects and onset of fatigue. Operators must not allow or cause a driver to work outside the operator's approved limits for periods that may endanger the safe operation of the vehicle and expose the driver, other road users and the environment to unacceptable levels of risk. An operator must demonstrate scheduling and rostering policies and techniques are being practiced and ensure accurate records of each drivers daily Time Working including rest activities, are kept.

Readiness for Duty. An operator must ensure all drivers are in a fit state to safely perform driving and non-driving duties.

Time Not Working. An operator must ensure a driver has sufficient continuous hours of Time Not Working to recover from the effects of fatigue caused by a period of Time Working and cumulative effects of fatigue caused by the extended periods of Time Working.

Health. The operator must ensure a health management and screening system is in place to best prevent and combat the onset and effects of fatigue and to address as a minimum such factors as medical history, sleep disorders, diet and substance abuse and provide preventative and remedial measures to assist drivers with the management of their health.

Management Practices. Management practices must ensure all drivers are suited to the transport task and that open lines of communication are fostered between management and drivers on matters that may enhance the safe operation of the business.

Workplace Conditions. The workplace conditions must provide environments which assist in the prevention of fatigue.

Vehicle Safety and Road Access Requirements. The operator must ensure that all vehicles owned and/or operated by them are safe to use on the road and that these vehicles do not expose other road users to unacceptable risk.

Driver Road Use Requirements. The operator must ensure drivers are licensed and authorised to drive the applicable category of vehicle safely on the road and in accordance with prescribed driving standards.

Training and Education. The operator must identify the fatigue management training and education needs of all employees and ensure that every staff member including managers are provided with training and education on the management of fatigue and the operator's fatigue management program.

Documented Policies and Procedures. The operator must prepare, implement and maintain documented policies and procedures that ensure the effective management, performance and verification of the fatigue management and accreditation requirements of the operator's FMP.

Responsibilities. The operator must assign and document the responsibilities and authorities of all positions involved in the management and operations of their FMP.

Management of Non-Compliance and Corrective Action. The operator must ensure all FMP non-compliance occurrences are reported, corrective action and preventative measures are taken in accordance with the level of risk identified, and an internal disciplinary system is in place to manage performance.

Records. The operator must ensure the identification, collection, storage and maintenance of records that demonstrate compliance with FMP standards. Records are to be kept for a minimum of 3 years.

Documentation Controls. The operator must operate a system to authorise, review and control all documents, including manuals, procedures, reference materials and legislation, required for the administration of a FMP.

Internal Audits. The operator must have an internal audit system to verify that all FMP activities and record keeping procedures comply with directions given in the relevant policy, procedures and instructions and any corrective action required, has been undertaken. This standard also requires the completion of a quarterly compliance statement.

CHANGES TO FMP

The FMP model has been further refined in anticipation of the commencement of Phase Two of the pilot project. The FMP is a live project that continues to develop as time goes on. Major enhancements to the FMP model include refinement of:

- the operating limits methodology; and
- the performance management model.

Operating Limits

When discussing the FMP, people are often interested in the maximum number of hours that are allowed to FMP operators and their drivers. As a society, we are so used to the formula approach to fatigue management that this question will inevitably come up.
So I will explain the FMP approach to this question for you today.

A set of working and rest parameters have been identified. These parameters are known collectively as 'operating limits' and include things such as:

- the maximum number of hours working in a 24 hour period;
- the minimum number of night time sleeps in a particular period (eg. 7 days); and
- the maximum number of aggregate work hours in a particular period (eg.14/28 days).

When preparing an FMP proposal, an operator will analyse their business and assess their fatigue risk factors to determine the set of operating limits that best meets their business

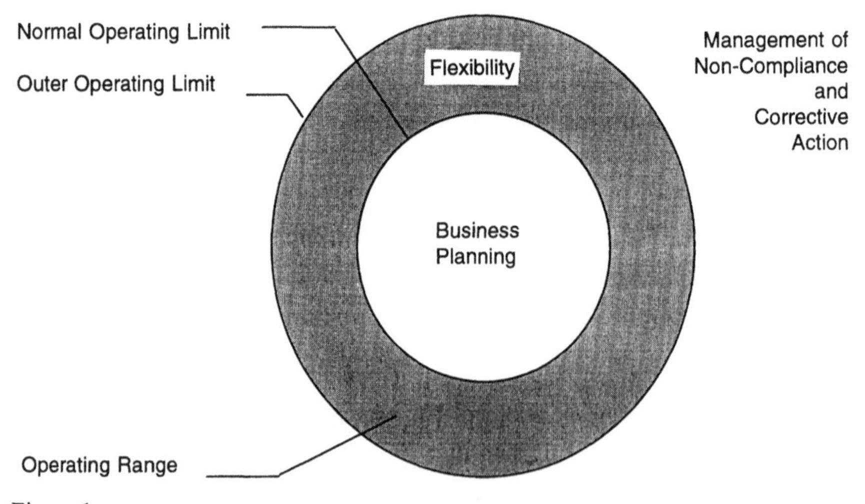

Figure 1

needs and minimises the risk of fatigue.

For each work and rest parameter, the operator will nominate a 'normal' limit and an 'outer' limit value as shown in Figure 1. The 'normal' limits provide the basis around which the business is planned and organised. These are the limits that are required to do the job in 'normal' circumstances. The 'outer' limit nominated is that limit past which the operator considers that the fatigue risk is significantly increased and some corrective action is required. It is at this point that the management system implemented by the operator kicks in, with the reporting of instances where an outer limit has been exceeded, a non-conformance is then raised, investigated and follow-up action is taken. The outer limit provides the mechanism to signal when countermeasures may be required, such as extra rest, to counteract the increased risk of fatigue.

The buffer zone between the normal and outer limits, the 'operational range', allows extra flexibility in work and rest limits to cater for unforeseen circumstances or for the driver to manage their fatigue more effectively. The operating limits model allows operators to take advantage of short term flexibility in their working arrangements whilst maintaining a reasonable average over the longer term. Each operating limit is assigned a 'frequency', which is the maximum number of times that the outer limit can be breached by an individual driver during a particular period (eg. once in 28 days). These frequencies will be valuable performance indicators, gauging the effectiveness of the operators fatigue management systems and procedures.

While each individual operator is allowed a different set of operating limits tailored to their business and fatigue management needs, there is no preset maximum number of hours under the FMP. It is up to each individual operator to justify the limits they are proposing and to have the countermeasures in place to effectively manage any increase in the risk of fatigue.

Performance Management

The FMP Project Team has strived to develop a performance management model that will be practical, effective and acceptable to all jurisdictions and industry.

The key components of the model are:

- audits;
- on-road enforcement; and
- sanctions.

Audits. As FMP is based on alternative compliance and quality assurance models, audits provide the primary method of monitoring and managing operator performance. Detailed audit procedures and checklists have been developed to assist the auditor. Several types of audit are employed in the program:

- The entry, or systems, audit is conducted prior to an organisation commencing operations under the FMP. The entry audit assesses whether the operator has implemented the management systems, policies and procedures necessary to meet the FMP Standards.

- The performance, or compliance, audit is conducted at three months and nine months after starting on the FMP. The performance audit assesses whether the operator has met the terms of the accreditation agreement and performance (measures against performance indicators, frequencies and standards). This type of audit can also be triggered to investigate an accident, incident or complaint.

- The surveillance audit is designed to be performed by enforcement officers to quickly investigate a particular incident or randomly check an operator's performance. The surveillance audit checks to ensure that essential components of an operator's FMP are working satisfactorily and to make recommendations to help determine whether a performance audit is required.

All entry and performance audits will be conducted by FMP accredited independent auditors during Phase Two of the pilot project. These auditors will assist the FMP Project Team in reviewing and evaluating the audit procedures to ensure the audits are effective and that operators get value for their money.

On-Road Enforcement. Even though most of the emphasis is on audits, on-road enforcement still has an important part to play in monitoring the performance of FMP drivers. Enforcement officers check to ensure that the driver meets the conditions of the program and are carrying the appropriate documentation. If not the driver will be breached under the regulations for not carrying and completing a logbook.

During the pilot project, enforcement officers complete an interception report for each interception, providing feedback on both good and bad incidents. These interception reports are forwarded to the operator for follow-up action if necessary. If serious problems are detected, the enforcement officer can recommend that a surveillance audit be conducted to investigate.

Like other alternative compliance schemes, it is intended that FMP drivers should be fast tracked through interceptions as compliance is assured through other mechanisms.

Sanctions. As FMP drivers are able to operate outside the regulated regime whilst they meet the conditions of the program, there are no associated logbook or driving hours offences. The FMP is designed to allow operators to report on and manage those instances where their outer limits may have been exceeded. However, if serious breaches of the accreditation agreement terms and conditions are detected, several options for managing the problem and bringing the operator into line are available.

Sanctions that can be exercised by the authority include:

- Corrective Action;
- Temporary Suspension;
- Increase Frequency of Audits;
- Variation of agreement terms or conditions; and
- Cancellation of FMP accreditation.

Many of these sanctions would result in significant disruption to the operator's business as well as increasing costs, thereby providing an appropriate incentive to maintain the required standards.

PROGRESS TO DATE

Phase One

To date, the three Phase One operators have travelled over 12 million kilometres without any fatigue related accidents or incidents. In the twelve months prior to commencing operations under FMP, one operator had experienced two fatigue-related accidents.

The operators' good records are also supported by the reports that have been received from enforcement officers. With over 480 interceptions of FMP drivers since the beginning of the pilot, only 15 interceptions have required any follow-up action due to the administrative problem of drivers not having their work diaries fully completed.

An evaluation of the Phase One operators and their drivers was compiled to assess their performance and to test the survey instruments to be used during Phase Two (IWTDC (a), 1996). Of the 32 drivers surveyed:

- 88% reported total control over where and when they stopped for rest;
- 40% felt tired on some or more trips;
- more than half claimed their driving was never impaired by fatigue; and
- 90% indicated they would tell their supervisor if they were too tired to drive.

The report also found that the FMP was found to have a beneficial impact on business efficiency. Due to the consultative process used to develop FMP within the companies, improvements have been experienced in management practices and communication with drivers. This has resulted in an increase in driver morale with the vast majority of drivers being very enthusiastic about the program. The FMP has provided the drivers with an opportunity to 'work within the law' and they can now truthfully record their activities in their FMP Work Diaries. The old logbook system was seen as unworkable. This increase in driver satisfaction with the work environment is expected to translate into substantial decreases in staff turnover.

The Phase One operators have also experienced an increase in vehicle utilisation and an improved ability to meet customer requirements. This improvement in business efficiency and the safety focus have allowed the companies to expand their businesses and move into new arenas. For example, we are beginning to see fatigue management programs being specified as a prerequisite for companies wishing to tender for large contracts. On the other hand, however, FMP operators have occasionally found themselves undercut by those unscrupulous operators willing to break the law and ignore the fatigue risks involved.

Whilst these results are very encouraging, it is difficult to determine how much of the outcome is attributable to the FMP as no comparable information is available prior to the companies commencing operations under FMP. This is why it is important that Phase Two of the pilot project commences.

Phase Two Operators

Unfortunately, the Phase Two operators have yet to commence operations under their individual FMP's. Several operators have put forward FMP applications and are very keen to begin trialing their programs. However, the jurisdictions have yet to reach a consensus on the conduct of the pilot project and this has essentially stalled the progress of the pilot project for over a year. Queensland believes that restrictions other than the FMP standards should not be placed on the program, as this will not give the approach the chance to reach its full potential and prove itself in an unbiased evaluation.

Information from the first wave of surveys to be completed by the Phase Two operators and their drivers have provided us with a benchmark to compare their performance in the next two waves of evaluation (ITWDC (b), 1997). The results of the benchmarking surveys and interviews have yet to be finalised, however, the preliminary findings indicate that the truck drivers were all male with an average age of 44 years with 21 years driving experience. The survey found that truck drivers drive an average of 51 hours per week and average 179,000 kilometres per year. With such a high presence on our roads, it is imperative that these drivers are fit to do the job.

Some interesting points to come out of the data include:

- 52% reported an insufficient knowledge about managing fatigue;
- 53% felt tired on some or more trips;
- 19% believed their driving was impaired by fatigue on some or more trips;
- 24% indicated they might not tell their supervisor if they were too tired to drive; and
- 15% reported the use of 'stay awake' pills.

Based on the experiences of the Phase One operators, after implementation of the FMP in the Phase Two organisations, we would expect to see the results of the next two waves of driver evaluation to show:

- a reduction in the number of fatigue incidents experienced by drivers;
- an increased awareness of fatigue and how it can be prevented;
- improved management practices (rostering, scheduling, communication etc); and
- a reduction in the use of 'stay awake' pills.

CHALLENGES FOR THE FUTURE

Queensland has learnt a lot from working with industry in developing the FMP. Most importantly, we have learnt that if we are going to make a real impact on fatigue, we need to use a performance-based approach and focus on the driver, not an assertoric hours formula. As we move into the future, several challenges have presented themselves to regulators:

Administrative Burden. We need to minimise the administrative burden of FMP by guiding operators with the development of practical systems with a minimal amount of paperwork while still meeting the safety requirements.

Reduce Costs. We need to reduce the costs of audits by having effective audit procedures, audit-friendly systems and reducing the number of audits. For example, conducting an FMP and TruckSafe audit at the same time.

Work with Industry. We need to work closely with industry to develop practical programs that work and to provide appropriate facilities, such as the strategic placement of rest stop areas.

Improve Accident Reporting. We need to improve the way we report accidents and their causes to allow us to accurately measure our performance and the effectiveness of our interventions.

Research. We need to encourage and support research, like that being conducted by many of the dedicated people present at this conference, in order to learn more about fatigue and how we might make our roads a safer place to work.

We need to move forward and learn from our past experiences. The regulated regime with its formula approach is a token effort which provides the community with a false sense of security rather than actually managing the many causes of fatigue.

With the encouragement of the success of the program to date, Queensland is currently considering other variations of the FMP principle to accelerate the move towards performance-based methods of fatigue management. During 1998 Queensland Transport expects to be able to offer more alternatives to managing fatigue that will help guide the industry towards a positive cultural change and the full FMP approach.

23

The Western Australian Strategy For Managing Fatigue In The Road Transport Industry

Lance Poore, Western Australian Department of Transport and Laurence R Hartley, Institute for Research in Safety and Transport, Murdoch University, Western Australia, 6150

Introduction

As elsewhere, there is concern in Australia about the part played by fatigue in truck crashes. Several Australian states restrict truck driving hours to 12 in any rolling 24 hour period in an attempt to limit fatigue as a causal factor in road crashes involving heavy goods vehicles. These regulations are enforced by requiring drivers to maintain a log book of their driving hours which must be produced on demand from a police officer. While fatigue has been found to be a significant contributor to road crashes the effectiveness of specific driving hours restrictions for the control of fatigue and resultant reduction of crashes is not clear (Arnold and Hartley, 1998).

Two Australian states, Western Australia (WA) and the Northern Territory do not enforce restrictions on driving hours. There is thus a regime of self regulation of the transport industry in these States. Nevertheless Western Australia is concerned to reduce fatigue in the road transport industry in order to minimise fatigue related crashes.

The research described earlier in this volume by Arnold and Hartley (*Its not just hours of duty: ask the drivers*) whilst certainly not suggesting that the problem of driver fatigue under the self regulation policy in Western Australia (WA) is more serious than in other Australian States with driving hours regulations, nevertheless recognised the problem does exist and therefore needs to be addressed in Western Australia as in other states.

The Second International Conference on Fatigue in Transport (1996) made the following recommendations for better management of fatigue.

- Driver fatigue arises not only from hours spent at the wheel but also from many other causes. Limiting driving hours does not address all the other causes of fatigue. Effective fatigue management will require that the other causes of fatigue are also addressed.
- Whilst technical means of detecting fatigue are under development, self assessment remains the only method currently available to detect fatigue. Education about the signs and dangers of impending fatigue is needed to support self-assessment.
- Government must play a central role in promoting research and evaluation of fatigue regulation measures.
- Government has a vital role in leading discussion and promoting education about fatigue at all levels of the community.
- Government has a key role in implementing cost-effective road-based countermeasures to fatigue.
- Government has a central role in leading the development of appropriate regulation and enforcement of fatigue countermeasures and of accident analyses.
- The transport industry must play a more active role in setting Occupational Safety and Health standards and adopting 'best practices' throughout the industry.
- The transport industry has a responsibility to the community for both a productive and also a safe transport environment and should take steps to ensure these expectations are met.
- The community should engage in education programs at all levels on the causes and countermeasures to fatigue. The costs and benefits of introducing safer work practices to better manage fatigue in the industry should be debated widely.

When considering its response to the call to introduce national uniform restrictions on driving hours as a means of combating fatigue the WA Department of Transport acknowledged the recommendations from the conference and the findings from the research and noted the following points:

- there was no evidence that prescriptive driving hours managed fatigue better than the existing self-regulatory practices in Western Australia, and the driving hours of the WA industry were little different to those jurisdictions with prescriptive driving hours;
- drivers recognised many other factors than long driving hours as causes of their fatigue, including night driving, other work, poor sleep and inadequate rest. These problems were not solved by the prescriptive driving hours of other states;
- there was some evidence that self regulation in WA permitted greater flexibility for drivers to take discretionary rest than did prescriptive driving hours;
- the widespread geographical layout of the centres of population in W.A. dictated that transport company schedules would sometimes entail extended hours of driving during the trip which would contravene any proposed prescriptive driving hours regime;

- although there was some support from drivers and the transport industry for government involvement in its regulation, there was uniform opposition to the introduction of log books as a means of doing so, thus effectively ruling out prescriptive hours as a regulatory strategy;
- the Australian state of Queensland has introduced an alternative to prescribed driving hours, the Fatigue Management Program (FMP), which permits a degree of flexibility for the industry and drivers comparable to self regulation practices in WA. The FMP approach requires companies to demonstrate adequate fatigue management practices and countermeasures in exchange for exemption from adherence to prescriptive driving hours. Failure to demonstrate adequate fatigue management practices results in withdrawal of this exemption. The FMP approach therefore relies on the existence of prescriptive driving hours legislation to underpin its enforcement;
- there is no prescriptive driving hours legislation in WA to underpin an FMP approach in WA;
- the Second International Conference on Fatigue in Transportation (1996) recommended that the existing Occupational Safety and Health legislation in WA could be used to improve standards of fatigue management in the transport industry.

Following review of these factors the WA Department of Transport considered that prescriptive driving hours had little to recommend them as a fatigue management system in Western Australia. What was required was a system flexible enough to accommodate the geography of WA; that required companies and drivers to demonstrate adequate fatigue management practices; and that could be enforced by reference to standards and guidelines. An approach that can provide this flexibility is by way of the Duty of Care provision as required under the Occupational Safety and Health Act.

An Overview Of The Occupational Safety And Health Act In Western Australia

The Occupational Safety and Health Act sets objectives to promote and improve occupational safety and health standards in Western Australia. The broad provisions of the Act are supported by the Occupational Safety and Health Regulations that detail minimum requirements for specific hazards and work practices, and guidance material such as approved Codes of Practice as described later.

The Act contains a general *Duty of Care* which describes the responsibilities of people in relation to safety and health at work. Employers must, so far as is practicable:

- provide a workplace and safe system of work so that, as far as practicable, employees are not exposed to hazards;
- provide employees with information, instruction, training and supervision to enable them to work in a safe manner;
- consult and cooperate with safety and health representatives in matters related to safety and health at work;
- provide adequate protective clothing and equipment where hazards cannot be eliminated; and
- ensure plant is installed or erected so it can be used safely.

Responsibility also extends to these other worker categories:
- employees are required to take reasonable care to ensure their own safety and health at work and the safety and health of others affected by their work;
- self-employed persons must take reasonable care to ensure their own safety and health at work and, as far as practicable, ensure their work does not affect the safety and health of others;
- designers, manufacturers, importers and suppliers of plant must ensure that plant intended for work use is safe to install, maintain and use at workplaces. Safety and health information must be provided when plant and substances are supplied for use at work; and
- designers or builders of a building or structure for use as a workplace must ensure, so far as is practicable, that persons constructing, maintaining, repairing, servicing or using the building or structure are not exposed to hazards.

Section 19 of the Act states:

An employer shall, so far as is practicable, provide and maintain a working environment in which his employees are not exposed to hazards and in particular, but without limiting the generality of the foregoing, an employer shall provide and maintain workplaces, plant, and systems of work such that, so far as is practicable, his employees are not exposed to hazards.

The Act requires employers to provide information to their employees, to alert them to areas where hazards may exist and to improve their understanding of safe work practices. A person who, at a workplace, is an employer or the main contractor must ensure that, as soon as practicable following a request from a person who works at the workplace, there is available for that person's perusal an up to date copy of the Act; these regulations; all Australian Standards, Australian/New Zealand Standards and National Occupational Health and Safety Commission documents or parts of those Standards or documents referred to in these regulations that apply to that workplace; all codes of practice approved under section 57 of the Act that apply to that workplace; and guidelines or forms of guidance referred to in section 14 of the Act the titles of which have been published in the Government Gazette.

Under this Act employers are required to provide and maintain a working environment where employees and other persons, and in the case of the road transport industry, other road users are not exposed to hazards. For a commercial driver a vehicle is a work place.

An important element of the Act is that safer systems of work are required where they are practicable. Practicable means that the cost of introducing a risk reduction work practice is less than the cost of the injury resulting from the hazard. Thus, the workplace hazards and risks must be assessed if practicable steps to improve safety are to be taken. Since fatigue related crashes are typically severe and costly for road users, a variety of measures to reduce fatigue can be justified.

The single most important defence to a charge of negligence is for employers to have in written form the practices and procedures that will lead to safe and efficient operations. This is a Fatigue Management System. To comply with the Duty of Care provision a Fatigue Management System should be in place; if the company does not have a documented fatigue management system it could not demonstrate it had a safe system of work and would be in breach of the Act. Under the act there is also a Duty of Care for employees to take reasonable care of their own safety and health in the workplace. Thus an employee also has a responsibility to report for work rested and fit for duty. For the purposes of the Act sub-contractors are treated as if they are employees.

THE COMPANY FATIGUE MANAGEMENT SYSTEM

Transport operators generally have a risk management program that includes components such as vehicle maintenance systems and systems for ensuring compliance with road access requirements. A driver Fatigue Management System is another component within the risk management program.

A Fatigue Management System identifies and targets specific risk factors and control measures involved in freight tasks. Managing driver fatigue requires effective management practices and office procedures including:

- maintaining open lines of communication between management and drivers;
- encouraging feedback from drivers;
- ensuring that the Fatigue Management System is included in driver induction programs and in other Human Resource procedures and practices; and

- appropriate documentation and record keeping practices.

Documentation of policies and procedures associated with the Fatigue Management System provides practical evidence that a system is in place and is actively working to manage driver fatigue. It also allows the effectiveness of the system to be measured. Documentation should be professionally administered and include numbered and dated systems in place for updating information. An example would be an update of a driver procedure manual where a new page is to be inserted. There should be a documented system that ensures all drivers receive the new information.

Record keeping is also important. Records provide the detail that the program is working and standards are being met. Records are an essential part of an overall risk management program as they provide a history of a particular driver or management activity. This information may be of vital importance in any legal action. Records should be kept for a minimum of three years.

A Code Of Practice

A Company's Fatigue Management System should address a number of key areas. To assist industry in meeting their duty in these areas guidance is required on appropriate standards for these areas. The provision of a Code of Practice, developed in conjunction with industry, is intended to offer this guidance. A Code of Practice also provides guidance to the authorities. In the event of a government agency or another competent authority investigating an incident involving driver fatigue a comparison will be made between the system of work and the recognised acceptable standard. The Code of Practice developed in W.A. is intended to provide a standard for transport industry operations in WA that is based on research, is comprehensive in its coverage of the causes of fatigue, and includes appropriate measures for excluding them from schedules and rosters and providing countermeasures.

The Code's general operating standards for work and rest in road transport

Transport operations must as far as practicable be conducted within the standards described below.

Operating standards	Time spent in the activity
• Maximum continuous *Active Work (Driving and Non Driving work time)*	5 hours

- Minimum *Short Break Time* within every $5^1/_2$ hours — at least 30 minutes
- Maximum average *Active Work* time per rolling 24 hours over 14 days — 14 hours
- Total *Time Not Working* in any 24 hours — 8 hours
- Minimum 24 hours continuous periods of *Time Not Working* in 14 days — 2 periods
- Minimum continuous *Time Not Working* after *Active Work* in any 24 hours for solo drivers — 6 hours
- Maximum consecutive periods of *Active Work* time exceeding 14 hours in 24 hours — 0
- Maximum *Active Work* time in any 14 Days — 168 hours

In order to provide for some flexibility in working hours the *Active Work* time of 14 hours in 24 hours may be exceeded only when it is not practicable to operate according to the standards. Active work cannot exceed 14 hours on two successive days. Active work may exceed 14 hours due to such circumstances as:

- delays resulting from accidents, traffic or weather;
- the need to allow for provision of improved rest facilities or environments; or
- the need to allow for improved night time sleep.

Definition of Terms

- ***Time Working*** means the total time spent in *Active Work* plus *Short Break Time.*
- ***Time Not Working*** means time off at home, away from the vehicle or, if on a trip with the vehicle, includes sleep in an appropriate sleeper berth and does not include driving and related work.
- ***Active Work*** means the total time spent in *Driving Work Time* plus *Non Driving Work Time* such as loading, servicing and repairing the vehicle while performing the freight task.
- ***Short Break Time*** means time provided at work for rest and meals, and does not include *Non Driving Work Time* or *Time Not Working.*
- ***Driving Work Time*** means the time spent driving a heavy vehicle each day and does not include loading, servicing and repairing the vehicle.
- ***Non Driving Work Time*** means time spent on all other duties such as loading, servicing, repairing the vehicle and completing documentation.

- ***Driver*** means a person who is in control of a vehicle and includes company employees, subcontractors and any other contractual relationship with a company.

Detailed Standards and Guidelines

In addition the Code contains detailed standards and guidelines covering incident reporting, record keeping, scheduling, rostering, time working, rest periods, fitness for duty, health, management practices, workplace conditions, training, policy and procedures, responsibilities, management of non-compliance with the fatigue management system, record keeping and documentation. For each standard a variety of fatigue control measures are also proposed for occasions when it is not practicable to adhere to a standard for reasons of obtaining better sleep; delays resulting from accidents, traffic or weather; or to allow for provision of improved rest facilities or environments. These are not detailed here but include using two-up or shared driving, calling on relief drivers, amending the trip schedule or driver roster.

Scheduling

A key factor in managing driver fatigue is how a company schedules or plans individual trips to meet a freight task. Where practicable and reasonable, scheduling practices should include appropriate pre-trip or forward planning to minimise fatigue. A driver should not be required to drive unreasonable distances in insufficient time and without sufficient notice and adequate rest. Scheduling practices should not put the delivery of a load before a driver's safety, health or welfare.

To meet the standards, scheduling should ensure that:

- a driver is given at least 24 hours' notice to prepare for *Time Working* of 14 hours or more;
- a driver is not required to exceed 168 hours of *Active Work* in 14 days;
- *Active Work* does not average more than 14 hours per 24 hours over 14 days;
- total *Time Not Working* in any 24 hours is at least 8 hours;

- a solo driver has the opportunity for at least 6 hours of continuous sleep in a 24 hour period;
- continuous periods of *Active Work* do not exceed 5 hours;
- minimum *Short Breaks* total 30 minutes in 5 $^1/_2$ hours;
- flexible schedules permit *Short Breaks* or discretionary sleep;
- minimise driving if solo driver does not have the opportunity for at least 6 hours of continuous sleep in 24 hours;
- maximum consecutive periods of *Active Work* time exceeding 14 hours in 24 hours is zero;
- where night shift operations occur, active hours of work are reduced to reflect the higher crash rate from fatigue between 1-6 am; and
- maximise opportunity for sleep to prepare for trip by minimising very early departures.

Rostering

Rosters are the driver's planned pattern of work and rest for a week or more. A driver's roster and workload should be arranged to maximise the opportunity for a driver to recover from the effects or onset of fatigue.

To meet the standards, rostering should ensure that:

- a driver does not exceeded 168 hours *Active Work* in 14 days;
- a driver has at least one day of *Time Not Working* in 7 days, or two in 14 days;
- minimise irregular or unfamiliar work rosters;
- minimise schedules and rosters which depart from day time operations when drivers return from leave. Drivers returning from leave require time to adapt to working long hours especially at night;

- total *Time Not Working* is at least 8 hours per 24 hours;

- minimum *Short Breaks* total 30 minutes in 5 $^{1}/_{2}$ hours; and

- a solo driver has opportunity for at least 6 hours of continuous sleep in 24 hours.

Readiness for duty

Drivers should be aware of the impact of activities such as a second job, recreational activities, sport, insufficient sleep, consumption of alcohol and drugs, prescribed or otherwise, and stressful situations on their well being and capacity to work effectively. These activities may affect their state of fatigue, especially cumulative fatigue, and capacity to drive safely.

To meet the standards, readiness for duty means:

- a driver must be in a fit state for work when presenting for duty.

Health

Poor health and fitness of a driver is a contributing factor to fatigue and its effective management is critical to the safe operation of a vehicle. A health management system should be developed and implemented to identify and assist those drivers who are at risk. The system should include medical history, sleep disorders, diet, alcohol, substance abuse or dependency and lifestyle. The system should also promote better health management.

To meet the standards, health management systems should ensure:

- medical examinations, at least every three years until 49 years of age and every year thereafter, in accordance with the National Road Transport Commission Medical Examination of Commercial Drivers or the Road Transport Forum (RTF) Driver Health Program;

- assessment of sleep disorders, other fatigue related conditions and health problems eg diabetes;

- provision of appropriate employee assistant programs where practicable;

- the provision of information and assistance to promote management of driver health;
- training for drivers on risk factors for poor health and the control measures;
- drivers are informed of benefits of good dietary intake and necessity for exercise to combat obesity which can result in fatigue. Obesity results from excessive food intake and the sedentary habits of long distance driving and is an important risk factor for the development of obstructive sleep apnoea, a common sleep disorder causing day time sleepiness;
- encourage a healthy lifestyle program in workplace; and
- encourage drivers to take healthy foods in vehicle to eat on trip and avoid excessive consumption of high calorie food, especially at one sitting, which may cause sleepiness.

Workplace conditions

Unsafe and unsuitable workplace conditions contribute to fatigue. The ergonomic design standards of a vehicle cabin are important if a driver is to operate a vehicle safely on a road. Unsuitable depot facilities may prevent drivers from reducing the effects of fatigue. Operators should ensure workplaces comply with the Occupational Safety and Health Act and relevant Australian Design Rule specifications.

To meet the standards, workplace conditions should ensure as far as practicable:

- they meet appropriate Australian standards for seating and sleeping accommodation;
- vehicles that are used for sleep during periods of *Time Not Working* should be fitted with, as a minimum standard:
    - in a truck - a sleeper berth which meets ADR42 (Sleeper berths); and
    - in a tour bus/coach - adequate sleeping accommodation as prescribed by legislation;
- a vehicle cabin meets the requirement of the Occupational Safety and Health Act and includes, as a minimum, ventilation in accordance with ADR 42.17 and seating suspension that is adjustable to driver's weight and height;

- depots provide safe and suitable fatigue management facilities that meet the requirement of the Occupational Health and Safety Act;

- vehicles and other accommodation provide suitable facilities for rest; and

- truck cabins are air conditioned where practicable, are comfortable and checked before trip.

Training and education

Training and education must ensure all employees, contractors and managers understand the meaning of fatigue and have the knowledge and skills to practice effective fatigue management and comply with the Fatigue Management System. Training should be structured and programmed to meet the training needs of the participants. All training and education provided should be documented and participation recorded.

To meet the standards, training and education must include:

- duties imposed by the Occupational Health and Safety Act;

- the penalties for failure to comply with the Occupational Health and Safety Act;

- induction training before commencing work;

- the causes of driver fatigue and symptoms;

- management of driver fatigue and strategies for making lifestyle changes; and

- training and education programs are documented and employee attendance is recorded.

Responsibilities

The success of a Fatigue Management System is dependent on the operator, clients and drivers knowing and practising their responsibilities and authorities to ensure policies, procedures and contingency actions are performed as required by the Fatigue Management System. Responsibilities included in the Fatigue Management System should be defined and encompassed in position and job descriptions which should be kept current.

To meet the standards, responsibilities should include:

- the operator should develop the Fatigue Management System in consultation with drivers and suppliers;

- duties of the operator and drivers under the Occupational Health and Safety Act;

- where appropriate the operator should delegate staff to implement the Fatigue Management System; and

- maintaining records of trip schedules, rosters, time working, and other information to show that the company is conforming with its Fatigue Management System.

Documentation and records

Employers should give consideration to keeping records of all regular and irregular trips, driver's schedules and rosters. These could be based upon trip sheets, pay records and delivery dockets. They must show sufficient information for an auditor to determine that the company and its drivers have conformed to the Fatigue Management System.

To meet the standards, documentation and records should include:

- a Fatigue Management System which documents how the company and its drivers address the agreed operating standards and if the standards are not met, how control measures are put in place;

- documents that record all actual regular and irregular trip time schedules, driver's schedules and rosters;

- in the event that an agreed standard is not met the control measure(s) which have been adopted should be recorded;

- these should include all trips performed, including details of any trip alterations; and

- personnel records, kept on a confidential basis, that include copies of current medical certificates and details of any work restrictions imposed and applicable rehabilitation programs.

Management of incidents

A fatigue management program should require all unsafe incidents to be recorded. This information should be used to target unsafe practices and prevent injuries and damage. Comprehensive and thorough reporting of all unsafe incidents at work is required in a Fatigue Management System.

To meet the standards, management of incidents should ensure:

- all unsafe incidents that may cause a hazard or potential injury or harm are reported;
- sufficient information is collected for action to be taken to prevent a future occurrence of the cause of the unsafe incident;
- procedures to prevent any further harm or injuries due to this cause;
- policies that promote and encourage all employees, sub-contractors and relief staff to report all unsafe incidents including those where there has been no injury or damage;
- procedures are in place to monitor, record and investigate all incidents and to take corrective action; and
- a review of the Fatigue Management System after each unsafe incident.

ENFORCEMENT OF THE CODE

The Code is enforceable under the Occupational Safety and Health act which provides for a work improvement or a work prohibition notice to be served on an individual or company when an inspector from the enforcement agency, the Department of Occupational Safety and Health, believes the duty of care principle has been breached. Inspection of the work place can be at the instigation of a complaint or at any time. The improvement notice will state a time by which the breach must be rectified. The prohibition notice will state that the work must cease. The notices may stipulate how the breach in duty of care is to be rectified, such as by adherence to the industry Code of Practice or the establishment of a fatigue management system in conformity with the Code. The individual or Company may appeal to the Commissioner for Health and Safety and then to a Health and Safety Magistrate. In the event that a corporate body is found guilty of the breach members of the body may also be accountable, such as company managers. Fines for an employer found guilty of a breach of the act are $100,000, and $200,000 if death or injury are caused. Fines for employees are $10,000 and $20,000 for these offences.

Criminal sanctions can also be applied to companies for safety breaches.

Training, Implementation And Evaluation

The industry Code of Practice has been submitted for comment to the industry and public in late 1997, and comments will be considered for incorporation. The Code will be field tested with transport companies during early 1998 for full implementation in late 1998. A training program has been developed to assist implementation. The evaluation strategy draws on a number of potential indicators of the impact of the Code including deaths and injuries, working hours and work practices in the industry, and the costs and benefits to the industry and community.

The incentives for companies to introduce fatigue management conforming to the Code of Practice are that they can demonstrate a safer system of work and avoid prosecution under the Occupational Safety and Health Act; they will obtain discounted insurance premiums; some customers require fatigue management systems in their transport companies and adherence to the Code will be a defence in the event of companies being prosecuted for unsafe practices. The Code can also be used by drivers to demand improved scheduling and rostering from their supervisors; and it can be used by companies to insist that their customers cease demanding impossible delivery schedules or face prosecution for doing so under the Act.

References

Arnold, P.K. and Hartley, L.R. (1998) Its not just hours of work: ask the drivers. In Fatigue and Transportation, (Ed). L.R. Hartley. London: Elsevier.

Recommendations of the Second International Conference on Fatigue (1996) Ed. L. R. Hartley, Institute for Research in Safety and Transport: Murdoch University, Western Australia.

24

U.S. TRUCKING INDUSTRY FATIGUE OUTREACH

William C. Rogers, Ph.D., Director of Research, ATA Foundation, Alexandria, Virginia, USA

INTRODUCTION

The ATA Foundation, the research and education arm of the American Trucking Associations, has participated in several research partnerships with the Federal Highway Administration's Office of Motor Carriers (OMC) designed to identify causes of, and begin to develop countermeasures for, truck driver fatigue. Such research, coupled with recent sleep and fatigue research in the United States and abroad, has pointed to the critical importance of education and training as a vital component in any program to reduce the likelihood of fatigue-related truck crashes.

Preliminary surveys, as well as continual requests from the industry for information on driver fatigue, sleep apnea, fitness for duty tests, shift work, etc., give every indication that management in the trucking industry is concerned about driver fatigue -- and often quite unaware of current knowledge of sleep and fatigue. If this is true for management, then the level of knowledge among truck drivers and their dispatchers is undoubtedly even lower.

However, to move beyond the anecdotal stage, and to cost effectively design and target the fatigue outreach program, the Foundation decided to survey a sample of truck drivers early in the program to assess their knowledge of the facts about fatigue and countermeasures to fatigue, and to identify some of the misconceptions they might hold on this subject. The intent was to develop materials that emphasized the important facts and principles that drivers know the least about, and to counter the most dangerous and widespread misconceptions. The initial survey would also serve as a baseline for follow-on surveys to evaluate the effectiveness of the outreach program. The mail survey of 25,000 truck drivers in the U.S. and Canada was conducted in 1996, with 5,000 drivers responding.

Using the results of the survey, as well as input from a group of safety and fatigue experts in the U.S. and Canada, the Foundation developed and initiated a Trucking Industry Fatigue Outreach Project with OMC in 1996, under two broad areas: public information and education. Through October, 1997, the Foundation has developed and disseminated the following public information and educational materials:

- Public Service Announcements (PSA) to 1000 radio stations that provide tips from truck drivers to help all drivers reduce the risk of drowsy driving. The PSAs were released just before Memorial Day 1996 and publicized by a radio tour with Tom Donohue, President and CEO of the American Trucking Associations. The PSAs were released again at Christmas 1996.

- In May 1997, three new PSAs were developed and released to 900 radio stations just before Memorial Day. The PSAs were publicized with a radio tour by the Chairman of the ATA, Charlie Ramorino. The estimated audience impressions from all the PSAs is over one billion.

- 700,000 *Awake at the Wheel* brochures were printed and distributed, with more than 100,000 going to OMC offices throughout the country. The ultimate goal is to place the brochure in the hands of every commercial vehicle driver in the country.

- A video/book package entitled, *The Alert Driver*, was developed and is now being distributed to 35,000 motor carriers. The package is aimed at truck drivers and their families and contains useful material on sleep, fatigue, and how families can help drivers maintain alertness through proper rest.

- A "train-the-trainer" course, *Fatigue and the Truck Driver*, was developed and training sessions are underway to provide in-depth courses to safety managers and truck driver training school staffs around the country. The instructors receive detailed information on the latest scientific information on sleep, fatigue, and countermeasures, as well as an Instructors Guide with PowerPoint slides, the *Alert Driver* video package, and student handouts that will enable them to teach drivers, dispatchers, and managers how to better manage fatigue. A similar program, *Fatigue Outreach for Managers,* has also been developed.

- In April 1997, an international conference -- jointly sponsored by the American

Trucking Associations, the Association of American Railroads, the Federal Highway Administration, the Federal Railroad Administration, the National Highway Traffic Safety Administration, and the National Transportation Safety Board -- was convened in Tampa, Florida. Drawing on the latest research, more than 130 leading scientists, government officials, and transportation managers from around the world met and agreed that the most current data clearly supports the application of new approaches to the management of operator fatigue in our 24-hour society. The conference speakers reiterated the importance of education about the causes and signs of fatigue, and that both industry and government should aggressively support programs to educate drivers, their families, and motor carriers on this issue. The proceedings from this conference have been widely disseminated throughout the U.S. and Canada.

This paper will discuss the driver fatigue survey, the public information and education materials developed partly in response to the survey, and the fatigue outreach evaluation plan.

Survey Of Truck Drivers' Knowledge And Beliefs Regarding Driver Fatigue

Background

In 1996, the Trucking Research Institute (TRI) of the ATA Foundation entered into a cooperative agreement with the Office of Motor Carriers (OMC) of the Federal Highway Administration (FHWA) of the U.S. Department of Transportation to develop and implement an outreach program to mitigate problems of driver fatigue by providing informational and educational programs and materials on fatigue-related topics to members of the trucking industry and the general public in the United States and to a certain extent, Canada. To cost effectively design and target the outreach program, TRI decided to survey a sample of truck drivers early in the program to assess their knowledge of the facts about fatigue and countermeasures to fatigue, and to identify some of the misconceptions they might hold on this subject.

The TRI subcontracted the survey to Star Mountain, Inc., and enlisted the support and participation of the National Private Truck Council (NPTC), the Owner Operator Independent Drivers Association (OOIDA), and the Canadian Trucking Association (CTA) to identify organizations and drivers willing to participate in the survey.

Survey Design

The study team had earlier identified a set of important fatigue-related topics and principles essential to the safe operation of commercial vehicles. These topics were:

- Alertness and safety
- Attention and performance
- Sleep requirements to maintain alertness
- Sleep deprivation and its effects
- Sleep apnea
- Circadian rhythms and their implications
- Work scheduling
- Napping
- Medications and other substances
- Countermeasures to fatigue

Survey items were designed to cover each of these topics, as were items to gather information on common misconceptions regarding fatigue, its effects on performance, and countermeasures, with these latter items phrased with alternatives to capture the misconceptions. A few demographic items were included to categorize respondents by age, experience as truck drivers, type of trucking operation, and work schedule.

There were 25 items on the survey. The first 23 items were 13 multiple-choice, four alternative-questions, and 10 true-false. There were also two open-ended items -- one dealing with individual drivers' strategies for avoiding or counteracting fatigue and the other with drivers' perceived need for information about fatigue. The Canadian version was identical except that because melatonin is not legally available without prescription in Canada, CTA omitted the item on the safety and efficacy of melatonin, leaving the Canadian survey at 24 items.

Survey Population and Sampling

The population the survey sought to sample was all truck drivers in the United States and Canada, with special emphasis on long-haul over-the-road drivers -- those who drive irregular routes and hours -- but it was not possible within the time and operational constraints of this task to conduct a random or rigorously systematic sampling process. Rather, the survey took advantage of the capability of TRI and other trucking industry associations to recruit member companies to distribute the surveys. The ATA and NPTC companies were recruited because of their membership in safety-oriented councils in the associations, with the notion that knowledge deficiencies in drivers in such premiere companies would be even more substantial with average drivers.

The sampling population for CTA was at the discretion of the provincial trucking associations, with instructions to seek a representative sample of companies. OOIDA, and association of drivers rather than motor carriers, used its list of 15,999 "active" members (out of over 30,000 total membership), and surveyed every tenth membership number.

In total, nearly 25,000 surveys were shipped to the distributing organizations, but because of the informal nature of the distribution process, the exact total number of surveys distributed to drivers is unknown. Thus, it is impossible to calculate a response rate for the survey based on the number *actually distributed* to drivers. A rate can be calculated on the number of forms *sent to the distributing organizations.*

Analysis Methodology

The surveys were analyzed by Star Mountain, Inc., using SPSS and SAS statistical packages, and included simple tabulation of responses to each alternative for each multiple-choice item, for the total sample and for each organizational sample. In addition, cross tabulations were performed for selected combinations of demographic items and knowledge items. For example, most items were examined for differences in responses by organization and by driver age. The responses to the two open-ended questions were coded into categories and tabulated.

Survey Responses

A total of 4833 completed surveys were received and analyzed, for an overall response rate of 20% based on the number of surveys sent to the organizations for distribution. While not a systematic sample, the nearly 5,000 responses are numerous enough that the results can be viewed with considerable confidence as representative of a large and diverse group of truck drivers in the United States and Canada.

Conclusions

In general, the drivers who responded to this survey performed moderately well on the fatigue knowledge items. Per cent correct ranged from just over 60% to over 95% on some items, demonstrating considerable understanding of many operator fatigue issues. However, the following fatigue-related topics were determined to require special emphasis in outreach and education programs.

- Sleep disorders, especially sleep apnea, seem to be poorly understood by the respondents, and there is some evidence that younger drivers are the least informed. Over 300 respondents specifically asked for more information on sleep disorders, and it would seem especially important to let drivers and others in the industry know that there is effective treatment available for apnea and most other sleep disorders -- without having to quit the truck driving profession.

- Napping is recognized as an effective short-term countermeasure to fatigue by fewer drivers than might be desired and it would be useful to provide more information on effective napping strategies, along with other sleep hygiene principles.

- Sleep requirements may be underestimated by many drivers because 25% of the respondents believe that 5 to 6 hours of sleep per night is enough to maintain alertness. Since sleep researchers now agree that few people are able to function well over the long term on this small amount of sleep, the importance of 8 hours of sleep in every 24 hours needs emphasis.

- Circadian rhythm effects are not fully understood by the drivers responding to the survey, especially effects on mood and performance efficiency. On the other hand, responses to the open-ended questions suggest that many drivers realize that the current hours of service regulations, combined with operational pressures, often require them to sleep and work out of phase with natural body rhythms.

- Caffeine and its effects on the body are probably not fully understood by drivers since a large proportion of respondents seem unaware of the limitations on caffeine's stimulant effects and possible side effects of caffeine overuse. About 900 respondents specifically mentioned using caffeine-containing food or beverages to combat fatigue, so it is important to teach drivers about what caffeine can and cannot do and the circumstances under which caffeine can be most useful.

- Melatonin may be another substance that drivers need to know more about since many of them did not respond to the question on melatonin safety and many others responded incorrectly. Given all the publicity about melatonin in the U.S., drivers need to be aware of the limitations of current scientific knowledge, especially recommended doses and timing as well as the long-term health effects from repeated use.

Fatigue Outreach Efforts

In addition to the truck driver fatigue survey discussed above, TRI relied on a curriculum advisory committee to help identify target audiences and training needs. Then, TRI staff, in conjunction with Star Mountain, Inc., and OMC staff, developed a wide variety of fatigue programs and materials that can be loosely categorized as education or information outreach.

Brochures

In 1995, the American Automobile Association (AAA) published a brochure entitled, *Wake Up*, to provide information to automobile drivers about sleep, drowsiness, and sleep disorders. The brochure was later revised to also target truck drivers. TRI, upon the advice of its curriculum advisory committee and the concurrence of AAA, revised the brochure, making its content more serious and changing its title to *Awake at the Wheel.* Since 1996, more than 700,000 brochures have been distributed to truck drivers throughout the United States, including 100,000 to OMC offices in all 50 states. The ultimate goal is to place the brochure in the hands of every commercial vehicle driver in the country as they receive or renew their commercial drivers licenses.

Public Service Announcements

Radio stations in the United States will broadcast Public Service Announcements (PSAs) free of charge, but there is tremendous competition for such free air time, and, therefore, the message and its delivery must first of all appeal to the radio stations. TRI developed its first set of PSAs in 1996 and distributed them to 1000 radio stations throughout the country. The compact disks contained three messages (in 30- and 60-second lengths) that provided tips from truck drivers to help all drivers reduce the risk of drowsy driving. The compact disks were forwarded with a letter from the Associate Administrator for Motor Carriers and the Managing Director of the ATA Foundation encouraging the radio stations to ensure the effectiveness of the *Awake at the Wheel* campaign. The PSAs were released just before Memorial Day (this holiday at the end of May is the first big driving holiday weekend of the year) and publicized by a radio tour with the CEO of the American Trucking Associations. The PSAs were released again at Christmas 1996. In all, an estimated 750 million impressions -- the number of people who could have heard the PSAs -- were made at a cost of US $36,000.

In May 1997, three new PSAs were developed and released to 900 radio stations and publicized with a radio tour by the Chairman of the American Trucking Associations. This series of PSAs focused on fatigue topics such as circadian rhythms and napping, areas that the truck driver fatigue survey indicated drivers were in need of more information. As in 1996, the messages were delivered as tips from professional truck drivers for all drivers, and have now made an estimated 1.2 billion impressions.

Video/Book Package

Continuing with the positive theme of alertness, a video/book package entitled, *The Alert Driver*, was developed and distributed to more than 25,000 trucking companies. The 19-minute video contains footage of two truck drivers, in over-the-road operations as well as at home, during which the drivers and their wives discuss how they cope with such issues as irregular schedules and sleeping during the day. The video also contains interviews with two leading fatigue and sleep researchers who offer the latest information on sleep, fatigue, and fatigue countermeasures. The 73-page book included in the package -- *A Trucker's Guide to Sleep, Fatigue, and Rest in our 24-Hour Society*, contains more than 50 pages on all aspects of sleep, while the remainder of the book covers strategies for the alert truck driver. The goal of both the video and book was to present a large amount of information on a complex subject in an understandable manner without talking down to the audience -- truck drivers and their families. A total of 35,000 packages were produced at a cost of US $135,000.

Train-the-Trainer Course

The cornerstone of the fatigue outreach program is the train-the-trainer course, a four-hour program of instruction begun in 1997 that is offered free of charge at sites around the country. The course is directed at safety managers and truck driver training school instructors, who, in turn, will train thousands of truck drivers, both novice and experienced, as well as dispatchers, about the latest information on sleep, fatigue, and alertness. The course, *Fatigue and the Truck Driver,* was modeled after the NASA Fatigue Module that has been used for several years to educate aviation personnel. As part of the course, the new trainers receive an extensive resource guide on sleep and fatigue, as well as an Instructors Guide with Power Point slides, the *Alert Driver* video package, and student handouts. A similar program, *Fatigue Outreach for Managers*, has also been developed. During 1997, 12 train-the-trainer courses were conducted, with almost 1000 instructors receiving the training, and one course for managers.

Tampa Fatigue Conference

In April 1997, the ATA Foundation convened an international conference on managing fatigue in transportation. This conference -- jointly sponsored by the American Trucking Associations, the Association of American Railroads, the Federal Highway Administration, the

Federal Railroad Administration, the National Highway Traffic Safety Administration, and the National Transportation Safety Board -- brought together more than 130 leading scientists, government officials, and transportation managers from around the world in Tampa, Florida.

The theme of the conference, drawing on the Second International Conference on Fatigue and Transportation, in Fremantle, Australia, was that in modern 24-hour societies, fatigue must be managed -- and that the most current scientific research clearly supports the application of new approaches to the management of fatigue in transportation operations. The conference speakers also reiterated the importance of education about the causes and signs of fatigue, and that both industry and government should aggressively support programs to educate operators, their families, and transportation managers on this issue. The proceedings from this conference have been widely disseminated throughout the United States and Canada, and have been essential to the deliberations for revising the U.S. motor carrier Hours of Service regulations, not to mention putting to rest the outmoded notion that hours on duty are the sole determinant of operator alertness.

Evaluation Plan For Fatigue Outreach

The fatigue outreach and education programs developed by the American Trucking Associations Foundation under the sponsorship of the Federal Highway Administrations's Office of Motor Carriers, have as their overall objective the reduction of truck driver fatigue. To reach this objective, it is necessary to change the behavior of drivers -- and the people and organizations that supervise, influence, motivate, or interact with them. Hence, the outreach and education programs are designed to provide information, education, and persuasion to various constituencies both within and without the trucking industry in order to bring about the desired behavior changes.

The evaluation of training materials and programs can be broken down into "formative" evaluation, conducted during the development and pilot testing of materials, and "summative" evaluation, conducted after the training has been implemented.

Formative Evaluation

The formative evaluation of the fatigue outreach materials was designed to ensure that the materials were:

- accurate;
- covered the required subject matter adequately;
- were appropriate in format, content, level of difficulty and detail, and presentation style for the intended audiences; and
- provided the supporting materials that instructors need to present the material well.

To evaluate these characteristics, it was first necessary to clearly define the intended audience(s), the material to be taught to each, and the intended instructional outcomes. It was also necessary to identify who should judge the instruction, and to design evaluation materials to facilitate the judgments and ensure that the feedback received would be useful in improving the instruction. The early tasks in the outreach development project defined the intended audiences for the instruction. These included trucking industry managers, risk managers, dispatchers and supervisors, company trainers, and, of course, truck drivers. It was decided that the instructional materials would be developed first for the audiences needing the most coverage and detail -- trainers -- and would then be adapted for other audiences. This permitted early formative evaluation to concentrate on the accuracy and completeness of the materials, with more attention later on format, presentation, and specific audience issues.

Early work also identified a set of fatigue-related topics that should be covered in the training materials, and reviewed existing materials for use as information sources, format models, or materials that might be used to supplement newly developed instruction, such as videos, brochures and booklets on specific topics (e.g. sleep and aging).

Expert review was selected as the method for the evaluation of the early phases of instructional design and development. The initial definitions of audiences and instructional topics were reviewed by contractor and government technical experts, and by fatigue and training experts associated with ATA, to ensure the training was appropriately targeted and that all important subjects were covered. Prototype training materials were then presented to a group of industry and government personnel for feedback. Contractor subject matter experts also reviewed these materials to ensure that they were accurate and clearly presented the teaching points that had been selected earlier. These reviews led to many revisions of the training materials.

Pilot presentation of the course materials with evaluation by participants was then used to evaluate the customized materials for specific audiences. They provided feedback on the appropriateness, usefulness, presentation, and sufficiency of the courses. To facilitate the process, evaluation forms were designed that included both structured responses on a five-point scale and open-ended responses. During this time, the Truck Driver Fatigue Survey discussed earlier was developed, disseminated, and evaluated to aid in the development of fatigue emphasis areas; a "before and after" test of learning was also developed. After each pilot presentation, the responses to structured the response evaluation items were tabulated and evaluated, and additional changes were made in the training materials and instructional methods.

Summative (Outcome) Evaluation

The on-going evaluation plan is not designed to rigorously evaluate the effectiveness of the myriad of training courses, books, pamphlets, videos, and other outreach materials. Basically, there are two lines of evaluation: learning outcomes from the formal fatigue courses, and fatigue knowledge changes among truck drivers.

Fatigue course outcome evaluation has used post-tests to estimate the level of knowledge of course participants at the end of the instructional period, and participants are also asked to complete evaluation forms as in the formative evaluation period, to provide ongoing feedback that will contribute to any future revisions of the instructional materials. It may also be possible to use follow-up surveys of course participants to gauge their retention of the information learned, perceived usefulness of the material, and self-reported changes in behavior resulting from the learning; however, such follow-up surveys will depend on the cooperation of the hundreds of trainers who present the courses to perform the data collection and transmit the evaluation forms to the ATA Foundation. In all cases where Foundation subcontractors have presented train-the-trainer courses, evaluations have been performed. After 25 courses have been taught, a summary evaluation will be conducted that will present descriptive statistics on participant post-test performance and on evaluation form responses, note any trends emerging over time, and identify problems.

Overall outreach effectiveness evaluation will rely on survey methods to gauge the program's effectiveness, at least partially, by its effects on the more immediate targets of the fatigue outreach -- truck drivers' knowledge of and attitude toward fatigue. Thus, in 1998, the Foundation will administer a follow-up driver survey, targeting the same organizations and trucking companies in the initial survey. Of course, there will be no way of attributing changes in the knowledge of a sample of drivers to the Foundation's fatigue education and outreach program itself -- there are many uncontrolled variables that can affect drivers' knowledge levels, including mass media coverage of related topics.

But, the Foundation may be able to use careful survey methods to increase the confidence level that any improvements found are attributable to its efforts. First, the survey will ask respondents about their exposure to the Foundation's information and outreach programs and materials. While recall and self-report may not be error-free, this information should be helpful in determining the extent to which the materials are being disseminated, as well as correlating changes in knowledge to exposure to the materials. Second, it may be possible to identify the trucking companies that are using the fatigue courses and selectively survey their drivers to compare knowledge levels between trained and untrained drivers. Third, it may be worthwhile to survey a sample of trucking executives and risk managers on: (1) their perceptions of the effectiveness of the outreach and education programs; (2) changes in company policy or practices as a result of new fatigue knowledge and awareness; and (3) and changes they see in the frequency or severity in fatigue related problems.

In summary, it should be possible at least to identify a reasonable body of evidence to support an evaluation of the effectiveness of the fatigue outreach and education efforts, and to use this evidence to reach defensible conclusions.

INDEX

CPSIA information can be obtained
at www.ICGtesting.com
Printed in the USA
BVHW040827271218
536512BV00001B/2/P